MATHEMATICS Mad

The Made Simple series
has been created
especially for self-education
but can equally well
be used as
an aid to group study.
However complex the subject,
the reader is taken,
step by step,
clearly and methodically
through the course. Each volume
has been prepared by
experts,
taking account of
modern educational requirements,
to ensure the most
effective way of
acquiring knowledge.

Accounting
Acting and Stagecraft
Additional Mathematics
Administration in Business
Advertising
Anthropology
Applied Economics
Applied Mathematics
Applied Mechanics
Art Appreciation
Art of Speaking
Art of Writing
Biology
Book-keeping
British Constitution
Business and Administrative
  Organisation
Business Calculations
Business Economics
Business Statistics and Accounting
Calculus
Chemistry
Childcare
Commerce
Company Law
Company Practice
Computer Programming
Computers and Microprocessors
Cookery
Cost and Management Accounting
Data Processing
Economic History
Economic and Social Geography
Economics
Effective Communication
Electricity
Electronic Computers
Electronics
English
English Literature
Financial Management
French
Geology
German
Housing, Tenancy and Planning
  Law

Human Anatomy
Human Biology
Italian
Journalism
Latin
Law
Management
Marketing
Mathematics
Metalwork
Modern Biology
Modern Electronics
Modern European History
Modern Mathematics
Modern World Affairs
Money and Banking
Music
New Mathematics
Office Administration
Office Practice
Organic Chemistry
Personnel Management
Philosophy
Photography
Physical Geography
Physics
Practical Typewriting
Psychiatry
Psychology
Public Relations
Public Sector Economics
Rapid Reading
Religious Studies
Russian
Salesmanship
Secretarial Practice
Social Services
Sociology
Spanish
Statistics
Teeline Shorthand
Twentieth-Century British
  History
Typing
Woodwork

MATHEMATICS   Made Simple

Abraham Sperling, PhD, and
Monroe Stuart

Advisory editor
Patrick Murphy, MSc, FIMA

Made Simple Books
HEINEMANN : London

© William Heinemann Ltd., 1981

Made and printed in Great Britain
by Richard Clay (The Chaucer Press), Ltd, Bungay, Suffolk
for the publishers William Heinemann Ltd.,
10 Upper Grosvenor Street, London W1X 9PA

First edition April 1967
Reprinted December 1967
Reprinted October 1969
Reprinted February 1971
Reprinted January 1973
Revised and reprinted August 1974
Reprinted March 1976
Revised and reprinted September 1977
Reprinted October 1978
Reprinted October 1979
Reprinted November 1980
Reprinted October 1981
Reprinted January 1984

British Library Cataloguing in Publication Data

Sperling, Abraham P.
    Mathematics made simple.—(Made simple
    books, ISSN 0265–0541)
    1. Mathematics—1961–
    I. Title   II. Stuart, Monroe
    III. Murphy, Patrick, 1925–
    510   QA37.Z

ISBN 0–434–98491–4

Foreword

Scientific and industrial progress in recent years has made Mathematics one of the most important subjects of our time. It is no longer fashionable to boast an inability to "work with figures"; such an admission, in some circles, is considered tantamount to an admission of illiteracy. Anyone these days who wants to make progress within his working organization has to become familiar with the language and activity of Mathematics. Anyone who simply takes an intelligent interest in the life around him will find that a knowledge of Mathematics will make possible new, fascinating fields of thought—whether it concerns filling in football pools or calculating the possibilities of space travel.

Here under one cover you will find the fundamentals of four common branches of Mathematics: Arithmetic, Algebra, Geometry, and Trigonometry. The book has been carefully planned so that every idea is clearly presented and explained, before moving on to the next. Illustrative examples and problems show how ideas are applied, and the reader's understanding of this is never obscured by tedious calculations; the more involved situation always comes at a secondary stage.

MATHEMATICS MADE SIMPLE contains a number of special features. Perhaps the most important is that throughout the Arithmetic work, not only are the monetary calculations discussed in terms of a decimal currency but the rest of the text is fully metricated in accordance with the recommended International System of Units (S.I.), the modern form of the metric system.

The discussion of Logarithms and Trigonometry is very straightforward and involves none of the usual mystery associated with these topics. The last two chapters offer an entertaining initiation into the Theory of Probability, a subject of increasing importance and endless fascination.

Among the tables in this book you will find: Decimal Equivalents of Sixty-Fourths; Measures, Money, Simple and Compound Interest; Squares and Square Roots, Cubes and Cube Roots; Common Logarithms, Sine, Tangent, Secant to 4-figures. As you become familiar with them you will realize just how far they can all be used to make Mathematics simple.

The care in presentation and detail of discussion makes this book invaluable as basic groundwork for all mathematical study, possibly as a companion reader to one of the recognized courses such as O-level G.C.E. or comparable examinations. It is also readably enjoyable for anyone working independently, whether seeking to recapture forgotten knowledge or studying Mathematics for the first time. All the equipment you need is pencil, paper and interest.

PATRICK MURPHY

Other Made Simple titles by the same author:

Additional Mathematics
Applied Mathematics
Modern Mathematics

As advisory editor

New Mathematics
Statistics

Table of Contents

TEST NO. 1

Check up your knowledge with this test before you begin reading.

1. The rudder of an aeroplane broke off. The part that broke off represented $\frac{2}{5}$ of the length. A piece 6 m long was left intact. What was the length of the part that broke off?
(A) 2 m ......   (B) 4 m ......   (C) 6 m ......   (D) 8 m ......

2. The crew of a boat was increased by $\frac{3}{7}$ of its original number. They then had 117 men. How many men did they have originally?
(A) 84 ......   (B) 77 ......   (C) 91 ......   (D) 105 ......

3. In order to reach 33 m, a fireman's ladder had to be increased by 32 per cent of its length. How long was it?
(A) 22 m ......   (B) 22·5 m ......   (C) 27 m ......   (D) 25 m ......

4. Train travel is $2\frac{1}{2}$ times as fast as boat travel. How long would it take to go 600 km by boat if it takes a train 10 h?
(A) $10\frac{1}{3}$ h ......   (B) 25 h ......   (C) $14\frac{1}{4}$ h ......   (D) 15 h ......

5. A factory has enough oil to last 20 days if 2 drums are used daily. How many drums less must be used daily to make the oil last 30 days?
(A) $\frac{1}{3}$ ......   (B) $\frac{1}{2}$ ......   (C) $\frac{2}{3}$ ......   (D) 1 ......

6. A man took a loan for 1 year and 4 months at 6 per cent interest. At the end of that time he paid £432, which included the loan plus the interest. How much did he originally borrow?
(A) £397·60 ......   (B) £398·00 ......   (C) £400·00 ......   (D) £406·08 ......

7. The formula $C = \frac{5}{9}(F - 32)$ gives Celsius temperature in terms of Fahrenheit. What is the Celsius equivalent for a temperature of 113° on the Fahrenheit scale?
(A) 144° ......   (B) 96° ......   (C) 81° ......   (D) 45° ......

8. The value of 36 coins, 5p and 10p only, is £3·30. Find the number of 10p coins.
(A) 16 ......   (B) 20 ......   (C) 30 ......   (D) 34 ......

9. If it takes 9 men 15 days to complete a construction job, how long would it take if 5 men worked on the job?
(A) 27 days ......   (B) $8\frac{1}{3}$ days ......   (C) 21 days ......   (D) 29 days ......

10. An aeroplane is to be built with a cowling 6 m in length, a tail as long as the cowling plus $\frac{1}{4}$ the length of the body, and a body as long as the cowling and tail together. What will be the overall length of the aeroplane?
(A) 16 m ......   (B) 32 m ......   (C) 48 m ......   (D) 60 m

(Pass mark 70 per cent.)

Compare your result with Test No. 3, p. 243 when you have completed the book.

1

WHOLE NUMBERS

Arithmetic is the science of numbers.

A **whole number** is a digit from 0 to 9, or a combination of digits, such as 17, 428, 1521. Thus it is distinguished from a division or part of a whole number, such as a fraction like $\frac{5}{7}$ or $\frac{16}{9}$.

ADDITION OF WHOLE NUMBERS

You should be able to add whole numbers rapidly. In order to do this you must add mentally. Here are sample tests used for classification purposes. Speed and accuracy count.

You should get a score of 22 out of 25 correct, and should not take longer than two and one-half minutes. If you are not up to this level, use the exercise on the following page for practice.

TEST NO. 2

MENTAL ADDITION

A

1. 12 + 3 = 15	2. 16 + 4 = 20	3. 11 + 7 = 18
4. 24 + 5 = 29	5. 25 + 7 = 32	6. 36 + 6 = 42
7. 74 + 9 = 83	8. 21 + 18 = 39	9. 14 + 15 = 29
10. 32 + 19 = 51	11. 53 + 13 = 66	12. 64 + 28 = 92
13. 59 + 17 = 76	14. 65 + 38 = 103	15. 118 + 48 = 166
16. 139 + 46 = 185	17. 178 + 57 = 235	18. 274 + 89 = 363
19. 457 + 76 = 533	20. 326 + 134 = 460	21. 495 + 179 = 674
22. 697 + 267 = 964	23. 673 + 568 = 1241	24. 878 + 595 = 1473
25. 1578 + 673 = 2251		

B

26. 11 + 4 = 15	27. 15 + 3 = 18	28. 13 + 6 = 19
29. 23 + 5 = 28	30. 25 + 8 = 33	31. 35 + 6 = 41
32. 64 + 9 = 73	33. 19 + 18 = 37	34. 13 + 19 = 32
35. 32 + 29 = 61	36. 63 + 16 = 79	37. 54 + 38 = 92
38. 69 + 27 = 96	39. 75 + 38 = 113	40. 118 + 58 = 176
41. 149 + 36 = 185	42. 178 + 67 = 245	43. 264 + 79 = 343
44. 467 + 66 = 533	45. 336 + 144 = 480	46. 479 + 195 = 674
47. 687 + 257 = 944	48. 693 + 578 = 1271	49. 888 + 585 = 1473
50. 1468 + 724 = 2192		

C

51. 13 + 5 = 18	52. 35 + 3 = 38	53. 43 + 4 = 47
54. 52 + 6 = 58	55. 35 + 7 = 42	56. 47 + 7 = 54
57. 74 + 9 = 83	58. 21 + 28 = 49	59. 15 + 17 = 32
60. 42 + 39 = 81	61. 63 + 18 = 81	62. 64 + 38 = 102
63. 79 + 27 = 106	64. 85 + 48 = 133	65. 116 + 38 = 154
66. 139 + 46 = 185	67. 168 + 47 = 215	68. 254 + 89 = 343
69. 346 + 74 = 420	70. 457 + 134 = 591	71. 579 + 115 = 694
72. 677 + 237 = 914	73. 683 + 568 = 1251	74. 878 + 595 = 1473
75. 1558 + 723 = 2281		

SPEED TEST

PRACTICE IN SIGHT ADDITION

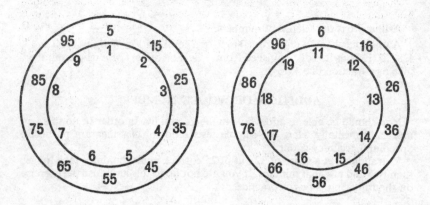

1. Add 1 to each figure in the outer circle; add 2 to each figure; add 3 to each figure; add 4, 5, 6, 7, 8, 9. Thus, mentally, you will say $1 + 5 = 6$, $1 + 15 = 16$, $1 + 25 = 26$, and so on going round the entire circle. Then add $2 + 5$, $2 + 15$, $2 + 25$, $2 + 35$, etc. Continue this until you have added every number from 1 to 9 to every number in the outer circle.

2. Add 11 to each figure in the outer circle. Thus, mentally, you will say $11 + 6 = 17$, $11 + 16 = 27$, $11 + 26 = 37$, $11 + 36 = 47$, and so on round the entire outer circle. Repeat this process for numbers from 12 through 19.

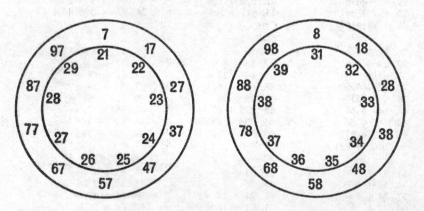

3. Follow same procedure as above using numbers from 21 through 29 as shown in the inner circle.

4. Follow same procedure as above using numbers from 31 through 39 as shown in the inner circle.

COLUMN ADDITION

Practice Exercise No. 1

This exercise is designed to present forty graded examples in column addition. Add Column I from A to B, then from B to C, then from C to D, then from D to E. Repeat for Columns II, III, and IV. Next add Column I from A to C, from B to D, from C to E. Repeat for the other columns. Add Column I from A to D, then from B to E. Repeat process. Finally, add each column from A to E. The complete answers will be found on p. 246.

	I	II	III	IV
A				
	6737	8956	6276	1712
	7726⎱ 10	7735⎱ 9	2985⎱ 10	1814
	2884⎰	6544⎰	4355⎰	2523
	8825	5459	5734	4411
	2201⎱ 10	4893⎱ 9	3756⎱ 9	4515
	4669⎰	6876⎰	7843⎰	5115
	1608	8574	2263⎱ 8	1220
	2599	8328	8545⎰	1418
B				
	5511	1681	9477	1541
	5522	6418	1668	1825
	8113	4527	6322	4236
	2037	2772	9755	1547
	8474	7858	4281	2625
	7745	6785	5727	1608
	3355	5274	2466	4838
	4505	4654	8515	1638
C				
	5754	5737	3594	1518
	2256	4862	5676	2417
	4445	6143	1229	3514
	6652	3688	8163	4656
	1868	6471	2223	2181
	6244	2423	7662	3435
	5471	1584	6141	1615
	4649	7845	8759	2344
D				
	6456	2417	3443	4011
	5554	7989	5682	9122
	3566	8016	1317	3517
	4273	5703	8831	1833
	8622	4298	4247	2328
	2488	1683	4042	4244
	4229	5316	1761	1613
	3698	6235	9278	8999
E				

Acquiring Speed

One way to acquire speed in column addition is to learn to group successive numbers at sight in such a way as to form larger numbers. Learn first to recognize successive numbers that make 10. In Column I of the preceding exercise are many combinations of two numbers adding up to 10. Practise again on this column, picking out these combinations as you go along.

Similarly, in Column II you will find many combinations making 9, while in Column III you will recognize groups adding to 10, 9, and 8. Practise such grouping with these columns also.

If your work calls for any considerable amount of column addition, learn to group numbers that add to other sums—11, 12, 13, 14, 15, etc., as well as any total at all that is less than 10. You need not limit yourself to groups of only two numbers. Learn to combine three or even more numbers at sight.

Persons who are exceptionally rapid at addition add two columns at a time (some do even three). Adding two columns at once is not as difficult as it may seem. Column IV of the preceding exercise has been specially designed as fairly easy practice in two-column addition. Try it! If you find that this method is not beyond your abilities practise using it as occasions arise.

Partial Totals

In actual work when you have long columns to add, write down your entire sum for each column separately instead of merely putting down a single digit and carrying the others. For instance, partial totals for Column I of the exercise would be set down thus:

$$\begin{array}{r} 166 \\ 1370 \\ 15200 \\ 142000 \\ \hline 158736 \end{array}$$

By this procedure you treat each column as an individual sum and to that extent simplify the work of checking.

Horizontal Addition

The method of partial totals is especially useful where the figures to be added are not arranged in column form—especially when they are on separate pieces of paper, such as invoices, ledger pages, etc.

In such a case you first go through the papers adding up only the figures in the units column; then you go through them again for the tens, for the hundreds, etc., setting down each successive partial total one place to the left. This procedure is usually very much quicker than the alternative one of listing the figures to be added.

Practice Exercise No. 2

Add the following, using the method of partial totals:
1. 67 + 28 + 24 + 12 + 55 + 82 + 87 + 34 =
2. 524 + 616 + 546 + 534 + 824 + 377 + 882 + 665 =
3. 551 + 473 + 572 + 468 + 246 + 455 + 264 + 455 =
4. 2642 + 6328 + 2060 + 9121 + 3745 + 5545 + 6474 + 5567 =
5. 2829 + 7645 + 1989 + 1237 + 4555 + 4652 + 8419 + 6463 =
6. 28 988 + 76 546 + 88 164 + 27 654 + 54 636 + 21 727 + 85 415 + 69 754 =

SUBTRACTION OF WHOLE NUMBERS

Subtraction is the process of finding the difference between two numbers. This is the same as finding out how much must be added to one number, called the **subtrahend**, to equal another, called the **minuend**.

For instance, subtracting 12 from 37 leaves a difference of 25 because we must add 25 to the subtrahend 12 to get the minuend 37. This may be written: 37 — 12 — 25; or:

$$37 \quad \text{(minuend)}$$
$$- 12 \quad \text{(subtrahend)}$$
$$\overline{25} \quad \text{(difference)}$$

The **minus sign** (—) indicates subtraction.

Here are several ways in which subtraction is indicated in verbal problems. They all mean the same as *subtract 4 from 16*.

(*a*) How much must be added to 4 to give 16?   Ans. 12
(*b*) How much more than 4 is 16?   Ans. 12
(*c*) How much less than 16 is 4?   Ans. 12
(*d*) What is the difference between 4 and 16?   Ans. 12

The circle arrangement on p. 4 may be used for practice in subtraction. In each case subtract the smaller number from the larger.

Different Methods of Subtraction

There are two different routines for subtraction in accepted use—the *borrow method* and the *carry method*.

Use whichever method you were taught at school, since that is the one which will come most easily to you. To understand the difference between the two methods consider their application to the example: from 9624 subtract 5846.

The **borrow method** proceeds thus:

$$8\,5\,1$$
$$9\cancel{6}24$$
$$5846$$
$$\overline{3778}$$

This may be read: subtracting 6 from 14 leaves 8, 4 from 11 leaves 7, 8 from 15 leaves 7, 5 from 8 leaves 3.

The **carry method** goes like this:

$$9624$$
$$5\cancel{8}\cancel{4}6$$
$$6\,9\,5$$
$$\overline{3778}$$

This may be read: subtracting 6 from 14 leaves 8, 5 from 12 leaves 7, 9 from 16 leaves 7, 6 from 9 leaves 3.

The *borrow method* is also known as the method of *decomposition* and the *carry method* as *equal addition*.

MULTIPLICATION OF WHOLE NUMBERS

Multiplication is a short method of adding a number to itself a given number of times.

The given number is called the **multiplicand.** The number of times it is to be added is called the **multiplier.** The result is called the **product.**

For instance, 4 times 15 means 15 + 15 + 15 + 15, or 60. This may be written: 15 × 4 = 60; or:

$$15 \quad \text{(multiplicand)}$$
$$\times \ 4 \quad \text{(multiplier)}$$
$$\overline{ 60} \quad \text{(product)}$$

The **sign of multiplication** is × ; it is read *multiply by.*

To multiply well and to use this ability in solving problems you must know by heart the product of any two numbers from 1 to 12. Below are tables of multiplication from 2 to 12; if you don't know all of them, memorize them now.

Table I. Multiplication Table from Two to Twelve

No.	×2	×3	×4	×5	×6	×7	×8	×9	×10	×11	×12
1	2	3	4	5	6	7	8	9	10	11	12
2	4	6	8	10	12	14	16	18	20	22	24
3	6	9	12	15	18	21	24	27	30	33	36
4	8	12	16	20	24	28	32	36	40	44	48
5	10	15	20	25	30	35	40	45	50	55	60
6	12	18	24	30	36	42	48	54	60	66	72
7	14	21	28	35	42	49	56	63	70	77	84
8	16	24	32	40	48	56	64	72	80	88	96
9	18	27	36	45	54	63	72	81	90	99	108
10	20	30	40	50	60	70	80	90	100	110	120
11	22	33	44	55	66	77	88	99	110	121	132
12	24	36	48	60	72	84	96	108	120	132	144

Short Cuts

There are so many short cuts in multiplication that they are an interesting study in themselves and throw a great deal of light on the general subject of the properties of numbers. It is beyond the plan of this book to treat this subject in detail, but the interested student may profitably experiment for himself with the devices presented on this page. More are considered in the chapters dealing with decimals and algebra.

Expanded multiplication table. Acquire a more extensive knowledge of the multiplication table. Aim to master it to 25 × 25. It is easy to learn ×13, ×15, ×20, and 25 × 25. Gradually include other numbers, as in Table IA.

Small multipliers of two places. Multiply 326 by 127:

```
        326
        127

       2282   instead of    2282
      39120                  6520
                            32600
      41402
                            41402
```

That is, multiply at once by 12, instead of multiplying first by 2 and then by 1.

Multipliers involving multiples of each other. Multiply 456 by 279.

```
        456
        279

       4104   instead of    4104
     123120                31920
                           91200
     127224
                           127224
```

Recognizing that 27 in the multiplier is 3 × 9, you here multiply the 4104 by 3 instead of doing two separate calculations of ×7 and ×2.

Multiplying by 11. Multiply 24 by 11. 2 + 4 = 6. Place a 6 between the 2 and the 4 and write the answer, 264. If the two numbers of the multiplicand add up to more than 10, add 1 to the hundreds figure in the answer. Multiply 48 × 11. 4 + 8 = 12. Make the first figure of the answer 5 and place the 2 of the 12 between 5 and 8 instead of between 4 and 8. Answer, 528.

Multiplying by 'near' figures. To multiply 36 × 49, recognize that 49 is *near* 50. Then since 36 × 50 = 1800 mentally, 36 × 49 must be only 36 less, or 1800 − 36 = 1764. Similarly, to multiply 3746 × 9988, recognize that 9988 is only 12 less than 10 000. Therefore, write down 37 460 000 and subtract 44 952 which is 3746 × 12, to get 37 415 048, the quick answer.

Practice Exercise No. 3

Use short-cut methods where applicable.
1. 32 × 47 =
2. 123 × 43 =
3. 182 × 52 =
4. 217 × 21 =
5. 136 × 24 =
6. 1112 × 893 =
7. 1457 × 369 =
8. 48 × 48 =
9. 83 × 53 =
10. 115 × 115 =
11. 4562 × 1211 =
12. 3765 × 648 =
13. 87 × 87 =
14. 96 × 46 =
15. 997 × 327 =

Mathematics Made Simple

Table IA. *Other Multiplication Tables*

No.	×13	×14	×15	×16	×19	×21	×24	×25
1	13	14	15	16	19	21	24	25
2	26	28	30	32	38	42	48	50
3	39	42	45	48	57	63	72	75
4	52	56	60	64	76	84	96	100
5	65	70	75	80	95	105	120	125
6	78	84	90	96	114	126	144	150
7	91	98	105	112	133	147	168	175
8	104	112	120	128	152	168	192	200
9	117	126	135	144	171	189	216	225
10	130	140	150	160	190	210	240	250
11	143	154	165	176	209	231	264	275
12	156	168	180	192	228	252	288	300
13	169	182	195	208	247	273	312	325
14	182	196	210	224	266	294	336	350
15	195	210	225	240	285	315	360	375
16	208	224	240	256	304	336	384	400
17	221	238	255	272	323	357	408	425
18	234	252	270	288	342	378	432	450
19	247	266	285	304	361	399	456	475
20	260	280	300	320	380	420	480	500
21	273	294	315	336	399	441	504	525
22	286	308	330	352	418	462	528	550
23	299	322	345	368	437	483	552	575
24	312	336	360	384	456	504	576	600
25	325	350	375	400	475	525	600	625

DIVISION OF WHOLE NUMBERS

Division is the process of finding how many times one number, called the **divisor**, goes into another number, called the **dividend**. Hence, division is 'multiplication in reverse', but its answer is called the **quotient**.

For instance, since 15 × 4 = 60, the dividend 60 divided by the divisor 4 produces the quotient 15. This may be written: 60 ÷ 4 = 15, or:

$$\begin{array}{r} 15 \text{ (quotient)} \\ \text{(divisor) } 4)\overline{60} \text{ (dividend)} \end{array}$$

The sign of division is ÷, and is read *divided by*.

If the answer in the above case were not at once obvious it could be obtained as follows: 6 ÷ 4 = 1 with 2 left over; write 1 above 6 and carry over the 2 to make the next partial dividend 20; then 20 ÷ 4 = 5, so write 5 over the 0 obtaining the exact quotient 15.

When a divisor does not thus divide into a dividend an exact number of times the last number left over is called the **remainder**. For instance, 63 ÷ 4 = 15 with the *remainder* 3.

The method of the above examples is called **short division** because the intermediate steps can be carried out mentally. The method called **long division** is exactly the same, but its intermediate steps are written out as follows:

$$\begin{array}{r} 279 \text{ (quotient)} \\ \text{(divisor) } 456)\overline{127229} \text{ (dividend)} \\ 912 \\ \hline 3602 \\ 3192 \\ \hline 4109 \\ 4104 \\ \hline 5 \text{ (remainder)} \end{array}$$

The last digits, 2 and 9, of the dividend in such an example are said to be 'brought down' in the intermediate steps.

Computing Averages

To find the average of several quantities, *divide their sum by the number of quantities.*

EXAMPLE: What was the average attendance at a church if the daily attendance from Monday to Friday was as follows: 462, 548, 675, 319, 521?

SOLUTION:

$$\begin{array}{r} 462 \\ 548 \\ 675 \\ 319 \\ 521 \\ \hline 2525 \end{array}$$

EXPLANATION: Add the quantities and divide the sum of 2525 by the number of days, 5. The average attendance is the quotient 505.

2525 ÷ 5 = 505. ANS.

Practice Exercise No. 4

1. $7258 \div 19 =$	2. $13\,440 \div 35 =$	3. $21\,492 \div 53 =$
4. $19\,758 \div 37 =$	5. $47\,085 \div 73 =$	6. $45\,522 \div 54 =$
7. $42\,201 \div 46 =$	8. $66\,822 \div 74 =$	9. $53\,963 \div 91 =$

CHECKING ANSWERS

Additions are checked by adding in the opposite direction.

Subtraction is checked by adding the subtrahend to the remainder. The sum should equal the minuend. In other words, in a completed subtraction example the sum of the middle and bottom figures should equal the top figure.

Simple multiplication may be quickly checked by reversing multiplicand and multiplier.

$$
\begin{array}{rcr}
\text{Prove} \quad 36 & \text{by} & 57 \\
57 & & 36 \\
\hline
252 & & 342 \\
1800 & & 1710 \\
\hline
2052 & & 2052 \\
\end{array}
$$

Simple division may be checked by multiplying divisor by quotient and adding the remainder, if any.

For long examples in multiplication, however, a good method of checking is that which is known as *casting out nines*. The same method is also applicable in principle to addition and subtraction, but nothing would be gained by using it for the latter, while for the former it would be cumbersome.

Casting Out Nines

This method of checking does not present an absolute proof of correctness but only a presumable one. It can fail, however, only if the solution of an example contains two errors that exactly offset one another. Since the chance of this happening is negligible, the method may be considered almost completely reliable.

The method of casting out nines is based on a peculiar property of the number 9. This is that—

The sum of the digits of a number (or the sum of these digits minus any multiple of 9) is equal to the remainder that is left after dividing the original number by nine.

No.	Sum of digits		Remainder after ÷ 9
21	3		3
32	5		5
62	8		8

	Sum of digits minus multiple of 9		
27	9	0	0
54	9	0	0
72	9	0	0
156	12	3	3
8765	26	8	8

In the cases of 27, 54, and 72, note that *when the nines have been cast out of the multiple of 9 the remainder is 0.*

In the case of 156 and 8765, note that you need only add the digits in the figure representing the sum of the original digits to arrive at the desired remainder.

In applying the method of casting out nines we are concerned only with the *remainders*. By the principles explained above check the remainders here given:

	Remainder		Remainder
24	6	1 466	8
36	0	16 975	1
58	4	206 534	2
138	3	7 898 875	7
257	5	56 879 876	2

The application of the method to multiplication and division is very simple and rapid, but before proceeding to explain this it will be well to do the following exercise.

Practice Exercise No. 5

What are the remainders after nines have been cast out of the following?

1. 35	2. 87	3. 126	4. 284
5. 982	6. 3465	7. 5624	8. 8750
9. 46 824	10. 65 448	11. 365 727	12. 584 977
13. 862 425	14. 7 629 866	15. 8 943 753	

To check multiplication, *multiply the remainders representing the original numbers, cast nines out of this product and compare the remainder with the remainder representing the answer.*

EXAMPLES:

Remainders

(1)

$$\begin{array}{r} 35 \\ \times\ 24 \\ \hline 140 \\ 700 \\ \hline 840 \end{array} \quad \begin{array}{r} -\ \ 8 \\ -\ \times\ 6 \\ \hline 48\ -\ 3 \\ \\ \hline -\ \ \ 3\ \checkmark \end{array}$$

(2)

$$\begin{array}{r} 54 \\ \times\ 38 \\ \hline 432 \\ 1\ 620 \\ \hline 2\ 052 \end{array} \quad \begin{array}{r} -\ \ 0 \\ -\ \times\ 2 \\ \hline 0\ -\ 0 \\ \\ \hline -\ \ \ 0\ \checkmark \end{array}$$

(3)

$$\begin{array}{r} 97\ 653 \\ 84\ 296 \\ \hline 585\ 918 \\ 8\ 788\ 770 \\ 19\ 530\ 600 \\ 390\ 612\ 000 \\ 7\ 812\ 240\ 000 \\ \hline 8\ 231\ 757\ 288 \end{array} \quad \begin{array}{r} -\ \ 3 \\ -\ \times\ 2 \\ \hline 6\ -\ 6 \\ \\ \\ \\ \\ \hline -\ \ \ 6\ \checkmark \end{array}$$

Note from the second of these examples that if *either* of the original numbers has a remainder of 0, the answer will have a remainder of 0.

The third example illustrates how easily the method may be applied to the most difficult multiplications.

If an answer is wrong, the method will also help you to find out quickly *where* the mistake has been made, since the casting out of nines can be applied to every step of the procedure. This is illustrated by the following example, which has been purposely made incorrect.

		Remainder should equal	
	358		7
	246		3
			—
	2148	7×6 or	6
wrong	14420	7×4 or	1
	71600	7×2 or	5
wrong	88168	7×3 or	3

To check division, *subtract the remainder from the dividend, then check the multiplication of quotient by divisor.*

EXAMPLE: Is $705\,776 \div 728 = 969\frac{344}{728}$ correct?

Remainders after
casting out 9's

$$705\,776$$
$$- \quad 344$$
$$\overline{705\,432} \quad — \qquad\qquad 3$$

$$969 \quad — \qquad 6$$
$$\times\ 728 \qquad \times\ 8$$
$$\overline{\qquad\qquad} \qquad \overline{\qquad}$$
$$48 \quad — \quad 3 \ \sqrt{}$$

The 3s check, therefore the answer may be considered correct.

Practice Exercise No. 6

Check by casting out nines whether each of the following is right or wrong.

1. $92 \times 61 = 5612$
2. $88 \times 72 = 6336$
3. $72 \times 37 = 2665$
4. $35 \times 99 = 3465$
5. $6284 \times 192 = 1\,236\,528$
6. $1938 \times 421 = 815\,898$
7. $664 \times 301 = 199\,864$
8. $736 \times 428 = 315\,008$
9. $893 \times 564 = 502\,652$
10. $1084 \times 839 = 892\,706$
11. $985 \times 916 = 902\,260$
12. $3241 \times 326 = 956\,566$
13. $47\,974 \div 83 = 578$
14. $21\,954 \div 67 = 327\frac{45}{67}$
15. $88\,445 \div 95 = 931$
16. $90\,159 \div 123 = 732\frac{64}{123}$
17. $229\,554 \div 234 = 981$
18. $307\,395 \div 345 = 890$
19. $59\,448 \div 96 = 619\frac{1}{4}$
20. $66\,822 \div 86 = 779$
21. $47\,320 \div 52 = 910$
22. $45\,414 \div 62 = 732\frac{30}{62}$
23. $78\,027 \div 93 = 839$
24. $31\,806 \div 38 = 839$

Practice Exercise No. 7

ADDITION, SUBTRACTION, MULTIPLICATION, AND DIVISION

Note: For each problem multiple answers are given, of which only one is correct. After you solve the problem check the answer that agrees with your solution.

1. A dealer bought 3 loads of coal of mass 6242 kg, 28 394 kg, and 143 686 kg. How much did he buy in all?
    (A) 76 324 ......   (B) 178 322 ......   (C) 268 422 ......   (D) 165 432 ......

2. If your Army pay is £152 a month, how much will you earn in a year?
    (A) £1800 ......   (B) £1884 ......   (C) £1824 ......   (D) £1956 ......

3. A company marched 48 km in 5 days. The first day they marched 12 km, the second day 9 km, the third day 7 km, the fourth day 9 km. How many kilometres did they march the last day?
    (A) 11 ......   (B) 8 ......   (C) 16 ......   (D) 20 ......

4. How many packets of cigarettes can you buy for £3·75 at the rate of 6 for £1·5?
    (A) 5 ......   (B) 10 ......   (C) 15 ......   (D) 20 ......

5. If a motor car travels 450 m in 15 seconds, how many metres does it go in $\frac{1}{3}$ of a second?
    (A) 30 ......   (B) 90 ......   (C) 60 ......   (D) 10 ......

6. Your grades on 5 tests were 80%, 90%, 70%, 60%, and 50%. What is the average of your 5 grades?
    (A) 80 ......   (B) 70 ......   (C) 75 ......   (D) 85 ......

7. A hangar is 100 m long, 50 m wide, and 10 m high. Estimate the cost of heating it at the rate of £25 per 1000 m³ per season.
    (A) £125 ......   (B) £250 ......   (C) £1250 ......   (D) £2250 ......

8. It takes 50 kg of cement to cover 10 m². How many kg of cement will be needed to cover a rectangular area 25 m by 10 m?
    (A) 250 ......   (B) 1500 ......   (C) 2000 ......   (D) 1250 ......
Note: In solving problems such as No. 8 *determine first what one unit will do.* In this case:
If 10 m² is covered by 50 kg then 1 m² is covered by 5 kg.

9. Before leaving his 142 km² estate, Mr. Curran sold 22 km² to Mr. Brown, 30 km² to Mr. Jones, 14 km² to Mr. Smith, and 16 km² to Mr. Ives. How many square kilometres did he have left?
    (A) 30 ......   (B) 40 ......   (C) 50 ......   (D) 60 ......

10. Two machinists operating the same lathe work 10 h each on a day- and on a night-shift respectively. One man turns out 400 pieces an hour, the other 600 pieces per hour. What will be the difference in their output at the end of 30 days?
    (A) 10 000 ......   (B) 6000 ......   (C) 60 000 ......   (D) 40 000 ......

11. You are given 12 days to drive to a destination 2400 km away. For the first 6 days you do 200 km a day. Due to an accident you can't drive for 2 days. What is the average number of kilometres per day that you have to drive to reach your destination on time?
    (A) 100 ......   (B) 200 ......   (C) 300 ......   (D) 400 ......

12. If, out of your annual income of £820, you pay £2 weekly on your car, send home £5 weekly, and pay a monthly insurance of £4, how much will that leave you to spend on a monthly basis?
    (A) 30 ......   (B) 32 ......   (C) 34 ......   (D) 37 ......

CHAPTER TWO

COMMON FRACTIONS

Although the product of any two whole numbers (Chapter One) is always another whole number, the quotient of two whole numbers may, or may not, be a whole number. For instance, $2 \times 3 = 6$, and $6 \div 3 = 2$; but $2 \div 3$ and $3 \div 2$ do not result in whole numbers. In these latter cases, therefore, we call the quotients **fractional numbers,** or, for short, **fractions.**

LANGUAGE OF FRACTIONS

More particularly, a **common fraction** is one in which the dividend, called the fraction's **numerator,** is written over the divisor, called the fraction's **denominator,** with a slanting or horizontal line between them to indicate the intended division. Thus, in common fraction form:

$$2 \div 3 = \tfrac{2}{3}$$

or with 2 as the numerator over 3 as the denominator.

From this example we see that $\tfrac{2}{3}$ by definition means $2 \div 3$, or '2 divided by 3'. Likewise, $\tfrac{3}{2}$ by definition means $3 \div 2$, or '3 divided by 2'. However, we shall soon see that, arithmetically:

$$\tfrac{2}{3} = \tfrac{1}{3} \times 2, \quad \text{and} \quad \tfrac{3}{2} = \tfrac{1}{2} \times 3$$

For this reason we commonly read the symbol '$\tfrac{2}{3}$' as '*two thirds*', and the symbol '$\tfrac{3}{2}$' as '*three halves*', etc.

A **proper fraction** has a value less than 1 (one) because, by definition, it has a numerator smaller than its denominator. Examples: $\tfrac{2}{3}, \tfrac{1}{4}, \tfrac{3}{5}$.

A so-called **improper fraction** has a value greater than 1 because, by definition, it has a numerator larger than its denominator. Examples: $\tfrac{3}{2}, \tfrac{7}{4}, \tfrac{31}{9}$. But it is quite 'proper' arithmetically to treat these fractions just like others.

A **mixed number** consists of a whole number and a fraction written together with the understanding that they are to be added to each other. Examples: $1\tfrac{3}{4}$ which means $1 + \tfrac{3}{4}$, and $2\tfrac{5}{7}$ which means $2 + \tfrac{5}{7}$.

A **simple fraction** is one in which both numerator and denominator are whole numbers, as in all the above examples of fractions.

A **complex fraction** is one in which either the numerator or the denominator is a fraction or a mixed number, or in which both the numerator and the denominator are fractions or mixed numbers. Examples are:

$$\frac{\tfrac{1}{4}}{2}, \frac{2}{\tfrac{2}{4}}, \frac{\tfrac{1}{4}}{\tfrac{1}{3}}, \frac{1\tfrac{1}{2}}{2}, \frac{2}{3\tfrac{1}{4}}, \frac{1\tfrac{1}{2}}{2\tfrac{3}{4}}$$

A FUNDAMENTAL RULE OF FRACTIONS

In much of our work with fractions we need to apply the **fundamental rule** that: *When the numerator and denominator of a fraction are both multiplied or divided by the same number the value of the fraction remains unchanged.* Examples are:

$$\frac{1}{2} = \frac{1 \times (2)}{2 \times (2)} = \frac{2}{4} = \frac{2 \times (25)}{4 \times (25)} = \frac{50}{100}, \text{ etc.;}$$

or

$$\frac{50}{100} = \frac{50 \div (10)}{100 \div (10)} = \frac{5}{10} = \frac{5 \div (5)}{10 \div (5)} = \frac{1}{2}, \text{ etc.}$$

From these examples we see that any common fraction can be written in as many different forms as we wish, provided always that the numerator, divided by the denominator, yields the same quotient. That particular form of a fraction which has the smallest possible whole numbers for its numerator and denominator is called **the fraction in its lowest terms**. Thus, the above fraction $\frac{50}{100}$, or $\frac{5}{10}$, or $\frac{2}{4}$, is $\frac{1}{2}$ in its lowest terms.

REDUCTION OF FRACTIONS

To change a fraction to its lowest terms, divide its numerator and its denominator by the largest whole number which will divide both exactly.

EXAMPLE: Reduce $\frac{12}{30}$ to its lowest terms

SOLUTION:
$$\frac{12}{30} = \frac{12 \div 6}{30 \div 6} = \frac{2}{5}, \quad \text{Ans.}$$

When you do not at once see the largest number which can be divided exactly into a large numerator and denominator, reduce the fraction by repeated steps, as follows:

EXAMPLE: Reduce $\frac{128}{288}$ to lowest terms.

SOLUTION:
$$\frac{128}{288} = \frac{128 \div 4}{288 \div 4} = \frac{32}{72} = \frac{32 \div 8}{72 \div 8}$$

$$= \frac{4}{9}, \quad \text{Ans.}$$

This is equivalent to:

$$\frac{128}{288} = \frac{128 \div 32}{288 \div 32} = \frac{4}{9}, \quad \text{Ans. (the same)}$$

When a fraction has been reduced to its lowest terms the numerator and the denominator of the fraction are said to be *prime to each other*.

Numbers are **prime to each other** when there is no other whole number that is contained exactly in both of them. Thus, 8 and 15 are *prime to each other* because there is no number that will divide both of them without a remainder.

A number that cannot be divided by any other number at all except 1 and itself is called a **prime number**. Thus, 1, 2, 3, 5, 7, 11, 13, 17, and 19 are prime numbers. But 4, 6, 8, 9, 10, 12, 14, 15, 16, 18, and 20 are not prime numbers because each of these can be divided by one or more smaller numbers.

A number that is contained exactly in two or more other numbers is called a *common divisor* or *common factor* of these numbers.

Greatest Common Divisor

The largest number that is contained exactly in two or more other numbers is called the **greatest common divisor (G.C.D.)** of these numbers. It is often called the **highest common factor (H.C.F.)**, but throughout this book the term **G.C.D.** will be used.

A knowledge of how to find the greatest common divisor of two or more numbers is necessary in order to perform various operations with fractions.

There are two methods of doing so—the **factoring method** and the **direct division method**. The *factoring method* is the handier where the numbers involved are small, and it has the added advantage that it can be applied at a single operation to more than two numbers.

To find the greatest common divisor by factoring, *arrange the numbers in a line and divide them by any prime number that is exactly contained in all of them; divide the quotients in the same way; continue thus to divide the quotients until a quotient is obtained which contains no common divisor; multiply the divisors that have been used to arrive at the desired greatest common divisor.*

EXAMPLE: Find the greatest common divisor of 42, 60, and 84.

SOLUTION:

$$
\begin{array}{r}
2)\overline{42,\ \ 60,\ \ 84} \\
3)\overline{21,\ \ 30,\ \ 42} \\
7,\ \ 10,\ \ 14 \\
2 \times 3 = \text{G.C.D.}
\end{array}
$$

EXPLANATION: The greatest common divisor of two or more numbers is equal to the product of all the *prime* factors (divisors) that are common to *all* of them. 7 and 14 are divisible by 7, and 10 and 14 by 2; but no number will divide all three of these quotients. Hence the only prime factors are 2 and 3.

The foregoing method is not readily applicable when the numbers are large and thus not subject to easy analysis. In such cases the following method is employed.

To find the greatest common divisor by direct division, *divide the larger number by the smaller, then divide the remainder into the smaller, and continue to divide remainders into previous divisors until no remainder is left; the last divisor used is the desired greatest common divisor.*

EXAMPLE: What is the greatest common divisor of 323 and 391?

SOLUTION:

$$
\begin{array}{l}
323)391(1 \\
\quad 323 \\
\quad \overline{} \\
\quad\quad 68)323(4 \\
\quad\quad\ \ 272 \\
\quad\quad\ \ \overline{} \\
\quad\quad\quad\quad 51)68(1 \\
\quad\quad\quad\quad\ \ 51 \\
\quad\quad\quad\quad\ \ \overline{} \\
\quad\quad\quad\quad\quad\quad 17)51(3 \\
\quad\quad\quad\quad\quad\quad\ \ 51 \quad\quad 17 = \text{G.C.D.} \\
\quad\quad\quad\quad\quad\quad\ \ \overline{}
\end{array}
$$

EXPLANATION: Since the greatest common divisor is contained in both 391 and 323, it must be contained in the difference between these numbers or 68; it must also be contained in any multiple (product) of 68 as well as in the difference between such a multiple and 323. Hence it must be contained in 51. Hence it must be contained in 68, 51 and in the difference between them or 17. Since no smaller number is contained in 17, this must be the required greatest common divisor.

If the successive divisions continue until a remainder of 1 is obtained this means that the original numbers have no common divisor.

If **more than two numbers are originally given**, *find the greatest common divisor of any two of them, then find the greatest common divisor of this result and another of the original numbers, continuing in this way until all the original numbers have been used. The last divisor is the required greatest common divisor.*

Practice Exercise No. 8

Find the G.C.D. of the following:

1. 12, 16, 28   2. 12, 72, 96   3. 14, 21, 35   4. 15, 45, 81
5. 32, 48, 80   6. 48, 60   7. 63, 99   8. 54, 234
9. 33, 165   10. 256, 608   11. 24, 32, 104   12. 36, 90, 153
13. 48, 120, 168   14. 64, 256, 400   15. 81, 117, 120

Practice Exercise No. 9

Reduce the following fractions to lowest terms:

1. $\frac{8}{12} =$ 2. $\frac{8}{20} =$ 3. $\frac{6}{15} =$ 4. $\frac{9}{15} =$
5. $\frac{12}{32} =$ 6. $\frac{16}{44} =$ 7. $\frac{10}{12} =$ 8. $\frac{13}{52} =$
9. $\frac{10}{16} =$ 10. $\frac{18}{54} =$ 11. $\frac{20}{36} =$ 12. $\frac{42}{126} =$
13. $\frac{15}{18} =$ 14. $\frac{144}{244} =$ 15. $\frac{8}{56} =$

To raise the denominator of a given fraction to a required denominator, *divide the denominator of the given fraction into the required denominator, then multiply both terms of the given fraction by the quotient.*

EXAMPLE: Change $\frac{1}{4}$ to sixty-fourths.

$$64 \div 4 = 16$$

$$\frac{1}{4} = \frac{1 \times 16}{4 \times 16} = \frac{16}{64}$$

EXPLANATION: 64 is the required denominator; 4 is the given denominator; 1 and 4 must each be multiplied by the quotient 16 to give the stepped-up fraction.

Practice Exercise No. 10

Change the following fractions to equivalent fractions having the indicated denominator.

1. $\frac{1}{2}$ to 8ths =   2. $\frac{1}{3}$ to 12ths =   3. $\frac{2}{5}$ to 20ths =   4. $\frac{4}{9}$ to 81sts =
5. $\frac{3}{6}$ to 48ths =   6. $\frac{2}{7}$ to 49ths =   7. $\frac{4}{32}$ to 64ths =   8. $\frac{5}{13}$ to 78ths =
9. $\frac{3}{8}$ to 24ths =   10. $\frac{2}{3}$ to 45ths =   11. $\frac{2}{6}$ to 36ths =   12. $\frac{11}{12}$ to 60ths =
13. $\frac{14}{25}$ to 75ths =   14. $\frac{9}{11}$ to 88ths =   15. $\frac{5}{12}$ to 96ths =   16. $\frac{7}{17}$ to 68ths =

To change an improper fraction to a whole or mixed number, *divide the numerator by the denominator and place the remainder over the denominator.*

EXAMPLE: Change $\frac{19}{5}$ to a mixed number.

$$\frac{19}{5} \text{ means } 19 \div 5 \qquad 5\overline{)19} \atop 3\frac{4}{5} \qquad \frac{19}{5} = 3\frac{4}{5}$$

Practice Exercise No. 11

Change to whole or mixed numbers:

1. $\frac{12}{5} =$ 2. $\frac{14}{7} =$ 3. $\frac{19}{12} =$ 4. $\frac{43}{5} =$ 5. $\frac{52}{8} =$
6. $\frac{114}{76} =$ 7. $\frac{32}{14} =$ 8. $\frac{28}{6} =$ 9. $\frac{19}{4} =$ 10. $\frac{21}{7} =$
11. $\frac{82}{41} =$ 12. $\frac{96}{6} =$

To change a mixed number to an improper fraction, *multiply the whole number by the denominator of the fraction, add the numerator to this product and place the sum over the denominator.*

EXAMPLE 1: $\quad 2\frac{7}{8} = \dfrac{8 \times 2 + 7}{8} = \dfrac{23}{8},$ ANS.

EXAMPLE 2: $\quad 4\frac{3}{5} = \dfrac{5 \times 4 + 3}{5} = \dfrac{20 + 3}{5}$

$$= \frac{23}{5}, \text{ ANS.}$$

Practice Exercise No. 12

Change to improper fractions:

1. $2\frac{3}{4} =$ 2. $3\frac{1}{4} =$ 3. $4\frac{4}{5} =$ 4. $5\frac{3}{8} =$ 5. $12\frac{2}{3} =$
6. $18\frac{3}{4} =$ 7. $19\frac{5}{6} =$ 8. $16\frac{1}{6} =$ 9. $12\frac{2}{7} =$ 10. $13\frac{3}{7} =$
11. $14\frac{1}{3} =$ 12. $22\frac{2}{5} =$

ADDITION AND SUBTRACTION OF FRACTIONS

Just as you cannot add or subtract numbers of feet and numbers of yards until you have first reduced both to a common unit such as feet, yards, or inches, you cannot add or subtract unlike fractions until you have first reduced them to a common denominator.

To add unlike fractions, *first change them to equivalent fractions with the same denominator. Then add the numerators and put the sum over the common denominator.*

EXAMPLE 1: Find $\frac{1}{2}$ plus $\frac{3}{4}$.

SOLUTION:

$$\frac{1}{2} + \frac{3}{4} = \frac{2}{4} + \frac{3}{4} = \frac{2+3}{4} = \frac{5}{4}, \text{ or } 1\frac{1}{4}, \quad \text{ANS.}$$

EXAMPLE 2: Find $\frac{1}{3}$ plus $\frac{3}{4}$.

SOLUTION:

$$\frac{1}{3} + \frac{3}{4} = \frac{4}{12} + \frac{9}{12} = \frac{4+9}{12} = \frac{13}{12}, \text{ or } 1\frac{1}{12}, \quad \text{ANS.}$$

EXPLANATION: The smallest number that contains both 3 and 4 is 12. We find that number by thinking of multiples of 4, the higher of the two denominators, as 4, 8, 12, 16. We stop as soon as we come to the first number that also contains 3. Then we work as follows:

$$\text{For } \frac{1}{3}: \quad 12 \div 3 = 4; \text{ so } \frac{1}{3} = \frac{4 \times 1}{4 \times 3} = \frac{4}{12}.$$

$$\text{For } \frac{3}{4}: \quad 12 \div 4 = 3; \text{ so } \frac{3}{4} = \frac{3 \times 3}{3 \times 4} = \frac{9}{12}, \text{ etc.}$$

Lowest Common Denominators

The 12 in the foregoing calculation is called the **lowest common denominator** of the fractions and is abbreviated **L.C.D.**

The term *lowest common denominator* is limited to use in connexion with fractions. Numerically it is identical with the **least common multiple (L.C.M.)** of the given denominators. The latter term has a more general use in mathematics. The *least common multiple* of two or more numbers is the smallest number that can be exactly divided by all the given numbers. Thus, 12 is the least common multiple of 3 and 4; 45 is the least common multiple of 9 and 15. When applied to fractions, 12 is the least common *denominator* of $\frac{1}{3}$ and $\frac{1}{4}$; 45 is the least common denominator of $\frac{1}{9}$ and $\frac{1}{15}$.

To find the least common multiple of two numbers, *first determine their greatest common divisor; divide the numbers by this, and multiply together the resulting quotients and the greatest common divisor.*

EXAMPLE 1: What is the least common multiple of 12 and 16?

$$
\begin{array}{r}
2)\underline{12, \ 16} \\
2)\underline{6, \ \ 8} \\
3, \ \ 4
\end{array}
$$
$$\text{G.C.D.} = 2 \times 2 = 4.$$
$$\text{L.C.M.} = 4 \times 3 \times 4 = 48.$$

EXAMPLE 2: What is the least common multiple of 54 and 81?

$$
\begin{array}{l}
54)81(1 \\
\ \ \underline{54} \\
27)54(2 \\
\ \ \ \underline{54} \qquad \text{G.C.D.} = 27. \\
\end{array}
$$

$$
\begin{array}{l}
27)\underline{54, \ 81} \\
\ \ \ 2, \ \ 3 \qquad \text{L.C.M.} = 27 \times 2 \times 3 = 162.
\end{array}
$$

The least common multiple of more than two numbers can be found by factoring, but in this case we must be careful not only to use the divisors that are contained in all the given numbers but also any divisors that may be contained in two or more of them.

EXAMPLE 3: What is the least common multiple of 6, 8, and 12?

$$
\begin{array}{r}
2)\underline{6, \ 8, \ 12} \\
2)\underline{3, \ 4, \ \ 6} \\
3)\underline{3, \ 2, \ \ 3} \\
1, \ 2, \ \ 1
\end{array}
$$
$$\text{L.C.M.} = 2 \times 2 \times 3 \times 2 = 24.$$

EXPLANATION: In the examples given above, which dealt with only two numbers, we multiplied the greatest common divisor by the quotients obtained by dividing the original numbers. To do so in this example would be wrong. The greatest common divisor of all the numbers is 2, but the quotients 3, 4, and 6 contain other common divisors that must be considered. 3 and 6 contain 3; 4 and 6 contain 2. We therefore continue the division, simply bringing down such numbers as cannot be divided by any given divisor, until no groups with a common divisor remain in the quotient.

This method may be stated as follows:

To find the least common multiple of more than two numbers by factoring, *divide all the numbers or any groups of two or more of them by such prime common divisors as may be contained in them and multiply together these divisors and the final quotients.*

To find the least common multiple of more than two numbers which cannot be readily factored, *find the least common multiple of two of them, then the least common multiple of this result and another of the given numbers; continue in this way until all the original numbers have been used.*

To subtract unlike fractions, *first change them to equivalent fractions with their L.C.D. Then find the difference of the new numerators.*

EXAMPLE: Find $\frac{3}{5} - \frac{1}{3}$.

SOLUTION: L.C.D. is 15. $\therefore \frac{3}{5} = \frac{9}{15}; \frac{1}{3} = \frac{5}{15}.$

Hence: $\frac{3}{5} - \frac{1}{3} = \frac{9}{15} - \frac{5}{15} = \frac{9 - 5}{15} = \frac{4}{15},$ ANS.

To find the L.C.D. when no two of the given denominators can be divided by the same number, *multiply the denominators by each other; the result is the L.C.D.*

EXAMPLE: Find $\frac{1}{2} + \frac{1}{3} + \frac{1}{5} + \frac{1}{7}$.

SOLUTION: $2 \times 3 \times 5 \times 7 = 210$ L.C.D.

$$\frac{1}{2} = \frac{105}{210}; \frac{1}{3} = \frac{70}{210}; \frac{1}{5} = \frac{42}{210}; \frac{1}{7} = \frac{30}{210};$$

$$\frac{105 + 70 + 42 + 30}{210} = \frac{247}{210} = 1\frac{37}{210}, \quad \text{ANS.}$$

To add and subtract mixed numbers, *treat the fractions separately; then add or subtract the results to or from the whole numbers.*

EXAMPLE: From $8\frac{1}{3}$ subtract $6\frac{3}{4}$.

SOLUTION: L.C.D. of $\frac{1}{3}$ and $\frac{3}{4} = 12$.

$$\frac{1}{3} = \frac{4}{12}; \frac{3}{4} = \frac{9}{12};$$

$$8\frac{4}{12} = 7 + \frac{12}{12} + \frac{4}{12} = 7\frac{16}{12};$$

$$\begin{array}{r} 7\frac{16}{12} \\ -6\frac{9}{12} \\ \hline 1\frac{7}{12} \quad \text{ANS.} \end{array}$$

EXPLANATION: Since $\frac{9}{12}$ is greater than $\frac{4}{12}$, it is necessary to use 1 from the 8, which becomes 7; then $7 + \frac{12}{12} + \frac{4}{12} = 7\frac{4}{12}$; subtracting $6\frac{9}{12}$ we get $1\frac{7}{12}$, ANS.

Practice Exercise No. 13

Do the following examples:

1. $\frac{7}{8} + \frac{3}{4}$ 2. $\frac{8}{9} - \frac{2}{3}$ 3. $\frac{7}{8} - \frac{3}{5}$ 4. $\frac{5}{6} + \frac{8}{9}$
5. $\frac{3}{4} + \frac{5}{12} - \frac{2}{3}$ 6. $\frac{7}{8} - \frac{1}{2} - \frac{1}{4}$ 7. $5\frac{1}{2} + 3\frac{3}{4}$ 8. $3\frac{2}{3} + 1\frac{5}{6} + 2\frac{1}{4}$
9. $15\frac{1}{9} + 8\frac{5}{6}$ 10. $12\frac{2}{3} - 6\frac{1}{3}$ 11. $9\frac{1}{3} - 7\frac{1}{4}$ 12. $16\frac{9}{8} - 9\frac{9}{6}$

MULTIPLICATION AND DIVISION OF FRACTIONS

To multiply a fraction by a whole number, *multiply the* NUMERATOR *by the whole number. The product will be the new numerator over the old denominator.*

$$\text{Thus: } 6 \times \frac{2}{3} = \frac{6 \times 2}{3} = \frac{12}{3} = 4, \text{ ANS.}$$

To divide a fraction by a whole number, *multiply the* DENOMINATOR *of the fraction by the whole number. The quotient will be the old numerator over the new denominator.*

$$\text{Thus: } \frac{2}{3} \div 5 = \frac{2}{3 \times 5} = \frac{2}{15}, \text{ ANS.}$$

$$\frac{1}{5} \div 2 = \frac{1}{5 \times 2} = \frac{1}{10}, \text{ ANS.}$$

To multiply one fraction by other fractions, *place the product of the numerators over the product of the denominators; then reduce.*

$$\frac{2}{3} \times \frac{1}{5} \times \frac{3}{6} = \frac{2 \times 1 \times 3}{3 \times 5 \times 6} = \frac{6}{90} = \frac{1}{15}.$$

To multiply mixed numbers, *change the mixed numbers to improper fractions.*

$$\text{Thus: } 1\frac{3}{4} \times 2\frac{2}{3} \times 1\frac{1}{2} = \frac{7}{4} \times \frac{8}{3} \times \frac{3}{2}$$

$$= \frac{7 \times 8 \times 3}{4 \times 3 \times 2} = \frac{168}{24} = 7, \text{ ANS.}$$

To divide a whole number or a fraction by a fraction, *invert the* DIVISOR *and* MULTIPLY.

Inverting a fraction means turning it upside down. Thus, $\frac{2}{3}$ inverted is $\frac{3}{2}$; $\frac{2}{5}$ inverted becomes $\frac{5}{2}$.

Inverting a whole number means putting a 1 above it. Thus, 2 becomes $\frac{1}{2}$; 16 becomes $\frac{1}{16}$.

EXAMPLE 1:
$$\frac{1}{2} \div \frac{2}{3} = \frac{1}{2} \times \frac{3}{2} = \frac{3}{4}, \text{ ANS.}$$

EXAMPLE 2:
$$\frac{4}{7} \div \frac{2}{5} = \frac{4}{7} \times \frac{5}{2} = \frac{20}{14} = 1\frac{6}{14}$$
$$= 1\frac{3}{7}, \text{ ANS.}$$

Cancellation

Cancellation is a short cut in the process of multiplication of fractions. It consists of taking out common factors in the numerator and denominator before dividing out.

$$\text{Thus, in the calculation } \quad \frac{\overset{1}{\cancel{2}}}{\underset{1}{\cancel{4}}} \times \frac{\overset{1}{\cancel{3}}}{\underset{3}{\cancel{6}}} \times \frac{\overset{1}{\cancel{4}}}{\underset{1}{\cancel{3}}} = \frac{1}{3}$$

the 4s cancel each other, the 3s cancel each other, and the 2 in the numerator is contained 3 times in the 6 in the denominator, leaving 3.

Cancellation can be applied only to multiplication and division of fractions; *never to addition and subtraction of fractions.*

Compare: $\frac{10}{25} \times \frac{4}{3} \times \frac{12}{8} =$

Long method: $\dfrac{10 \times 4 \times 12}{25 \times 3 \times 8} = \dfrac{480}{600} = \dfrac{48}{60} = \dfrac{4}{5}$

$$\textit{Cancellation: } \quad \frac{\overset{2}{\cancel{10}}}{\underset{5}{\cancel{25}}} \times \frac{\overset{1}{\cancel{4}}}{\underset{1}{\cancel{3}}} \times \frac{\overset{4}{\cancel{12}}}{\underset{2}{\cancel{8}}} = \frac{4}{5}$$

Practice Exercise No. 14

Do the following examples:

1. $\frac{3}{7} \times \frac{3}{5} =$
2. $\frac{3}{8} \times \frac{2}{3} =$
3. $\frac{2}{15} \times \frac{4}{21} \times \frac{7}{8} \times 5 =$
4. $\frac{14}{15} \times \frac{9}{28} =$
5. $\frac{3}{8} \times 12 =$
6. $\frac{15}{16} \times \frac{7}{90} =$
7. $\frac{5}{6} \times \frac{5}{9} =$
8. $\frac{7}{27} \times \frac{9}{14} =$
9. $18 \div \frac{1}{2} =$
10. $\frac{3}{5} \div \frac{1}{15} =$
11. $\frac{3}{8} \div \frac{1}{2} =$
12. $\frac{4}{7} \div \frac{1}{8} =$
13. $1\frac{2}{3} \times \frac{3}{4} =$
14. $2\frac{1}{2} \times 1\frac{2}{4} =$
15. $3\frac{1}{2} \div \frac{1}{4} =$
16. $1\frac{1}{8} \div \frac{3}{16} =$
17. $\dfrac{\frac{2}{3} + \frac{1}{4} + \frac{1}{2}}{\frac{5}{6} - \frac{1}{6} - \frac{1}{4}} =$
18. $\dfrac{\frac{2}{3} \div \frac{1}{3}}{\frac{1}{4} \times \frac{1}{3}} =$
19. $\dfrac{2\frac{1}{2} + 3\frac{1}{3}}{4 + \frac{2}{3}} =$
20. $\dfrac{\frac{1}{4} \text{ of } 8}{1\frac{1}{2} \text{ of } 3} =$
21. $\dfrac{\frac{1}{5} + \frac{8}{14} + \frac{3}{7}}{\frac{3}{7} \div 10} =$
22. $\dfrac{\frac{1}{6} + \frac{2}{3} + \frac{3}{4} + \frac{1}{3}}{\frac{7}{12} - \frac{1}{6} + \frac{1}{3}} =$

To simplify complex fractions, *convert the numerators and denominators to simple fractions, then follow the rules for adding, subtracting, multiplying, or dividing simple fractions.*

Thus:

$$\frac{\frac{1}{2} + \frac{3}{4}}{2 + \frac{1}{2}} = \frac{\frac{5}{4}}{\frac{5}{2}} = \frac{\overset{1}{\cancel{5}}}{\underset{2}{\cancel{4}}} \times \frac{\overset{1}{\cancel{2}}}{\underset{1}{\cancel{5}}} = \frac{1}{2}, \text{ Ans.}$$

$$\frac{1\frac{2}{3} \times 2\frac{1}{4}}{\frac{1}{2} \div \frac{2}{3}} = \frac{\overset{5}{\cancel{\frac{5}{3}}} \times \frac{9}{4}}{\frac{1}{2} \times \frac{3}{2}} = \frac{\frac{45}{12}}{\frac{3}{4}} = \frac{\overset{15}{\cancel{45}}}{\underset{1}{\cancel{12}}} \times \frac{\overset{1}{\cancel{4}}}{\cancel{3}} = 5, \text{Ans.}$$

In performing cancellation, when the quotient of any division is 1, this figure may be written down as in the preceding examples; but this is not strictly necessary.

Larger Numbers Involving Fractions

To multiply a whole number and a mixed number together, *perform separate multiplications and add the results.*

EXAMPLE 1: Multiply 17 by $6\frac{3}{4}$.

$$\begin{array}{r} 17 \\ 6\frac{3}{4} \\ \hline 102 \\ 12\frac{3}{4} \\ \hline 114\frac{3}{4}, \text{ Ans.} \end{array}$$

EXPLANATION: We multiply 17 first by 6, the whole-number part of the multiplier, and then by the fractional part, $\frac{3}{4}$; this is simply taking $\frac{3}{4}$ of it. Finally, we add the results.

EXAMPLE 2: Multiply $17\frac{3}{5}$ by 4.

$$\begin{array}{r} 17\frac{3}{5} \\ 4 \\ \hline 2\frac{2}{5} \\ 68 \\ \hline 70\frac{2}{5}, \text{ Ans.} \end{array}$$

EXPLANATION: We first multiply $\frac{3}{5}$ in the multiplicand by 4, the multiplier; thus 4 times $\frac{3}{5}$ is $\frac{12}{5}$, equal to $2\frac{2}{5}$, which is in effect taking $\frac{3}{5}$ of the multiplier 4. We then multiply the whole-number part, 17, by 4. Finally, we add the two products.

To divide a mixed number by a whole number, *divide the whole-number part of the quotient, reduce any remainder to a single fraction and divide this by the divisor.*

EXAMPLE: Divide $17\frac{3}{8}$ by 6.

$$6)17\frac{3}{8}$$
2 with $5\frac{3}{8} = \frac{43}{8}$, remainder.

Next: $\frac{43}{8} \times \frac{1}{6} = \frac{43}{48}$.

Then: $2 + \frac{43}{48} = 2\frac{43}{48}$, Ans.

EXPLANATION: Having divided the whole number as in simple division, we have a remainder of $5\frac{3}{8}$, which we reduce to an improper fraction and divide by the divisor. Annexing this result to the quotient 2, we obtain $2\frac{43}{48}$ for the answer.

To divide a whole number by a mixed number, *reduce the divisor and dividend to equivalent fractions having the same denominator; then divide the numerator of the dividend by the numerator of the divisor.*

EXAMPLE: Divide 25 by $4\frac{3}{5}$.

$$
\begin{array}{c|c}
4\frac{3}{5} & 25 \\
5 & 5 \\
\hline
\multicolumn{2}{c}{23)125(5\frac{10}{23}} \\
\multicolumn{2}{c}{115} \\
\hline
\multicolumn{2}{c}{10}
\end{array}
$$

EXPLANATION: We first reduce the divisor and dividend to fifths, and then divide as in whole numbers. By multiplying divisor and dividend by the same number, 5, their relation to each other for the purpose of division is the same as before, and the quotient is not changed. The answer here is $\frac{125}{23} = 5$ and $\frac{10}{23}$.

Practice Exercise No. 15

1. $9\frac{3}{8} \times 5 =$	2. $12\frac{3}{5} \times 7 =$	3. $9 \times 8\frac{11}{12} =$	4. $10 \times 7\frac{1}{8} =$
5. $11\frac{6}{7} \times 8 =$	6. $7\frac{6}{11} \times 5 =$	7. $23\frac{7}{12} \times 6 =$	8. $8\frac{3}{8} \times 5 =$
9. $9 \times 6\frac{5}{8} =$	10. $12 \times 637\frac{1}{2} =$	11. $17\frac{3}{5} \div 7 =$	12. $18\frac{3}{7} \div 8 =$
13. $27\frac{11}{12} \div 9 =$	14. $31\frac{1}{10} \div 11 =$	15. $78\frac{4}{5} \div 12 =$	16. $36 \div 9\frac{7}{8} =$
17. $97 \div 13\frac{11}{12} =$	18. $342 \div 14\frac{47}{131} =$	19. $113 \div 21\frac{1}{7} =$	20. $19 \div 2\frac{3}{7} =$

To find what part one number is of another, *make the first number the numerator and the second the denominator of a fraction; reduce to lower terms if possible.*

Thus, 5 is $\frac{5}{8}$ of 8, because 1 is $\frac{1}{8}$ of 8, and 5 is 5 times 1.

Again, 2 is $\frac{4}{7}$ of $3\frac{1}{2}$, because if we write the relation as a complex fraction we get $\dfrac{2}{3\frac{1}{2}}$, and multiplying both numerator and denominator by 2 gives us $\frac{4}{7}$.

Similarly, to find what part $3\frac{1}{2}$ is of $5\frac{1}{6}$, we reduce both to twelfths. $3\frac{1}{2} = \frac{42}{12}$; $5\frac{1}{6} = \frac{62}{12}$; $\frac{42}{62} = \frac{21}{31}$.

To find a number when a specified fractional part is given, *divide the given number by the specified fraction.*

EXAMPLE: 360 is $\frac{5}{6}$ of what number?

SOLUTION: $360 \times \frac{6}{5} = 432$, ANS.

EXPLANATION: Since 360 is $\frac{5}{6}$ of the number, $\frac{1}{6}$ will be determined by dividing 360 by 5, and the whole number will be found by multiplying this sixth part by 6. The necessary calculations are performed at once by simply inverting the specified fraction and then multiplying.

Practice Exercise No. 15a

1. What part of 11 is 8?	2. What part of 16 is $\frac{3}{8}$?	3. What part of $5\frac{1}{4}$ is 4?
4. What part of $3\frac{1}{4}$ is $2\frac{1}{4}$?	5. 160 is $\frac{4}{5}$ of?	6. 144 is $\frac{9}{10}$ of?
7. 143 is $\frac{13}{14}$ of?	8. $\frac{15}{16}$ is $\frac{5}{7}$ of?	

Practice Exercise No. 16

1. How many sheets of metal $\frac{1}{32}$ mm thick are there in a pile $25\frac{1}{2}$ mm high?
   (A) 550 ......   (B) 105 ......   (C) 408 ......   (D) 816 ......

2. If a man can do a piece of work in 16 days, how much of it can he do in $\frac{1}{2}$ day?
   (A) $\frac{1}{2}$ ......   (B) $\frac{1}{32}$ ......   (C) $\frac{1}{8}$ ......   (D) $\frac{1}{16}$ ......

3. Find the area in square metres of the figure represented by the diagram below in which each measurement is a fraction of a metre.

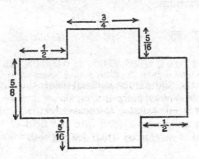

   (A) $1\frac{1}{2}$ ......   (B) $2\frac{1}{32}$ ......   (C) $1\frac{9}{16}$ ......   (D) $6\frac{2}{3}$ ......

4. If $\frac{1}{3}$ the length of a beam is 10 m, how long is the entire beam?
   (A) 3 ......   (B) 30 ......   (C) $3\frac{1}{3}$ ......   (D) 9 ......

Note: **To find the value of the whole when the fractional part is given,** *invert the fraction and multiply it by the given part.*

5. If $\frac{1}{3}$ of a machine's daily output is equal to $\frac{1}{2}$ of another machine's daily output, and if the total day's output is 600 parts, how many parts were produced by the more efficient machine?
   (A) 300 ......   (B) 360 ......   (C) 200 ......   (D) 440 ......

6. If it takes 5 hours to do $\frac{2}{3}$ of a job, how long will it take to complete the job?
   (A) $7\frac{1}{2}$ ......   (B) $3\frac{1}{2}$ ......   (C) 7 ......   (D) 10 ......

7. Two machines turn out an equal amount of parts daily. One day they both break down; the second to break down turns out $\frac{5}{6}$ of its usual amount while the first turns out $\frac{1}{2}$ of what the second does. What fraction of their usual combined total is lost because of the break-down?
   (A) $\frac{1}{2}$ ......   (B) $\frac{3}{8}$ ......   (C) $\frac{3}{4}$ ......   (D) $\frac{1}{6}$ ......

8. An aviator made 3 flights. The first was 432 km, the second was only $\frac{1}{2}$ that distance; and the third $\frac{1}{3}$ of the original distance. How far would he have to go on a fourth flight to equal $\frac{1}{2}$ the distance covered by the second and third trips?
   (A) 180 ......   (B) 450 ......   (C) 360 ......   (D) 275 ......

9. It takes 24 days to complete the first $\frac{1}{4}$ of a ship; 31 days to complete the next $\frac{3}{8}$ of it. If the average rate of speed for this much of the job were maintained, how long would it take to build the entire ship?
   (A) 96 ......   (B) 80 ......   (C) 33 ......   (D) 88 ......

10. The distance between two cities is 3000 kilometres. Two trains leave the two cities at the same time. One train travels at the rate of $62\frac{3}{5}$ kilometres per hour, the other at $69\frac{4}{5}$ kilometres per hour. How far apart will the two trains be at the end of 5 hours?
   (A) 662 ......   (B) 2338 ......   (C) 2220 ......   (D) 1842 ......

CHAPTER THREE

DECIMAL FRACTIONS

Decimal fractions are a special way of writing proper fractions that have denominators beginning with 1 and ending with one or more zeros. *Thus*, when written as decimal fractions,

$$\frac{1}{10}, \quad \frac{2}{100}, \quad \frac{3}{1000}, \quad \frac{4}{10\,000}, \quad \frac{5}{100\,000}, \text{ etc.}$$

become

$$0·1, \quad 0·02, \quad 0·003, \quad 0·000\,4, \quad 0·000\,05, \text{ etc.}$$

The period before the digits is the **decimal point**; the digits following it are said *to stand in certain decimal places*.

The word *decimal* means *relating to the number ten*, and to calculate fractions by decimals is simply to extend into the field of fractions the same method of counting that we employ when dealing with whole numbers.

READING DECIMAL FRACTIONS

Read the number after the decimal point as a whole number and give it the name of its last decimal place.

Thus,

0·135 is read as *one hundred thirty-five thousandths*.

4·18 is read as *four and eighteen hundredths*.

Another way to read

0·135 is *point, one-three-five*.

4·18 may be read *four, point, one-eight*.

Practice Exercise No. 17

Read each number aloud and write it as a common fraction or as a mixed number.

1. $0·01 =$ 2. $0·5 =$ 3. $0·625 =$ 4. $2·10 =$ 5. $23·450 =$
6. $0·000\,8 =$ 7. $0·060\,8 =$ 8. $0·234\,1 =$ 9. $0·043\,29 =$ 10. $18·020 =$

In examples 11 to 20 write the numbers as decimals.

11. $\frac{3}{10} =$ 12. $\frac{5}{100} =$ 13. $\frac{321}{1000} =$ 14. $12\frac{1}{100} =$ 15. $124\frac{3}{10000} =$
16. $18\frac{7}{10} =$ 17. $\frac{300}{1000} =$ 18. $\frac{145}{100} =$ 19. $\frac{223}{10} =$ 20. $\frac{4330}{1000} =$

Place of digit	How to read it	Example		
First decimal place	Tenths	0·3	is	$\frac{3}{10}$
Second decimal place	Hundredths	0·03	is	$\frac{3}{100}$
Third decimal place	Thousandths	0·003	is	$\frac{3}{1000}$

Place of digit	How to read it	Example
Fourth decimal place	Ten thousandths	$0{\cdot}000\,3$ is $\dfrac{3}{10\,000}$
Fifth decimal place	Hundred thousandths	$0{\cdot}000\,03$ is $\dfrac{3}{100\,000}$

Table II. *Decimal Equivalents of Sixty-Fourths*

Fraction	Decimal	Fraction	Decimal
$\frac{1}{64}$	0·015 625	$\frac{33}{64}$	0·515 625
$\frac{1}{32}$	0·031 25	$\frac{17}{32}$	0·531 25
$\frac{3}{64}$	0·046 875	$\frac{35}{64}$	0·546 875
$\frac{1}{16}$	0·062 5	$\frac{9}{16}$	0·562 5
$\frac{5}{64}$	0·078 125	$\frac{37}{64}$	0·578 125
$\frac{3}{32}$	0·093 75	$\frac{19}{32}$	0·593 75
$\frac{7}{64}$	0·109 375	$\frac{39}{64}$	0·609 375
$\frac{1}{8}$	0·125	$\frac{5}{8}$	0·625
$\frac{9}{64}$	0·140 625	$\frac{41}{64}$	0·640 625
$\frac{5}{32}$	0·156 25	$\frac{21}{32}$	0·656 25
$\frac{11}{64}$	0·171 875	$\frac{43}{64}$	0·671 875
$\frac{3}{16}$	0·187 5	$\frac{11}{16}$	0·687 5
$\frac{13}{64}$	0·203 125	$\frac{45}{64}$	0·703 125
$\frac{7}{32}$	0·218 75	$\frac{23}{32}$	0·718 75
$\frac{15}{64}$	0·234 375	$\frac{47}{64}$	0·734 375
$\frac{1}{4}$	0·25	$\frac{3}{4}$	0·75
$\frac{17}{64}$	0·265 625	$\frac{49}{64}$	0·765 625
$\frac{9}{32}$	0·281 25	$\frac{25}{32}$	0·781 25
$\frac{19}{64}$	0·296 875	$\frac{51}{64}$	0·796 875
$\frac{5}{16}$	0·312 5	$\frac{13}{16}$	0·812 5
$\frac{21}{64}$	0·328 125	$\frac{53}{64}$	0·828 125
$\frac{11}{32}$	0·343 75	$\frac{27}{32}$	0·843 75
$\frac{23}{64}$	0·359 375	$\frac{55}{64}$	0·859 375
$\frac{3}{8}$	0·375	$\frac{7}{8}$	0·875
$\frac{25}{64}$	0·390 625	$\frac{57}{64}$	0·890 625
$\frac{13}{32}$	0·406 25	$\frac{29}{32}$	0·906 25
$\frac{27}{64}$	0·421 875	$\frac{59}{64}$	0·921 875
$\frac{7}{16}$	0·437 5	$\frac{15}{16}$	0·937 5
$\frac{29}{64}$	0·453 125	$\frac{61}{64}$	0·953 125
$\frac{15}{32}$	0·468 75	$\frac{31}{32}$	0·968 75
$\frac{31}{64}$	0·484 375	$\frac{63}{64}$	0·984 375
$\frac{1}{2}$	0·5	1	1·0

In the last part of Exercise No. 17 you converted fractions to decimals by placing the decimal point and the correct number of ciphers (0's) before the numerator and eliminating the denominator. You could do this because all the denominators were 10s or some multiple of ten such as 100, 1000, 10 000, etc. It is not possible, however, to do this in all cases. Hence—

To change any common fraction into decimals, *divide the numerator by the denominator and write the quotient in decimal form.*

EXAMPLE 1: Change ⅗ to a decimal.

SOLUTION: 5)3·0
      ‾‾‾‾
      0·6 ANS.

EXAMPLE 2: Change ⅜ to a decimal.

SOLUTION: 8)3·000
      ‾‾‾‾‾‾
      0·375 ANS.

Above you will find a table in which the decimal equivalents are worked out for fractions up to 64ths. All of these are frequently used in various types of technical work. Similarly, conversions of decimals to common fractions are often necessary in shop practice.

Practice Exercise No. 18

Using the foregoing table, find the decimal equivalents to the nearest thousandths of the following fractions:

1. $\frac{1}{2}$ =   2. $\frac{3}{4}$ =   3. $\frac{3}{8}$ =   4. $\frac{5}{16}$ =   5. $\frac{9}{16}$ =
6. $\frac{17}{32}$ =   7. $\frac{28}{32}$ =   8. $\frac{14}{16}$ =   9. $\frac{22}{32}$ =   10. $\frac{56}{64}$ =

ADDITION AND SUBTRACTION OF DECIMALS

To add or subtract decimals, *place the numbers in a column with the decimal points in a column. Add or subtract as for whole numbers, placing the decimal point in the result in the column of decimal points.*

EXAMPLE 1: Find the sum of 2·43, 1·485, 0·3, 12·02, and 0·074.

SOLUTION:

	2·43	or	2·430
	1·485		1·485
	0·3		0·300
	12·02		12·020
	0·074		0·074
	‾‾‾‾‾		‾‾‾‾‾
	16·309		16·309

EXPLANATION: Since 1·485 and 0·074 are three-place numbers, we write zeros after 2·43, 9·3, and 12·02. This does not change the value but helps to avoid errors.

EXAMPLE 2: Find the difference of 17·29 and 6·147.

SOLUTION:

	17·29	or	17·290
	−6·147		−6·147
	‾‾‾‾‾		‾‾‾‾‾
	11·143		11·143

EXPLANATION: As above, we add a zero to 17·29 to make it a three-place number. This does not change the value, and is not strictly necessary, but helps to avoid errors.

Note: *The value of a decimal fraction is not changed if zeros are written at the right end of it.*

Practice Exercise No. 19

Do the following examples:

1. $0.2 + 0.07 + 0.5 =$
2. $2.6 + 22.4 + 0.03 =$
3. $22.8 + 5.099 + 613.2 =$
4. $0.005 + 5 + 16.2 + 0.96 =$
5. $15.4 + 22 + 0.01 + 1.48 =$
6. $28.74 - 16.32 =$
7. $0.005 - 0.0005 =$
8. $1.431 - 0.562 =$
9. $1.002 - 0.2 =$
10. $8.04 - 7.96 =$
11. $72.306 + 18.45 - 27.202 -$
12. $14 \quad 6.3 \mid 2.739 =$
13. $27.65 + 18.402 - 2.39 + 7.63 =$
14. $18.000\,6 + 14.005 + 12.34 =$
15. $93.8 - 16.432\,7 - 20.009 =$
16. $14.29 - 6.305 - 3.472\,65 =$

MULTIPLICATION OF DECIMALS

To multiply decimals, *proceed as in multiplication of whole numbers. But in the product, beginning at the right, point off as many decimal places as there are in the multiplier and in the multiplicand combined.*

EXAMPLE 1: Multiply 3·12 by 0·42.

SOLUTION: 3·12 (Multiplicand—has two decimal places.)
0·42 (Multiplier—has two decimal places.)

624
12480

1·3104 ANS. (Product has two plus two, or four decimal places.)

EXPLANATION: Since there is a total of four decimal places when we add together those in the multiplier and in the multiplicand, we start at the right and count four places; hence we put the decimal point off to the left of the 3, which marks the fourth place counted off.

EXAMPLE 2: Multiply 0·214 by 0·303.

SOLUTION: 0·214
0·303

642
64200

0·?64842 = 0·064 842, ANS.

EXPLANATION: There are a total of six places in the multiplier and in the multiplicand, but there are only five numbers in the product; therefore we prefix a zero at the left end, and place our decimal point before it to give the required six decimal places. If we needed eight places and the answer came out to five places we would prefix three zeros and place the decimal point to the left of them.

To multiply a decimal by any multiple of ten, *move the decimal point as many places to the* RIGHT *as there are zeros in the multiplier*

Thus, $0.31 \times 100 = 31$; $0.021 \times 100 = 02.1 = 2.1$ (here we drop the zero since before a whole number it is meaningless), $0.31 \times 1000 = 310$ (we add a cipher to make the third place).

Reciprocally: **to divide a decimal or a whole number by 10 or by a multiple of 10,** *we move the decimal point as many places to the* LEFT *as there are zeros in the divisor.*

Thus, $42 \div 10 = 4 \cdot 2$; $15 \cdot 6 \div 100 = 0 \cdot 156$; $61 \div 1000 = 0 \cdot 061$ (prefixing zero to give the required number of decimal places).

Practice Exercise No. 20

Do the following examples:

1. $18 \cdot 5 \times 4 =$	2. $3 \cdot 9 \times 2 \cdot 4 =$	3. $45 \times 0 \cdot 72 =$
4. $143 \times 0 \cdot 214 =$	5. $0 \cdot 56 \times 0 \cdot 74 =$	6. $0 \cdot 224 \times 0 \cdot 302 =$
7. $7 \cdot 43 \times 0 \cdot 132 =$	8. $0 \cdot 021 \times 0 \cdot 204 =$	9. $0 \cdot 601 \times 0 \cdot 003 =$
10. $0 \cdot 014 \times 0 \cdot 006 \, 4 =$	11. $13 \cdot 2 \times 2 \cdot 475 =$	12. $0 \cdot 132 \times 2 \cdot 475 =$
13. $0 \cdot 236 \times 12 \cdot 13 =$	14. $9 \cdot 06 \times 0 \cdot 045 =$	15. $0 \cdot 008 \times 751 \cdot 1 =$
16. $8 \cdot 7 \times 10 =$	17. $0 \cdot 006 \, 9 \times 10 =$	18. $95 \cdot 6 \times 100 =$
19. $0 \cdot 045 \, 3 \times 100 =$	20. $4 \cdot 069 \times 1000 =$	21. $0 \cdot 000 \, 094 \times 10 \, 000 =$
22. $9 \cdot 2 \times 10 =$	23. $7 \cdot 49 \times 100 =$	24. $534 \cdot 79 \div 100 =$
25. $492 \cdot 568 \div 1000 =$	26. $24 \cdot 965 \, 3 \div 1000 =$	27. $5 \cdot 980 \div 100 =$
28. $0 \cdot 071 \, 56 \div 1000 =$	29. $4956 \cdot 74 \div 10 \, 000 =$	30. $0 \cdot 038 \, 649 \div 100 \, 000 =$

DIVISION OF DECIMALS

Law of division: *A quotient is not changed when the dividend and divisor are both multiplied by the same number.*

EXAMPLE 1: $7 \cdot 2 \div 0 \cdot 9$.

$7 \cdot 2 \times 10 = 72$ Thus, multiplying dividend and divisor by ten gives
$0 \cdot 9 \times 10 = 9$ $\qquad\qquad 72 \div 9 = 8$. ANS.
$72 \div 9 = 8$

To check: $8 \times 9 = 72$, and $8 \times 0 \cdot 9 = 7 \cdot 2$.

To divide a decimal by a whole number, *proceed as with whole numbers, but place the decimal point in the quotient directly above the decimal point in the dividend.*

EXAMPLE 2: $20 \cdot 46 \div 66$.

SOLUTION:

$$\begin{array}{r} 0 \cdot 31 \text{ ANS.} \\ 66\overline{)20 \cdot 46} \\ 198 \\ \hline 66 \\ 66 \\ \hline 0 \end{array}$$

EXPLANATION: Dividing as with whole numbers, simply place the decimal point in the quotient directly above the decimal point in the dividend. Check the answer by multiplying the quotient by the divisor.

EXAMPLE 3: How many yards in $165 \cdot 6$ in.?

SOLUTION:

$$\begin{array}{r} 4 \cdot 6 \text{ ANS.} \\ 36\overline{)165 \cdot 6} \\ 144 \\ \hline 216 \\ 216 \\ \hline 0 \end{array}$$

EXPLANATION: Since there are 36 in. in one yard, we divide the number of inches by 36, pointing off the quotient decimally as in the previous example.

To divide a decimal by a decimal, *move the decimal point of the divisor to the right until it becomes a whole number* (i.e. *multiply it by ten or a multiple of ten*). *Next move the decimal point of the dividend the same number of places to the right, adding zeros if necessary.* (Multiplying divisor and dividend by the same number does not change the quotient.) *Then proceed to divide as in Example 2, above.*

EXAMPLE 4: $131 \cdot 88 \div 4 \cdot 2$.

SOLUTION:

$$
\begin{array}{r}
31 \cdot 4 \quad \text{Ans.} \\
4_{\times}2 \cdot \overline{)131_{\times}8 \cdot 8} \\
126 \\
\hline
58 \\
42 \\
\hline
168 \\
168 \\
\hline
0
\end{array}
$$

EXPLANATION: Division of a decimal by a decimal is simplified if the divisor is made a whole number. In this case the divisor 4·2 was made a whole number by moving the decimal point one place; therefore we also moved the decimal point one place in the dividend. Then placing the decimal point in the quotient directly above the decimal point in the dividend, we proceed as for division of whole numbers.

Check the answer by multiplying the quotient by the divisor or by casting out nines.

It is a corollary of the foregoing that when we are working with two numbers which contain the same number of decimal places the decimal points may be disregarded and the two numbers treated like whole numbers. This is often useful to remember when working against time.

To carry out a decimal quotient to a given number of places, *add zeros to the right of the dividend until the dividend contains the required number of places.*

EXAMPLE 5: Find $0 \cdot 3 \div 0 \cdot 7$ to the nearest thousandth.

SOLUTION:

$$7 \overline{)3 \cdot 0000}$$

$0 \cdot 4285$, ANS.—But change to $0 \cdot 429$ as the answer requested.

EXPLANATION: In Examples 1, 2, and 3 the quotients had no remainder. Often division problems do not come out exactly, as in Example 5. We then add zeros to the right of the dividend in order to carry out the division to the number of decimal places required by the work.

As a general rule carry out the division to one more decimal place than is needed. If the last figure is five or more drop it and add one to the figure in the preceding place. (This was done in Example 5, above.) If the last figure is less than five, just drop it entirely.

Averages

In finding averages of several quantities you should round off the result to the smallest part stated in the problem. Thus, if a problem is stated in decimal thousandths the result should be given in thousandths.

Rule: To find the average of several decimal quantities, *divide their sum by the number of quantities.*

EXAMPLE: Find the average of the following dimensions: 1·734, 1·748, 1·640, and 1·802

SOLUTION:

```
1·734    4)6·924
1·748    1·731 ANS.
1·640
1·802
─────
6·924
```

EXPLANATION: Add the quantities and divide the sum, 6·924, by the number of quantities, 4.

Practice Exercise No. 21

Carry out answers only as far as three places.

1. 0·34 ÷ 2   2. 0·35 ÷ 7   3. 5·4 ÷ 9   4. 47·3 ÷ 10
5. 4·2 ÷ 0·01   6. 1·11 ÷ ·3   7. 0·987 ÷ 21   8. 0·254 6 ÷ 0·38
9. 2·83 ÷ 0·007 10. 0·081 ÷ 0·002 2

ALIQUOT PARTS

An **aliquot part** of a number is a part that exactly divides the number. The term is used especially of the decimal values of exact fractional divisions of basic currency. Everybody should be familiar with the following values.

Aliquot Parts of Basic Currencies

Fractional Part	£ sterling Gt. Britain	$ Australian or American	R (rand) S. Africa
$\frac{1}{10}$	10p	10c	10c
$\frac{1}{8}$	12½p	12½c	12½c
$\frac{1}{6}$	16⅔p	16⅔c	16⅔c
$\frac{1}{4}$	25p	25c	25c
$\frac{1}{3}$	33⅓p	33⅓c	33⅓c
$\frac{3}{8}$	37½p	37½c	37½c
$\frac{1}{2}$	50p	50c	50c
$\frac{2}{3}$	66⅔p	66⅔c	66⅔c
1	£1	$1	R1

Note: This table only gives fractional parts. It *does not* say that £1 is worth $1 or R1. Furthermore it does not imply that coins of value ½p, ⅔c, etc. actually exist.

Problems involving any of these values will usually be simplified if they are worked with common fractions instead of decimals.

To find the total cost when the price involves an aliquot part, *multiply the quantity by the price expressed as a common fraction.*

EXAMPLE 1: What will 1751 castings cost at $12\frac{1}{2}$p each?

SOLUTION: $1751 \times \frac{1}{8} = 218\frac{7}{8} = £218 \cdot 87\frac{1}{2}$p, ANS.

EXAMPLE 2: What would 15 500 litres of wine cost at $2 \cdot 75$ per litre?

SOLUTION: $15\ 500 \times \dfrac{11}{4} = \dfrac{170\ 500}{4}$

$$= \$42\ 625, \text{ ANS.}$$

To find the quantity when the total cost is given and the price is an aliquot part, *divide the cost by the price expressed as a common fraction.*

EXAMPLE: How many metres of cloth at $83\frac{1}{3}$p per metre can be bought for £1000?

SOLUTION: $1000 \div \frac{5}{6} = 1000 \times \frac{6}{5} = \frac{6000}{5}$

$$= 1200 \text{ metres, ANS.}$$

Practice Exercise No. 22

1. 276 metres at $16\frac{2}{3}$p =
2. 344 litres at $62\frac{1}{2}$p =
3. 267 kilogrammes at $83\frac{1}{3}$c =
4. 1900 litres at R1·25 —
5. 2514 barrels at £5·83⅓ =
6. $6200 buys how many tons at $7.75?
7. £1000 buys how many litres at £1·33⅓?
8. £500 buys how many metres at 37½p?
9. £10 buys how many articles at $8\frac{1}{3}$p?
10. $1 buys how many grammes at 62½c?

Practice Exercise No. 23

PROBLEMS INVOLVING DECIMALS

1. A labourer mixed 250 kg of mortar. If 0·84 of it was sand how many kilogrammes of sand were used?
  (A) 166 ...... (B) 14 ...... (C) 210 ...... (D) 50 ......

2. A square picture frame has an outside perimeter of 8·32 m. If the sides are 0·04 m wide, what is the perimeter of the inside?
  (A) 8·16 m ...... (B) 8·48 m ...... (C) 8·24 m ...... (D) 8 m ......

3. A link pin is supposed to have a diameter of 0·675 mm, but one of these was made 0·000 7 mm too large. What is its diameter?
  (A) 0·674 3 mm...... (B) 0·675 7 mm...... (C) 0·767 5 mm...... (D) 0·472 5 mm......

4. Mr. Brown received £44·20 for his eggs. How many dozen did he sell if the price he received was £0·20 per dozen?
  (A) 200 ...... (B) 221 ...... (C) 210 ...... (D) 231 ......

5. A 16 section road is 158·72 metres long. How long are the first six sections when joined together?
  (A) 9·92 m ...... (B) 26·45 m ...... (C) 66 m ...... (D) 59·52 m ......

6. What would you consider to be the probable length of an object if you measured it four times and obtained the following readings: 0·641 m, 0·642 m, 0·646 m, 0·647 m?

    (A) 0·643 m ......   (B) 0·644 m ......   (C) 0·645 m ......   (D) 0·646 m ......

7. A certain engine cylinder is 33·75 mm in diameter. The piston which is to fit into it must have 0·045 mm clearance. What should be the diameter of the piston?

    (A) 38·75 mm ......   (B) 33·695 mm ......   (C) 33·66 mm ......   (D) 3·420 mm ......

PERCENTAGE

(*Discount, Interest, Commission, Profit, and Loss*)

Percentage is a term used in arithmetic to denote that a whole quantity divided into 100 equal parts is taken as the standard of measure. Percentage is indicated by the **per cent sign** (%).

Thus *per cent*, or %, means a number of parts of one hundred (100). For example, 4% may be written as $\frac{4}{100}$ or 0·04. Notice that $\frac{4}{100}$ reduces to $\frac{1}{25}$.

Percentages may be added, subtracted, multiplied, or divided, just as other specific denominations are treated.

$$\textit{Thus:} \quad \begin{aligned} 6\% + 8\% &= 14\% \\ 18\% - 12\% &= 6\% \\ 18\% \div 9\% &= 2 \\ 7\% \times 5 &= 35\% \end{aligned}$$

In the actual working of a problem, when the % sign is not used the percentages must be changed to a common fraction or a decimal.

To change a percentage to a fraction, *divide the percentage quantity by 100 and reduce to lowest terms.*

$$\textit{Thus:} \quad \begin{aligned} 8\% &= \tfrac{8}{100} = \tfrac{2}{25} \\ 75\% &= \tfrac{75}{100} = \tfrac{3}{4} \\ 80\% &= \text{?} \end{aligned}$$

Table III.   Fractional Equivalents of Percentages

$10\% = \frac{1}{10}$	$12\frac{1}{2}\% = \frac{1}{8}$	$8\frac{1}{3}\% = \frac{1}{12}$
$20\% = \frac{1}{5}$	$25\% = \frac{1}{4}$	$16\frac{2}{3}\% = \frac{1}{6}$
$40\% = \frac{2}{5}$	$37\frac{1}{2}\% = \frac{3}{8}$	$33\frac{1}{3}\% = \frac{1}{3}$
$50\% = \frac{1}{2}$	$62\frac{1}{2}\% = \frac{5}{8}$	$66\frac{2}{3}\% = \frac{2}{3}$
$60\% = \frac{3}{5}$	$87\frac{1}{2}\% = \frac{7}{8}$	$83\frac{1}{3}\% = \frac{5}{6}$

Practice Exercise No. 24

Change to fractions.

1. 1% =	2. 2% =	3. 4% =	4. 7% =
5. $\frac{1}{2}$% =	6. $6\frac{1}{4}$% =	7. $6\frac{2}{3}$% =	8. $7\frac{1}{2}$% =
9. $\frac{1}{3}$% =	10. $\frac{3}{4}$% =	11. $1\frac{1}{2}$% =	12. $3\frac{1}{4}$% =

To change a percentage to a decimal, *remove the per cent sign and move the decimal point two places to the left.*

EXAMPLE 1: Change 25% to a decimal.

    25% = 0·25. Moving decimal point two places to the left.

EXAMPLE 2: Change 1·5% to a decimal.

1·5% = 0·015.   To move the decimal point two places to the left, one zero had to be prefixed.

To change a decimal to a percentage, *move the decimal point two places to the right and add a per cent sign.*

EXAMPLE 1: Change 0·24 to a percentage.

0·24 = 24%.   Moving the decimal point two places to the right and adding the % sign.

EXAMPLE 2: Change 0·0043 to a percentage.

0·0043 = 0·43%.   Note that this is less than 1%.

EXAMPLE 3: Change 2·45 to a percentage.

2·45 = 245%.   Note that any whole number greater than 1 which designates a percentage is more than 100%.

The terms commonly used in percentage are **rate** (*R*), **base** (*B*), **percentage** (*P*)**, and amount** (*A*).

Rate (*R*) or **rate per cent** is the fractional part in hundredths that is to be found.

For example, in 4% of 50 = 2,
4%, $\frac{1}{25}$ or 0·04 is the *rate*.

The **base** (*B*) is the whole quantity of which some percentage is to be found. In 4% of 50 = 2, 50 is the *base* or whole quantity.

The **percentage** (*P*) is the result obtained by taking a given hundredth part of the base.

In 4% of 50 = 2, 2 is the *percentage* or the part taken.

The **amount** (*A*) is the sum of the base and the percentage. In 4% of 50 = 2, the *amount* is 50 + 2, or 52.

The Three Types of Percentage Problems

A. **Finding a percentage of a number:** given the base and the rate, to find the percentage.

EXAMPLE 1: Find 14% of £300.

$$
\begin{array}{r}
300 \text{ base} \\
\times\, 0{\cdot}14 \text{ rate} \\
\hline
1200 \\
3000 \\
\hline
£42, \quad \text{ANS.}
\end{array}
$$

Percentage = Base × Rate; or *P* = *B* × *R*.

EXPLANATION: 14% is equal to 0·14. Multiplying 300 by 0·14, the product is £42 of 14% of £300.

B. **Finding what percentage one number is of another:** given the base and the percentage to find the rate.

EXAMPLE 2; 120 is what percentage of 240?

$$\tfrac{120}{240} = \tfrac{1}{2} = 50\%, \quad \text{ANS.}$$
$$120 = \text{the percentage.}$$
$$240 = \text{the base.}$$
$$50\% = \text{the rate.}$$

$$\text{Rate} = \text{Percentage} \div \text{Base; or } R = \frac{P}{B}.$$

EXPLANATION: This is another way of saying what fractional part of 240 is 120. Change the answer to percentage.

C. Finding a number when a percentage of that number is known: given the rate and the percentage, to find the base.

EXAMPLE 3: 225 is 25% of what amount?

$25\% = \tfrac{1}{4}, 225 \div \tfrac{1}{4} = 225 \times 4/1 = 900,$ ANS.

Alternatively: $25\% = 0.25, 225 \div 0.25 = 900,$ ANS.

$225 =$ the percentage. $25\% =$ the rate. $900 =$ the base.

$$\text{Base} = \text{Percentage} \div \text{Rate; or } B = \frac{P}{R}.$$

EXPLANATION: This is another way of saying $\tfrac{1}{4}$ of a number equals 225, and asking what does the whole number equal.

The formulae given above make the solution of percentage problems easy if you learn to identify the base, the rate, and the percentage.

Practice Exercise No. 25

1. Find 30% of 620.
2. Find $12\tfrac{1}{2}\%$ of 96 men.
3. What is 4% of 250 kg?
4. How much is $62\tfrac{1}{2}\%$ of £80?
5. Find $\tfrac{1}{2}$ of 1% of 190 kg.
6. 8 is what percentage of 32?
7. $7\tfrac{1}{2}$ m is what percentage of 15 m?
8. £14 is what percentage of £200?
9. $\tfrac{1}{4}$ is what percentage of $\tfrac{4}{5}$?
10. $\tfrac{1}{2}$ is what percentage of $\tfrac{3}{4}$?
11. $12 = 25\%$ of what number?
12. 10 is 20% of what number?
13. 8 is $2\tfrac{1}{2}\%$ of what number?
14. 16% of a sum $= 128$. What is the sum?
15. What number increased by 25% of itself equals 120?

16. A loaded truck has a mass of 20 000 kg. If 80% of this represents the load, what is the mass of the truck?

    (A) 2000 kg ......   (B) 8000 kg ......   (c) 4000 kg ......   (D) 16 000 kg ......

17. A bin has a mass which is 8% of its contents. If the mass of the contents is 275 kg, what is the mass of the bin?

    (A) 22 ......   (B) 34·75 ......   (c) 22·15 ......   (D) 207 ......

18. A brass bar of mass 75 kg is made of 45% zinc and the balance of copper. How many kilogrammes of copper does it contain?

    (A) 55 ......   (B) 41·25 ......   (c) 33·75 ......   (D) 0·6 ......

19. What number increased by 75% of itself is equal to 140?
(A) 105 ......   (B) 215 ......   (C) 46⅔ ......   (D) 80 ......

20. One man in a crew receives a bonus of £5·25, which is 17·5% of the total bonus allowed for the job. What is the total bonus?
(A) £93·65 ......   (B) £433·12 ......   (C) £30·00 ......   (D) £31·50 ......

21. A truck carrying 6750 kg of coal has a total mass of 9000 kg. What percentage of the total mass was due to the mass of the truck?
(A) 15% ......   (B) 20% ......   (C) 25% ......   (D) 30% ......

22. A bronze statue with a wooden base has a mass of 28 kg. The base has a mass of 3·5 kg. What percentage of the total mass was bronze?
(A) 87½% ......   (B) 12½% ......   (C) 9·8% ......   (D) 83⅛% ......

23. In a factory 8% of the machines broke down. They were replaced by new ones. How many machines are there in the shop if 144 machines were replaced?
(A) 1728 ......   (B) 1800 ......   (C) 1440 ......   (D) 360 ......

24. If the voltage loss of an electrical transmitting line is 5%, what will be the voltage at the end of the line if the generator voltage is 140?
(A) 147 ......   (B) 7 ......   (C) 28 ......   (D) 133 ......

25. 33⅓% of a machinist's daily output is equal to 50% of another man's output. The slower man turns out 1500 machine screws daily. What is the faster man's output?
(A) 500 ......   (B) 1000 ......   (C) 2000 ......   (D) 2250 ......

CHAPTER FOUR

DISCOUNT, COMMISSION, INTEREST, PROFIT AND LOSS

The arithmetical principles learned in studying the topic of percentage can be applied directly in solving problems in Discount, Commission, Interest, and Profit and Loss. In order to understand problems in those fields it is first necessary to become familiar with the meanings of the terms used in each branch. Here is a table of terms used in connexion with these topics, matched according to their equivalence to each other and to those used in dealing with the general subject of percentage.

Table IV.

Percentage	Commission	Discount	Interest	Profit and Loss
Base (B)	Cost or Selling price	List price	Principal	Cost
Rate per cent (R)	Rate of commission	Rate	Rate	Per cent gain or loss
Percentage (P)	Commission	Discount	Interest	Gain or loss in money
Amount	Total cost or Net proceeds	Net price	Amount	Selling price

You will notice that the same basic relationship that holds true for percentage ($P = B \times R$), holds true for all the others.

DISCOUNT

Discounts in Trading

A **discount** is a reduction in the selling price of an article or in the amount of a bill. Discounts are expressed as rates or percentages.

The **list price** or **marked price** of an article is the original price.

The **net price** is the list price minus the discount, i.e. the price actually paid by the customer.

To find the amount of a discount, *multiply the list price, or base, by the rate of discount.*

Formula: $P = B \times R$, indicating the first type of percentage problem.

EXAMPLE 1: A used car listed at £925 was sold at a discount of 10%. Find the discount.

$$B = £925, R = 10\%, P = ?$$
$$P = B \times R; P = £925 \times 0.10 = £92.50, \text{ Ans.}$$

To find the net price, *subtract the amount of the discount from the original or list price.*

EXAMPLE 2: A filing cabinet is listed at £32.50. Find the net price if the discount is 12%.

SOLUTION: $N = ? P = ? R = 12\%. B = £32.50.$

$$P = B \times R = £32.50 \times 0.12 = £3.90$$
$$N = B - P = £32.50 - £3.90 = £28.60, \text{ Ans.}$$

ALTERNATIVE SOLUTION: Instead of multiplying the list price by the per cent discount and deducting the result, a short method is to deduct the per cent discount from 100% and multiply the remainder by the list price to get the net price directly. Thus, in the same example we have:

$$100\% - 12\% = 88\%.$$
$$£32.50 \times 0.88 = £28.60, \text{ Ans. as before.}$$

Chain discounts is the term used when more than one discount is given. Thus, when two or more discounts are given, the second discount is taken off what remains after the first discount has been deducted from the list price. The same process is continued in cases where three or more discounts are allowed.

EXAMPLE 3: A second-hand car is listed at £400 with discounts of 20%, 10%, and 5%. What is the net price?

SOLUTION:
$$£400 \times \tfrac{1}{5} = £80$$
$$- 80$$
$$\overline{}$$
$$£320 \times \tfrac{1}{10} = £32$$
$$- 32$$
$$\overline{}$$
$$£288 \times \tfrac{1}{20} = £14.4$$
$$14.4$$
$$\overline{}$$
$$£273.60 \text{ net price, Ans.}$$

A SHORTER WAY to do this example: Discounts of 20%, 10%, and 5% mean selling at 80%, 90%, and 95% of the list and net prices.

> *Thus,* £400 × 0·8 × 0·9 × 0·95 is the final selling price
> i.e. £400 × 0·684 = £273·60, ANS. as before.

The *method* here shown is generally used in business, but it is important to realize that there are several different *types* of discount.

Trade discount. A wholesaler supplies a retailer with a list of his goods. Usually the stated prices are the prices the retailer charges the customer (*list price*). The retailer is therefore entitled to a discount from this list price. This is called a **trade discount**, since it is only available for people selling goods at retail prices.

Retail discount. A retailer may offer the goods to the public at less than recommended retail prices (*marked price*). This **retail discount** is often called a *markdown*, especially in 'Sales'. There are also quantity discounts, cash discounts, and so on.

Practice Exercise No. 26

1. What is the selling price of a £30 chair if it is sold at a discount of 25%?
   (A) £22 ......   (B) £20 ......   (C) £22·5 ......   (D) £24·0 ......

2. How much can be saved by buying at a discount of 66⅔% a table marked at £54?
   (A) £33 ......   (B) £30 ......   (C) £36 ......   (D) £35 ......

3. A decorator buys 60 rolls of wallpaper. The list price is £0·40 a roll. He gets a discount of 25% and an additional 2½% for cash. How much did he pay for the wallpaper?

4. During three successive weeks of a sale a coat originally priced at £10 receives markdowns of 40%, 10%, and 20%. What is the final price?

5. The insurance premium for a car is £40. If there are chain discounts of 50% (no-claim bonus) and 17½% (for the first £10 of damage to be paid by owner), what does the owner pay for his insurance?

COMMISSION OR BROKERAGE

Commission or brokerage is the amount of money paid to an agent or broker for buying or selling goods. When the commission is some per cent of the value of the services performed that rate per cent is called the rate of commission.

Gross proceeds or **selling price** is the money received by the agent for his employer. In commission or brokerage calculations this is called the **base**.

Net proceeds is the gross proceeds minus the commission.

To find the commission, *multiply the principal amount, or the base, by the rate of commission.*

The **formula** is:

$$\text{Commission } (P) = \text{Base } (B) \times \text{Rate } (R),$$
or $$P = B \times R.$$

EXAMPLE 1: An agent obtained sales of £5000. His commission is 2½%. How much does he receive?

SOLUTION: $P = B \times R.$
$B = £5000, R = 2\frac{1}{2}\%, P = ?$
$£5000 \times 0.025 = £125$ commission,   ANS.

EXAMPLE 2: A broker received £200 for selling a piece of property. His rate of commission was 5%. What was the selling price?

SOLUTION: $\dfrac{P}{R} = B.$

$P = £200, R = 5\%, B = ?$

$\dfrac{£200}{0.05} = £4000$ selling price,   ANS.

Or: $£200 \div \frac{1}{20} = £200 \times \frac{20}{1} = £4000$,   ANS.

Practice Exercise No. 27

1. An agent sells £25 000 worth of property at $1\frac{1}{2}\%$ commission. How much does he receive?
     (A) £3500 ......   (B) £375 ......   (C) £166 ......   (D) £750 ......

2. A salesman sells 100 sets of silverware at £165 per set. At 16% what would his commission be?
     (A) £26·40 ......   (B) £2700 ......   (C) £2640 ......   (D) £2000 ......

3. An agent earned £400 commission at a rate of 5% for selling a farm. What was the selling price of the farm?
     (A) £8000 ......   (B) £3800 ......   (C) £2000 ......   (D) £6000 ......

4. For a very difficult sale a commission agent received $37\frac{1}{2}\%$. His fee was £600. What was the amount of the sale?
     (A) £4800 ......   (B) £1600 ......   (C) £1200 ......   (D) £2000 ......

INTEREST

Interest problems employ the rules of percentage problems but include the additional factor of *time*.

The **interest** (*I*) is the amount of money paid for the use of money.

The **principal** (*P*) is the *base* or the money for the use of which interest is paid.

The **rate** (*R*) is the percentage charged on the basis of one year's use of the money.

The **time** (*T*) is the number of years, months, and days over which the money is used.

The **amount** (*A*) is the sum of the principal and the interest.

To find the interest for any given period of time, *multiply the principal by the rate by the time. Interest calculated in this manner is called simple interest.*

Formula: $I = P \times R \times T.$

EXAMPLE 1: Find the simple interest on £900 for 2 years at 6%.

SOLUTION: $I = ?, \quad P = £900, \quad R = 6\%, \quad T = 2.$
$£900 \times 0.06 = £54, £54 \times 2 = £108$ interest.   ANS.

To find the amount, *add the interest (I) to the principal (P).*
Formula: $A = P + I.$

EXAMPLE 2: Find the interest and amount of £400 for 3 years, 3 months at 6%.
Note: When the order is stated in this form, 3 months is $\frac{1}{4}$ year, i.e. it does
not matter how many days are in each month.

SOLUTION: $I = P \times R \times T$, and $A = P + I.$
            £400 × 0·06 × 3 =   £72·00
            £400 × 0·06 × $\frac{1}{4}$ =   6·00
                              £78 interest ⎫
        $A = P + I$ ⎬ ANS.
          = £400 + £78 = £478 amount ⎭

Indirect Cases of Interest

To find the rate when the principal, interest, and time are given, *divide the total
interest by the time to get the amount of the interest for one year; then divide
this quotient by the principal.*

EXAMPLE 3: What must be the rate of interest on £400 to produce £25 in 6 months?

SOLUTION: £25 ÷ $\frac{1}{2}$ = £25 × $\frac{2}{1}$ = £50 interest for 1 year.
            £50 ÷ 400 = $\frac{1}{8}$ or 12$\frac{1}{2}$% rate of interest, ANS.

To find the time when the principal, interest, and rate per cent are given,
*multiply the principal by the rate to obtain the amount of interest for one year;
then divide the total interest by the interest for one year.*

EXAMPLE 4: How long will it take for £600 to yield £40 in interest at a rate of 4%?

SOLUTION: £600 × 0·04 = £24·00 interest for 1 year.
            $£\frac{40}{24} = 1\frac{16}{24} = 1\frac{2}{3}$ years, ANS.

**To find the principal when the interest, the rate per cent, and the time are
given,** *divide the interest by the time to get the interest for one year, then divide
this by the rate.*

EXAMPLE 5: How much money will you have to lend to get £24 interest at 6%, if
you lend it for 6 months?

SOLUTION: £24 ÷ $\frac{1}{2}$ = 24 × $\frac{2}{1}$ = £48 interest for 1 year.
        $\frac{£48}{0·06}$ = £800 or $\frac{£48}{\frac{6}{100}}$ = £48 × $\frac{100}{6}$ = £800,   ANS.

Note: There are alternative methods for solving interest problems of the above
type. See if you can work them out.

To find the principal when the amount, rate per cent, and time are given,
divide the given amount by the amount of £1 for the given time at the given rate.

EXAMPLE: How much would you have to deposit at 4% in order to withdraw £819
at the end of 1 year and 3 months?

SOLUTION:
£1 at 4% for 1 year will amount to £1·04.
£1 at 4% for 1$\frac{1}{4}$ years will amount to £1·05.
∴ The number of pounds that will amount to £819 is £819 ÷ £1·05 or £780,
ANS.

ACCURATE INTEREST

Two different rules are in use for the purpose of calculating interest. They are,

(i) A day is counted as one-365th of a year, e.g. British Banking, U.S. Government and Federal Banks.

(ii) A day is counted as one-360th of a year, e.g. Fairly common practice in the U.S.A. and Continental business circles. This rule is further extended to considering the year as twelve 30-day months.

In the case of rule (ii) we have to note that when dates are given for an interest-bearing period the time from a given date in one month to the same date in any other month is calculated as even months of 30 days each. Thus, February 5 to May 5 is 90 days. Similarly, February 5 to March 6 is 31 days, but February 5 to March 4 is only 27 days.

EXAMPLE: Using rule (ii), what is the interest on £150 at 5% from February 5 to March 11?

SOLUTION: There are 36 interest-bearing days between these dates.

$$\therefore \text{ Interest} = £150 \times \frac{5}{100} \times \frac{36}{360} = £0.75, \quad \text{ANS.}$$

Interest calculated under rule (i) is sometimes called **accurate interest** or **exact interest**.

To compute exact interest time between dates in different years:

1. *Determine the number of days remaining in the earlier of the given years after the given date, and to this add the number of days in the later of the given years up to the later date.*

2. *Find the number of full years remaining in the interval and add to these the number of days previously determined.*

EXAMPLE 2: For purposes of exact interest, what is the time between October 15, 1937, and March 11, 1945?

SOLUTION:

1937	October	16	days
	November	30	„
	December	31	„
1945	January	31	„
	February	28	„
	March	11	„
		147	„

1938 to 1944 inclusive = 7 years
7 years 147 days, ANS.

EXPLANATION: Starting with October 1937 there are 16 days left in October, 30 days in November, and 31 in December. Then from the beginning of 1945 there are 31 days in January, 28 in February, and 11 in March up to the 11th day of March. All these days come to a total of 147, and these are added to the 7 full years occurring in the given interval.

A **leap year** of 366 days occurs every four years, with February containing 29 days. To be a leap year the date must be divisible by 4, except that the final year of each century is only a leap year when divisible by 400. Thus, 2000 will be a leap year, as was 1600; but 1800 and 1900 were not.

To find the exact number of days between dates in different years:

1. *Find the number of years and days by the method used to determine time for exact interest.*

2. *Multiply the number of years by 365 and add to this the separate number of days as previously found.*

3. *Add one day for every leap year occurring in the full years included in the interval.*

EXAMPLE 3: What is the exact number of days between October 15, 1937, and March 11, 1945?

SOLUTION: By Example 2 we found that this interval contained 7 years and 147 days. Multiplying 365 by 7 and adding 147 we get 2702 days. But 2 leap years occurred among the full years (1940 and 1944). Hence we add two days to make a total of 2704 days.

Bank Discount

When a bank lends money against a promissory note it collects the interest charges in advance by deducting them from the full value or *face* of the note. This is called **discounting** the note. The amount thus deducted is called the **discount**. The remainder left when the *discount* has been deducted from the *face* of the note is called the **net proceeds** and is what the borrower receives. When the note falls due at **maturity** the borrower pays back to the bank its full face value.

EXAMPLE 1: A 3-month note for £400 at 6% is discounted when made. Find the discount and the net proceeds.

SOLUTION: Interest for 3 months = £6·00 discount $\Big\}$ ANS.
£400 − £6 = £394 net proceeds

EXAMPLE 2: A 3-month note for £400 at 6%, dated April 15, is discounted May 15. Find the net proceeds.

SOLUTION: Term of discount = 3 months − 1 month = 2 months.
Interest for 2 months = £4·00 discount.
£400 − £4 = £396 net proceeds, ANS.

Compound Interest

Compound interest is interest which for each successive interest period is figured on a base that represents the original principal plus all the interest that has accrued in previous interest periods.

To compute compound interest, *add the interest for each period to the principal before calculating the interest for the next period.*

EXAMPLE: What is the compound interest on £5000 for 4 years at 6% compounded annually?

SOLUTION: £5000 original principal
0·06 rate
                _____
300·00 interest for 1st year
5000·00
                _____

5300·00 new principal, 2nd year
  0·06
_____
 318·00 interest for 2nd year
5300·00
_____
5618·00 new principal, 3rd year
  0·06
_____
 337·08 interest for 3rd year
5618·00
_____
5955·08 new principal, 4th year
  0·06
_____
 357·30 interest for 4th year
5955·08
_____
6312·38 amount at end of 4th year
5000·00 original principal
_____
£1312·38 compound interest for 4 years, Ans.

INTEREST TABLES

Simple Interest

Showing the interest on £1000 at various rates; based on a 30-day month and a 360-day year.

Time	2½%	3%	3½%	4%	4½%	5%	5½%	6%
1 day	0·069	0·083	0·097	0·111	0·125	0·139	0·153	0·167
2 days	0·139	0·167	0·194	0·222	0·250	0·278	0·306	0·333
3 days	0·208	0·250	0·292	0·333	0·375	0·417	0·458	0·500
4 days	0·278	0·333	0·389	0·444	0·500	0·556	0·611	0·667
5 days	0·347	0·417	0·486	0·556	0·625	0·694	0·764	0·833
6 days	0·417	0·500	0·583	0·667	0·750	0·833	0·917	1·000
1 month, $\frac{1}{12}$ year	2·083	2·500	2·917	3·333	3·750	4·167	4·583	5·000
2 months, $\frac{1}{6}$ year	4·167	5·000	5·833	6·667	7·500	8·333	9·167	10·000
3 months, $\frac{1}{4}$ year	6·250	7·500	8·750	10·000	11·250	12·500	13·750	15·000
6 months, $\frac{1}{2}$ year	12·500	15·000	17·500	20·000	22·500	25·000	27·500	30·000
1 year	25·000	30·000	35·000	40·000	45·000	50·000	55·000	60·000

The last 5 rows of this table are exactly the same for simple interest calculated on the basis of a 365-day year.

Compound Interest

Showing the amount of a basic unit of currency at various rates. Thus £1, $1, R1 become £, $, R1·4071 after 7 years at 5%.

Yr.	2%	2½%	3%	3½%	4%	4½%	5%	5½%	6%	7%
1	1·02000	1·02500	1·03000	1·03500	1·04000	1·04500	1·05000	1·05500	1·06000	1·07000
2	1·04040	1·05063	1·06090	1·07123	1·08160	1·09203	1·10250	1·11303	1·12360	1·14490
3	1·06121	1·07689	1·09273	1·10872	1·12486	1·14117	1·15763	1·17424	1·19102	1·22504
4	1·08243	1·10381	1·12551	1·14752	1·16986	1·19252	1·21551	1·23882	1·26248	1·31080
5	1·10408	1·13141	1·15927	1·18769	1·21665	1·24618	1·27628	1·30696	1·33823	1·40255
6	1·12616	1·15969	1·19405	1·22926	1·26532	1·30226	1·34010	1·37884	1·41852	1·50073
7	1·14869	1·18869	1·22987	1·27228	1·31593	1·36086	1·40710	1·45468	1·50363	1·60578
8	1·17166	1·21840	1·26677	1·31681	1·36857	1·42210	1·47746	1·53469	1·59385	1·71819
9	1·19509	1·24886	1·30477	1·36290	1·42331	1·48610	1·55133	1·61909	1·68948	1·83846
10	1·21899	1·28009	1·34392	1·41060	1·48024	1·55297	1·62889	1·70814	1·79085	1·96715
11	1·24337	1·31209	1·38423	1·45997	1·53945	1·62285	1·71034	1·80209	1·89830	2·10485
12	1·26824	1·34489	1·42576	1·51107	1·60103	1·69588	1·79586	1·90121	2·01220	2·25219
13	1·29361	1·37851	1·46853	1·56396	1·66507	1·77220	1·88565	2·00577	2·13293	2·40984
14	1·31948	1·41297	1·51259	1·61870	1·73168	1·85194	1·97993	2·11609	2·26090	2·57853
15	1·34587	1·44830	1·55797	1·67535	1·80094	1·93528	2·07893	2·23248	2·39656	2·75903
16	1·37279	1·48451	1·60471	1·73399	1·87298	2·02237	2·18287	2·35526	2·54035	2·95216
17	1·40024	1·52162	1·65285	1·79468	1·94790	2·11338	2·29202	2·48480	2·69277	3·15882
18	1·42825	1·55966	1·70243	1·85749	2·02582	2·20848	2·40662	2·62147	2·85434	3·37993
19	1·45681	1·59865	1·75351	1·92250	2·10685	2·30786	2·52695	2·76565	3·02560	3·61653
20	1·48595	1·63862	1·80611	1·98979	2·19112	2·41171	2·65330	2·91776	3·20714	3·86968

Money will double itself at *compound interest* in 35·003 years at 2%, in 23·450 years at 3%, in 17·675 years at 4%, in 14·207 years at 5%, in 11·896 years at 6%, in 10·245 years at 7%. These figures are based on annual compounding of interest. If interest is compounded semi-annually or quarterly the periods will be only very slightly shorter. At *simple interest* money doubles itself in 25·000 years at 4%, in 20 years at 5%, in 16·667 years at 6%, in 14·286 years at 7%.

EXPLANATION (of last example): We first find the interest on the original principal (£5000) for one year. This amounts to £300. We add this to £5000, making the new principal at the start of the second year £5300. One year's interest on £5300 comes to £318, which we add to £5300 to arrive at £5618 as the new principal at the start of the third year. We continue in this way until we reach the end of the fourth year, when we find that principal and interest together amount to £6312·38. From this we subtract the original principal of £5000 leaving £1312·38 as the compound interest for the entire period of 4 years.

Alternatively we may solve the problem by using the above table for Compound Interest as follows.

The table shows that when £1 is invested for 4 years at 6% compound interest, paid annually, it will amount to £1·26248. This means that with the same time and rate of interest, £1000 will amount to £1·26248 × 1000 = £1262·48.

Similarly, £5000 will amount to £1·26248 × 5000 = £6312·40, so that in 4 years the £5000 has earned compound interest equal to £6312·40 − £5000 = £1312·40, the insignificant difference of 2p in the answers being caused by restricting the accuracy of the table to five decimal places.

These tables effect a considerable simplification of these problems as we shall now see.

EXAMPLE 1: Calculate which of the following 5-year investments will yield the greater interest
      (i) £4000 at 4½% compound interest, paid annually.
      (ii) £4000 at 6% simple interest.

SOLUTION: (i) The compound interest table shows that in 5 years at $4\frac{1}{2}\%$ compound interest, paid annually, an investment of £1 will amount to £1·24618.

∴ £4000 will amount to £1·24618 × 4000 = £4984·72

That is, £4000 invested under these terms has yielded an interest of £984·72.

(ii) The simple interest table shows that in 1 year at 6% simple interest an investment of £1000 will yield £60 interest.

∴ £4000 will yield £60 × 4 = £240 in 1 year.

In 5 years the total simple interest obtained will be £240 × 5 = £1200. Thus investment (ii) will yield £1200 − £984·72 = £215·28 more than investment (i), ANS.

EXAMPLE 2: A parent invests £100 at 4% compound interest, paid annually, on each of a child's first five birthdays. Calculate how much this investment amounts to on the child's twenty-first birthday.

SOLUTION: The first £100 invested will be in the bank for 20 years. The second £100 invested will be in the bank for 19 years and so on. Using the results given in the compound interest table we have the following amounts:

£100 at 4% for 20 years amounts to £219·112
£100 at 4% for 19 years amounts to £210·685
£100 at 4% for 18 years amounts to £202·582
£100 at 4% for 17 years amounts to £194·790
£100 at 4% for 16 years amounts to £187·298

Total	£1014·467 ANS.

We may also use the compound interest table to calculate the principal which needs to be invested at a particular rate of interest so as to yield a required amount in a number of years time.

To illustrate the method consider the fact (to be found in the table) that £1 invested at 6% compound interest will amount to £2·0122 at the end of 12 years. Some of the results which arise from this fact are tabulated below.

Thus, £1 amounts to £2·0122
£2 amounts to £4·0244
£15 amounts to £2·0122 × 15 = £30·1830

Just as we obtain these results by multiplication so we may obtain other results by division.

Thus: £1 ÷ 2   = £0·50 amounts to £2·0122 ÷ 2 = £1·0061
£1 ÷ 5   = £0·20 amounts to £2·0122 ÷ 5 = £0·40244
£1 ÷ 2·0122 = £0·49697 amounts to £2·0122 ÷ 2·0122 = £1

We have now found that £0·49697 invested at 6% compound interest will amount to £1 in 12 years.

Clearly £0·49697 × 100 = £49·697 will amount to £100 in 12 years. In this sense we may say that £49·697 is the present worth of £100 in 12 years time at 6% compound interest.

EXAMPLE 3: Calculate the principal which needs to be invested at 4% compound interest paid annually in order to realize an amount of £1000 in 9 years.

SOLUTION: From the table we see that in 9 years a principal of £1 at 4% compound interest paid annually will amount to £1·42331.

$$\therefore \quad \frac{£1}{1\cdot42331} \text{ amounts to } \frac{£1\cdot42331}{1\cdot42331} = £1$$

$$\therefore \frac{£1000}{1\cdot42331} \text{ amounts to } £1000$$

By long division £1000 ÷ 1·42331 = £702·59, Ans.

Note: We could anticipate the work of Chapter Twelve and use logarithms to obtain the last result. However, in combining four-figure logarithms with the six-figure compound interest table some approximations are necessary. Thus we could only consider 1000 ÷ 1·423 as follows:

$$
\begin{array}{rl}
\log 1000 = & 3\cdot000 \\
\log 1\cdot423 = & 0\cdot1532 \\
\hline
\text{difference} & 2\cdot8468
\end{array}
$$

∴. Quotient = antilog 2·8468 = £702·8, a difference of only 21p between the two answers.

True Discount

True discount (not to be confused with *bank* discount) is the deduction made from the face value of an obligation payable at some future date in order to determine what this obligation is worth at the present time. **Present worth** is the sum of money which, if invested at the same rate as that which applies to the given obligation, will equal the face value of the debt when it becomes due.

Rule: 1. To find present worth, *divide the face value of the indebtedness by the amount of £1 for the given time at the given rate.*

2. To find true discount, *subtract present worth from face value.*

True discount may apply to obligations bearing either simple or compound interest.

Since a sound bond on the day when it matures is worth neither more nor less than its face value, it follows that as such a bond approaches maturity its value on the stock market tends more to coincide with *present worth* as determined by the principles of true discount.

EXAMPLE: What is the present worth of an obligation for £500 payable 2 years and 6 months from now and bearing interest at 6% not compounded. Also, what is the true discount?

SOLUTION: The face value is £500.
0·06 × 2½ = 0·15 interest rate for 2½ years.
£1·00 × 0·15 = £0·15 interest on £1 for 2½ years.
£1·00 + £0·15 = £1·15 amount of £1 for 2½ years.
£500 ÷ £1·15 = £434·78 present worth ⎱ ANS.
£500 − £434·78 = £65·22 true discount ⎰

EXPLANATION: Since £1 would grow to £1·15 in 2 years and 6 months at 6% simple interest, to find out how many pounds would grow to £500 we divide £500 by £1·15 and get £434·78 as the required amount. The difference between this and £500 is the true discount. This example may be checked by multiplying £434·78 by 0·15 to arrive at £65·22 interest at 6% for 2½ years, corresponding with the same amount calculated as true discount.

RATES

A local authority (county, county borough, borough or urban district council) requires revenue to meet the cost of providing various services for its residents. These services include Education, Housing, Public Health, Town and Country Planning, Libraries, Fire Brigades, Police, Care for the Aged, etc. Part of the cost is met by the national Treasury, but the major part is raised by the Local Authority by taxes known as the **rates**.

The Authority assesses the value of each house, land, buildings, etc., in its areas as a basis for levying the rate. This assessment is called the **rateable value**. It will depend upon the size, position, use of the property under consideration.

E.g. A house may be given a rateable value of £150, so that if the Authority levy a rate of £0·50 in the £ the owner will be required to pay £75 during the course of the year; usually in two instalments £37·50 each. The £75 is called the owner's *rates*.

The total of all these rateable values is the rateable value of the borough (Authority).

The rate to be levied is $\dfrac{\text{Estimated expenditure}}{\text{Rateable value of borough}}$.

E.g. Suppose the estimated expenditure for the coming year is £3 000 000 and the rateable value of the borough is £4 000 000.

The rate for the coming year will be $\dfrac{3\,000\,000}{4\,000\,000}$, i.e. $\frac{3}{4}$ or £0·75 in the £.

The owner of the above house will now have to pay £150 $\times \frac{3}{4} = $ £112·50.

It is as well to point out that very often the estimated expenditure exceeds the rateable value of the borough.

Consider a borough with estimated expenditure of £4 500 000 and rateable value of £4 000 000.

The rate will be $\dfrac{4\,500\,000}{4\,000\,000} = \dfrac{9}{8}$ or £1·12½ in the £.

The owner with a house of rateable value £150 would have to pay £150 $\times \frac{9}{8}$ = £168·75.

The yield of a penny-rate is sometimes useful to make a quick estimate of the effects of any suggested improvements in the amenities of the borough on its rates.

The yield of a penny-rate is, of course, the amount of money raised by levying a rate of a penny in the £. If the rateable value of a borough is £3 600 000, then a penny-rate will raise 3 600 000p or £36 000. Suppose this borough proposes to erect a new swimming bath and social centre at a cost of £306 000. It may be seen at once that 306 000 ÷ 36 000p will have to be added to the rate to cover this expenditure, i.e. an increase of 8½p in the £.

Consequently, our owner with a house of rateable value £150 will be called upon to pay an extra 8½p $\times$ 150 = £12·75.

EXAMPLE 1 : The rateable value of a house is £120. The rates are 62½p in the £. How much does the owner pay this coming year?

SOLUTION: The rate is $\dfrac{62\frac{1}{2}}{100} = \dfrac{5}{8}$

∴ The owner pays £120 $\times \frac{5}{8} = $ £75 rates, ANS.

EXAMPLE 2: The yield of a penny-rate in a borough is £11 000. The cost of suggested improvements in Health services together with a pay rise for the Police is £38 500. What is the rateable value of the borough and what is the necessary increase in rate to cover these improvements?

SOLUTION: A rate of £1 in the £ will yield the rateable value. ∴ The rateable value is £11 000 $\times$ 100 = £1 100 000, ANS.

The increase in rate will be 38 500 ÷ 11 000p = 3½p, ANS.

WATER RATE

The other tax paid by house owners is the charge for water. This is in the form of a levy on the rateable value of the house supplied. There are extra charges for the use of garden hoses, sprinklers, etc.

Basic water rate is at present 3% in most parts of the country, i.e. the owner of a house of rateable value £170 must pay 3% of £170 = £5·10 per year for his water supply.

Practice Exercise No. 28

Note: Unless otherwise stated interest is to be considered as simple interest rather than compounded.

1. What is the interest on £188·60 for one year at $4\frac{1}{2}$%?
   (A) £9·43 ......   (B) £18·11 ......   (C) £8·49 ......   (D) £9·8 ......

2. Find the interest on £1850 for 2 years and 6 months at 4% a year.
   (A) £74·00 ......   (B) £185·00 ......   (C) £111·00 ......   (D) £277·50 ......

3. How much interest will be charged for a loan of £275 for 3 months if the rate is $4\frac{1}{2}$% a year?
   (A) £3·09 ......   (B) £12·37 ......   (C) £4·12 ......   (D) £3·45 ......

4. I lend £60 000 for 2 years. On $\frac{2}{3}$ of it I receive 4% interest, on the balance I receive 5% interest. What is the total amount of interest money I receive?
   (A) £2600 ......   (B) £5200 ......   (C) £520 ......   (D) £5600 ......

5. At what rate will £300 yield £67·50 interest in 4 years and 6 months?
   (A) 2% ......   (B) £3% ......   (C) 4% ......   (D) 5% ......

6. At what rate will £150 double itself in 14 years?
   (A) $6\frac{1}{4}$ ......   (B) $7\frac{1}{2}$ ......   (C) $4\frac{1}{6}$ ......   (D) $8\frac{1}{6}$ ......

7. What principal will yield £36 in 3 months at 3% interest?
   (A) £300 ......   (B) £4800 ......   (C) £2400 ......   (D) £1200 ......

8. How long will it take to earn £80·00 interest with a deposit of £800 at 6%?
   (A) $1\frac{1}{2}$ years ......   (B) $1\frac{2}{3}$ years ......   (C) $1\frac{3}{4}$ years ......   (D) 2 years ......

9. Using the compound interest table on page 47 calculate the amounts of the following investments: (i) £100 at 3% for 4 years; (ii) £1000 at 5% for 7 years; (iii) £10 at 7% for 3 years.

10. Calculate the principal which needs to be invested at 5% compound interest, paid annually, in order to realize an amount of £1000 in 8 years.

11. The rateable value of a house is £400. The rates are 57p in the £. How much does the owner pay this coming year?

12. The yield of a penny-rate in a borough is £24 000. Calculate the rateable value of the borough and the sum raised by increasing the rates by 2p.

PROFIT AND LOSS

To find profit or gain, *subtract the cost from the selling price.*

EXAMPLE 1: A man bought a motor car for £600 and sold it for £800. How much profit did he make?

SOLUTION: Selling price − cost = profit.
             £800 − £600 = £200, ANS.

To find the loss, *subtract the selling price from the cost.*

EXAMPLE 2: A dealer paid £800 for a machine and sold it the following year for £200. How much did he lose?

SOLUTION: Cost − selling price = loss.
             £800 − £200 = £600, ANS.

Note: In all profit and loss problems the words *gain* and *profit* have the same meaning and are used interchangeably.

When a percentage of profit or loss is given it is understood, unless stated to the contrary, that this percentage is based on the *cost*. *Thus*, if a man states simply that he sold something at a profit of 10% he is understood to mean that it was sold for an amount equal to its cost plus 10% of its cost. In modern business, however, it is quite customary to figure profit and loss as a percentage of *selling price*. This is because commissions, discounts, certain taxes, and other items of expense are commonly based on selling price, and in a complicated business it makes for simplicity in accounting to base profit and loss also on selling price.

There are, accordingly, two distinct kinds of profit and loss problems— those in which profit or loss is based on cost, and those in which profit or loss is based on selling price. Before such a problem can be solved it must be known in which of these classes it belongs. Bear in mind, though, that profit or loss is always to be considered as based on cost unless it is stated or otherwise known that it is based on selling price.

Profit and Loss Based on Cost

To find the percentage gain or loss, *divide the amount gained or lost by the cost.*

EXAMPLE 3: A wrench that cost 80 cents is sold at a profit of 20 cents. Find the percentage or rate of profit.

SOLUTION: Gain ÷ Cost = % profit.
$\frac{20}{80} = \frac{1}{4}$ or 25%,   ANS.

EXAMPLE 4: A book that cost 20p is sold for 8p. Find the percentage loss.

SOLUTION: Cost − Selling price = Loss.
20p − 8p = 12p loss.
Loss ÷ Cost = % loss.
$\frac{12}{20} = \frac{3}{5}$ or 60%,   ANS.

To find the gain and the selling price when the cost and the percentage gain are given, *multiply the cost by the percentage gain and add the result to the cost.*

EXAMPLE 5: Find the gain if the selling price of a desk that cost £48 represents a profit of $16\frac{2}{3}$%.

SOLUTION: Cost × Percentage gain = Gain.
£48 × $\frac{1}{6}$ = £8,   ANS.

To find the loss and the selling price when the cost and the percentage loss are given, *multiply the cost by the percentage loss and subtract the product from the cost.*

EXAMPLE 6: A damaged chair that cost £110 was sold at a loss of 10%. Find the loss and the selling price.

SOLUTION: Cost × Percentage loss = Loss.
£110 × $\frac{1}{10}$ = £11 loss,   ANS.
Cost − Loss = Selling price.
£110 − £11 = £99 selling price,   ANS.

To find the cost when the profit and the percentage profit are given or to find the cost when the loss and the percentage loss are given, *divide the profit or loss by the percentage profit or loss.*

EXAMPLE 7: A house was sold at a profit of £9000. The rate of profit was $37\frac{1}{2}\%$. What was the cost of the house?

SOLUTION: $37\frac{1}{2}\%$ or $\frac{3}{8}$ of the cost $= £9000$.

$$\text{Cost} = £9000 \div \tfrac{3}{8} = £\overset{£3000}{\cancel{9000}} \times \frac{8}{\cancel{3}}$$
$$= £24\,000, \quad \text{ANS.}$$

ALTERNATIVE SOLUTION:
$37\frac{1}{2}\% = \frac{3}{8}$
If $\frac{3}{8}$ of the cost $= £9000$, $\frac{1}{8}$ of the cost $= £3000$, and $\frac{8}{8} = 8 \times £3000 = £24\,000$ cost, ANS.

When the rate is reducible to a decimal fraction, *use decimal multiplication or division.*

EXAMPLE 8: A boat was sold at a loss of £120, representing a loss of 6%. What was the cost of the boat?

SOLUTION: 6% of the cost $= £120$
100% or total cost $= £120 \div 0.06$
$= £2000$ cost, ANS.

To find the cost when the selling price and the percentage profit are given, *divide the selling price by 1 plus the percentage profit.*

EXAMPLE 9: A table was sold for £80. This figure included a profit of 25%. Find the cost.

SOLUTION: Profit $= 25\%$ or $\frac{1}{4}$ of cost.
Total cost $= \frac{4}{4}$.
∴ Selling price $= 1 + \frac{1}{4} = \frac{5}{4}$ of cost.
$\frac{5}{4}$ of cost $= £80$.
∴ $£80 \div \frac{5}{4} = £64$ cost, ANS.

To find the cost when the selling price and the percentage loss are given, *divide the selling price by 1 minus the percentage loss.*

EXAMPLE 10: A silver set was sold for £168; the loss was 4%. How much did the silver set cost?

SOLUTION: Loss $= 4\%$ of cost. Total cost $= 100\%$.
Selling price $= 100\% - 4\% = 96\%$.
96% of cost $= £168$.
Cost $= £168 \div 0.96 = £175$, ANS.

Profit and Loss Based on Selling Price

Modern accounting practice favours the basing of profit and loss on selling price rather than on cost. This is because commissions and other selling expenses are calculated as percentages of selling price, and it simplifies accounting to base profit and loss on selling price also.

To find the percentage profit or loss, *divide the amount gained or lost by the selling price.*

EXAMPLE 1: A box of chocolates sells for $1.50 at a profit of 50c. What percentage of profit on selling price does this represent?

SOLUTION: Gain ÷ Selling price = % profit.
      $0.50 ÷ $1.50 = 0·33⅓ or 33⅓% profit, Ans.

EXAMPLE 2: On every radio selling for £40 a merchant lost £8. What was his rate of loss on selling price?

SOLUTION: Loss ÷ Selling price = % loss.
      £8 ÷ £40 = 0·20, or 20% loss, Ans.

To find the profit and the cost when the selling price and the percentage profit are given, *multiply the selling price by the percentage profit and subtract the result from the selling price.*

EXAMPLE 3: Eggs selling for £3 a gross carry a profit of 15% of selling price. Separate this selling price into cost and profit.

SOLUTION: Selling price × % profit = Profit.
      Selling price − Profit = Cost.
      £3 × 0·15 = £0·45 profit.
      £3 − £0·45 = £2·55 cost.

To find the loss and the cost when the selling price and the percentage loss are given, *multiply the selling price by the percentage loss and add the result to the selling price.*

EXAMPLE 4: At a clearance sale shoes selling at £6·00 are sold at a loss of 25% of selling price. What is the loss and the original cost?

SOLUTION: Selling price × % loss = Loss.
      Selling price + Loss = Cost.
      £6·00 × 0·25 = £1·50 loss   ⎫
      £6·00 + £1·50 = £7·50 cost ⎭ Ans.

To find the selling price when the profit and the percentage profit are given, or to find the selling price when the loss and the percentage loss are given, *divide the profit or loss by the percentage profit or loss.*

Note: This rule should be compared with the one under *Profit and Loss Based on Cost.* The two rules are exactly similar except that in one case 100% represents cost, while in the other case 100% represents selling price.

EXAMPLE 5: A kind of tape is selling at a profit of 12% on selling price, equal to 6p per metre. What is the tape selling at?

SOLUTION: Profit ÷ % profit = Selling price.
      6 ÷ 0·12 = 50p selling price, Ans.

EXAMPLE 6: A loss of £13·50 on a floor-waxing machine represents 15% loss on the selling price. What is the selling price?

SOLUTION: Loss ÷ % loss = Selling price.
      £13·50 ÷ 0·15 = £90·00 selling price, Ans.

To find the selling price when the cost and the percentage profit are given, *subtract the percentage profit from 100% and divide the cost by the remainder.*

EXAMPLE 7: A machine costing £90 to produce is sold at a profit of 40% of selling price. What is the selling price?

SOLUTION: Cost ÷ (100% − % profit) = Selling price.
      £90 ÷ 0·60 = £150 selling price, Ans.

To find the selling price when the cost and the percentage loss are given, *add the percentage loss to* 100% *and divide the cost by this sum.*

EXAMPLE 8: Socks that cost £0·25 per pair were sold at a loss of 25% of selling price. What did they sell for?

SOLUTION: Cost ÷ (100% + % loss) = Selling price.
£0·25 ÷ 1·25 = £0·20 selling price, ANS.

When profits are based on cost, profit is commonly referred to as **mark-up** over selling price, and the percentage profit on cost is called **percentage mark-up** to distinguish it from *percentage profit* (i.e. on selling price).

To reduce percentage profit on selling price to percentage mark-up (percentage profit on cost), *divide profit on selling price by* 100% *minus percentage profit on selling price.*

EXAMPLE 9: 40% profit on selling price is what percentage mark-up (percentage profit on cost)?

SOLUTION:
% profit on selling price ÷ (100% − % profit on selling price)
= % profit on cost.
0·40 ÷ 0·60 = 0·66⅔ or 66⅔% profit on cost, ANS.

To reduce percentage mark-up (percentage profit on cost) to percentage profit on selling price, *divide percentage mark-up by* 100% *plus percentage mark-up.*

EXAMPLE 10: A cricket bat marked up 25% carries what percentage of profit on selling price?

SOLUTION: % profit on cost ÷ (100% + % profit on cost)
= % profit on selling price.
0·25 ÷ 1·25 = 0·20 or 20% profit on selling price, ANS.

To reduce percentage loss on selling price to percentage loss on cost, *divide percentage loss on selling price by* 100% *plus percentage loss on selling price.*

EXAMPLE 11: 20% loss on selling price is what percentage loss on cost?

SOLUTION:
% loss on selling price ÷ (100% + % loss on selling price)
= % loss on cost.
0·20 ÷ 1·20 = 0·16⅔ or 16⅔% loss on cost, ANS.

To reduce percentage loss on cost to percentage loss on selling price, *divide percentage loss on cost by* 100% *minus percentage loss on cost.*

EXAMPLE 12: 33⅓% loss on cost is what percentage loss on selling price?

SOLUTION: % loss on cost ÷ (100% − % loss on cost)
= % loss on selling price.
0·33⅓ ÷ 66⅔ = ½
= 0·50 = 50% loss on selling price, ANS.

The four foregoing rules may appear confusingly difficult to distinguish, but they will be easily remembered if the student will ask himself whether the percentage required in the answer is to be larger or smaller than the one that is given. Where profits are concerned, the percentage of selling price is smaller than the percentage of cost. Where losses are concerned, the percentage of selling price is larger than the percentage of cost. If a smaller percentage is required in the answer, the given percentage is divided by 100% (or 1) plus the given percentage. If a larger percentage is required in the answer the given percentage is divided by 100% minus the given percentage. It should also be noted in these examples that when the given percentage is added to 100% the sum represents a cost or a selling price corresponding exactly with the *kind* of percentage wanted in the answer.

Practice Exercise No. 29

1. Apples cost 5c each. They are sold at a 20% profit. Find the selling price of one dozen apples.

    (A) 12c ......   (B) $1.00 ......   (C) 72c ......   (D) 60c ......

2. The Army bought 10 000 pairs of boots. They cost £2·50 a pair to manufacture and were sold to the Army at a 6% profit. What was the total selling price?

    (A) £26 500 ......   (B) £25 000 ......   (C) £1500 ......   (D) £40 000 ......

3. A dealer paid £200 for 10 books. He sold $\frac{2}{3}$ of them at £30 apiece and the remainder at £25 apiece. What was his per cent profit?

    (A) 70% ......   (B) 40% ......   (C) 35% ......   (D) 80% ......

4. A dealer made a profit of 25% in selling a boat for £500. How much did the boat cost him?

    (A) £300 ......   (B) £380 ......   (C) £620 ......   (D) £400 ......

5. A man having to get rid of his car in a hurry sold it at a 15% loss, which meant that he lost £75. How much did the car cost him originally?

    (A) £575 ......   (B) £425 ......   (C) £500 ......   (D) £600 ......

6. A department store sold a damaged bedroom suite at a $37\frac{1}{2}$% loss. The amount they lost was £90. What was the selling price of the bedroom suite?

    (A) £240 ......   (B) £150 ......   (C) £90 ......   (D) £330 ......

7. A farmer had to sell an inferior crop at a loss of $12\frac{1}{2}$%. He sold it for £259. What did the crop cost him to produce?

    (A) £296 ......   (B) £281·38 ......   (C) £295 ......   (D) £291·38 ......

8. A furniture dealer sold a piano for £806 at a 24% mark-up. How much did it cost him?

    (A) £871 ......   (B) £650 ......   (C) £612·56 ......   (D) £604·33 ......

9. If you bought two farms for £10 000 each, and sold one of them at a gain of 18%, and the other at a loss of 18%, how much did you make?

    (A) £1800 ......   (B) £00 ......   (C) £3600 ......   (D) £900 ......

10. You sold two building lots at £3600 each. On one your rate of profit was 20%, on the other your rate of loss was 20%. How much did you lose on the total transaction?

    (A) £00 ......   (B) £900 ......   (C) £300 ......   (D) £600 ......

11. The £4 which a dealer receives for a fountain pen includes $37\frac{1}{2}$% profit. What did it cost him? (*This style of wording indicates that profit is based on selling price.*)

    (A) £3·00 ......   (B) £2·00 ......   (C) £3·37$\frac{1}{2}$ ......   (D) £2·50 ......

12. A grocer sells $1000 worth of tea at a profit of $12\frac{1}{2}$% on the selling price. What did it cost him?

    (A) $900 ......   (B) $875 ......   (C) $1125 ......   (D) $800 ......

13. A hardware merchant sells nails at a mark-up of 15%. What percentage of the selling price is profit?

    (A) 13·04% ......   (B) 17·64% ......   (C) 30% ......   (D) 20% ......

14. 15% loss on selling price is what percentage loss on cost?

    (A) 13·04% ......   (B) 17·65% ......   (C) 15% ......   (D) 85% ......

15. A and B live in two different boroughs. A owns a house of rateable value £170 and pays a rate of £0·62½ in the £. B owns a house of rateable value £150 but pays a rate of £0·77½ in the £. Who pays the greater tax, and by how much?

    (A) A £6 more ...   (B) B £10 more ...   (C) B £5 more ...   (D) A £20 more ...

CHAPTER FIVE

DENOMINATE NUMBERS

A **denominate number** is one that refers to a unit of measurement which has been established by law or by general usage. Examples are: 1 *metre*, 8 *kilogrammes*, 3 *seconds*.

A **compound denominate number** is one that consists of two or more units of the same kind. Examples are: 1 *metre* 3 *millimetres*, 2 *hours* 15 *minutes*, 1 *kilogramme* 14 *grammes*.

Denominate numbers are used to express measurements of many kinds, such as:

(*a*) Linear (length)   (*b*) Square (area)   (*c*) Cubic (volume)
(*d*) Weight (newtons)   (*e*) Time (seconds)   (*f*) Angular (degrees)

This classification is by no means complete. Systems of currency (dollars and cents, pounds sterling and pence, etc.) would, for instance, be considered denominate numbers, and the various foreign systems of weights and measures would, of course, come under the same head, though they are beyond the scope of this book.

To gain facility in working out arithmetic problems involving denominate numbers it is necessary to know the most common tables of measures, such as are given here for reference. Fewer of these tables will be required in the future as we progress towards metrication, but, during the transition period many of us will need the reassurance which such tables can bring by translating the unfamiliar back to the familiar. Note the abbreviations used, since these are in accordance with the manner in which the values are usually written.

TABLE OF MEASUREMENTS

1. Length or Linear Measure

Linear units are used to measure distances along straight lines.

U.S. OR ENGLISH SYSTEM (with metric equivalent)

12 inches (in)   = 1 foot (ft) = 0·304 8 metres (m)
3 feet or 36 inches = 1 yard (yd) = 0·914 4 metres
1760 yards   = 1 mile (There is no abbreviation for mile.)
5280 feet   = 1 mile   = 1·609 34 kilometres (km)

NAUTICAL MEASURE

6080 feet = 1 UK nautical mile = 1·853 18 kilometres
1 fathom = 6 feet (of depth)   = 1·828 8 metres

The present UK nautical mile of 6080 feet is approximately 0·06 per cent greater than the International nautical mile of 1852 metres which is accepted by all other countries.

METRIC SYSTEM (with English equivalent)

	Unit		*Metres*		
	1 millimetre (mm)	=	0·001	=	0·039 37 in
10 millimetres	= 1 centimetre (cm)	=	0·01	=	0·393 7 in
1000 millimetres	= 1 metre (m)	=	1·0	=	39·370 1 in
1000 metres	= 1 kilometre (km)	=	1000·0	=	0·621 371 mile

2. Square Measure

This is used to measure the area of a surface; it involves two dimensions, length and width.

SQUARE AREA OF MEASURE

144 square inches (in²) = 1 square foot (ft²) = 0·092 903 m²
9 square feet   = 1 square yard (yd²) = 0·836 127 m²
640 acres   = 1 square mile   = 2·589 99 km²

METRIC SQUARE MEASURE

100 square millimetres (mm²)	= 1 square centimetre (cm²)	=	0·155 in²
100 square centimetres (cm²)	= 1 square decimetre (dm²)	=	15·5 in²
100 square decimetres (dm²)	= 1 square metre (m²)	=	1·195 99 yd²
10⁶ square millimetres	= 1 square metre		
100 square metres	= 1 are (a)	=	119·599 yd²
100 ares	= 1 hectare (ha)	=	2·471 05 acres
100 hectares	= 1 square kilometre (km²)	=	247·105 acres
10⁶ square metres	= 1 square kilometre		

3. Cubic Measure

This is used to find the volume or amount of space within the boundaries of three-dimensional figures. It is sometimes referred to as *capacity*.

CUBIC OR VOLUME MEASURE

1728 cubic inches (in³) = 1 cubic foot (ft³) = 0·028 316 8 m³
27 cubic feet (ft³) = 1 cubic yard (yd³) = 0·764 555 m³
1 cubic inch (in³) = 16·387 1 cm³

LIQUID MEASURE OF CAPACITY

$$
\begin{aligned}
4 \text{ gills (gi.)} &= 1 \text{ pint (pt.)} &&= 0\cdot568 \text{ litre (l)} \\
2 \text{ pints} &= 1 \text{ quart (qt.)} &&= 1\cdot137 \text{ litre} \\
4 \text{ quarts} &= 1 \text{ gallon (gal.)} &&= 4\cdot546 \text{ litre}
\end{aligned}
$$

The Imperial gallon is used in the United Kingdom.
1 Imperial gallon = 1·200 94 U.S. gallon

METRIC MEASURE OF CAPACITY

$$
\begin{aligned}
1000 \text{ cubic millimetres (mm}^3) &= 1 \text{ cubic centimetre (cm}^3) &&= 0\cdot061\,023\,7 \text{ in}^3 \\
1000 \text{ cubic centimetres} &= 1 \text{ cubic decimetres (dm}^3) &&= 0\cdot035\,314\,7 \text{ ft}^3 \\
1 \text{ cubic decimetre} &= 1 \text{ litre (l)} \\
1000 \text{ cubic decimetres} &= 1 \text{ cubic metre (m}^3) &&= 1\cdot307\,95 \quad \text{yd}^3 \\
10^9 \text{ cubic millimetres} &= 1 \text{ cubic metre}
\end{aligned}
$$

Note: All measurements of volume should be given in the basic units of m³, dm³. cm³, and mm³. A litre should be regarded as a special name for 1 dm³.

4. Measures of Mass

These are used to determine the quantity of matter a body contains.

ENGLISH SYSTEM

$$
\begin{aligned}
16 \text{ ounces (oz)} &= 1 \text{ pound (lb)} &&= 0\cdot453\,592\,37 \text{ kg} \\
14 \text{ pounds} &= 1 \text{ stone (st)} &&= 6\cdot350\,29 \text{ kg} \\
2 \text{ stone} &= 1 \text{ quarter (qtr)} &&= 12\cdot700\,6 \text{ kg} \\
112 \text{ pounds (lb)} &= 1 \text{ hundredweight (cwt)} &&= 50\cdot802\,3 \text{ kg} \\
2240 \text{ pounds (lb)} &= 1 \text{ ton} &&= 1016\cdot05 \text{ kg}
\end{aligned}
$$

METRIC SYSTEM

$$
\begin{aligned}
1000 \text{ milligrammes (mg)} &= 1 \text{ gramme (g)} &&= 0\cdot035\,274 \text{ oz} \\
1000 \text{ grammes} &= 1 \text{ kilogramme (kg)} &&= 2\cdot204\,62 \text{ lb} \\
1000 \text{ kilogrammes} &= 1 \text{ megagramme (Mg)} &&= 2204\cdot62 \text{ lb} \\
1000 \text{ kilogrammes} &= 1 \text{ tonne (t)}
\end{aligned}
$$

5. Measures of Time

$$
\begin{aligned}
1000 \text{ milliseconds} &= 1 \text{ second (s)} \\
60 \text{ seconds} &= 1 \text{ minute (min)} \\
1000 \text{ seconds} &= 1 \text{ kilosecond (ks)} \\
60 \text{ minutes} &= 1 \text{ hour (h)} \\
24 \text{ hours} &= 1 \text{ day (d)} \\
7 \text{ days} &= 1 \text{ week} \\
2 \text{ weeks} &= 1 \text{ fortnight} \\
365 \text{ days} &= 1 \text{ common year} \\
366 \text{ days} &= 1 \text{ leap year} \\
12 \text{ calendar months} &= 1 \text{ year} \\
10 \text{ years} &= 1 \text{ decade} \\
100 \text{ years} &= 1 \text{ century (C.)}
\end{aligned}
$$

6. Angular or Circular Measure

ANGULAR (∠) OR CIRCULAR (○) MEASURE

60 seconds (″)	=	1 minute (′)
60 minutes	=	1 degree (°)
90 degrees	=	1 right angle
360 angle degrees	=	4 right angles

7. Money

UNITED STATES MONEY   ENGLISH MONEY after Feb. 1971

10 mills (m)	= 1 cent (c., ct. or ¢)		2 new half pennies	=	1 new penny (p)
10 cents	= 1 dime (d.)		100 new pence	=	1 pound (£)
10 dimes	= 1 dollar ($)				
10 dollars	= 1 eagle (E.)				

REDUCTIONS INVOLVING COMPOUND UNITS

Reduction Descending

EXAMPLE: Reduce $3\frac{1}{4}$ days to units of lower denomination.

SOLUTION: $3\frac{1}{4}$ days $= \dfrac{13}{4} \times 24 = 13 \times 6 = 78$ h, ANS.

EXAMPLE: Reduce 2·98 hectares to square metres.

SOLUTION: 2·98 ha = 2·98 × 100 × 100 = 29 800 m², ANS.

Reduction Ascending

EXAMPLE: How many right angles in 470 degrees?

SOLUTION: 470 degrees $= \dfrac{470}{90} = \dfrac{47}{9} = 5\frac{2}{9}$ right angles, ANS.

EXAMPLE: How many tonnes in 4379 kg?

SOLUTION: 4379 kg $= \dfrac{4379}{1000} = 4\cdot379$ t, ANS.

Practice Exercise No. 30

1. How many millimetres in 1·5 metres?
2. How many metres in 1·53 kilometres?
3. How many grammes in 1·53 kilogrammes?
4. How many seconds in 1 hour?
5. How many minutes in 1 hour 15 minutes?
6. How many minutes in 0·5 right angles?
7. How many hours in 2 weeks and 3 days?
8. How many square metres in 3·98 ares?
9. How many square metres in 2·3 hectares?

10. How many square kilometres in 238·4 hectares?
11. How many hectares in 4·6 square kilometres?
12. How many cubic metres in 100 000 cubic decimetres?
13. How many right angles in 315 degrees?
14. How many metres in 1750 mm?
15. How many tonnes in 5600 kg?
16. How many ares in 2347 m²?
17. How many hectares in 56722 m²?

THE FOUR FUNDAMENTAL OPERATIONS APPLIED TO DENOMINATE NUMBERS

Addition

EXAMPLE: Add 3 days 12 h 36 min, 2 days 23 h 42 min, 1 day 14 h 34 min.

SOLUTION:

days	h	min
3	12	36
2	23	42
1	14	34
8	2	52 ANS.

EXPLANATION: List like units under one another. Add the minutes, take out the number of hours contained in them (1) and put down the remainder. Carry 1 to the hour column and proceed as before.

Subtraction

EXAMPLE: From 38° 4′ 38″ take 27° 16′ 49″

SOLUTION:

degrees	minutes	seconds
38	4	38
27	16	49
10	47	49 ANS.

EXPLANATION: To make the subtraction of seconds possible we convert one of the 4 minutes into 60 seconds to yield the result 98 − 49 = 49. To make the subtraction of minutes possible we convert one of the 38 degrees into 60 minutes to yield the result 63 − 16 = 47. The final result now follows. (We have here used the method of decomposition.)

The advantage of metric units is that the conversion factors are always multiples of ten so that each of the quantities concerned may be expressed as a simple decimal number as we now illustrate.

Multiplication

EXAMPLE: Multiply 2 hectares 53 ares 45 square metres by 5.

SOLUTION: We may express the given area in hectares. Thus 45 m² = 0·45 a so that we now have 53·45 a. But, 53·45 a = 0·5345 ha so that we now have 2·5345 ha as the given area.
The multiplication is now 2·5345 × 5 = 12·6725 ha,   ANS.
We may return this answer to the original form as 12 ha 67 a 25 m² if required.

EXAMPLE: Multiply 4 tonnes 750 kilogrammes 586 grammes by 8.

SOLUTION: This total mass may be expressed as 4·750 586 t.
Therefore 4·750 586 × 8 = 38·004 688 t,   ANS.
As in the previous example we may return to the original form quite easily, 38 t 4 kg 688 g.

Division

EXAMPLE: Divide 12 ha 84 a 36 m² by 8.

SOLUTION: We begin by expressing the given area as a decimal number. 36 m² = 0·36 a so that we now have 84·36 a which is 0·8436 ha, giving a final area of 12·8436 ha to be divided by 8, and yield 1·605 45 ha for our answer. On the other hand we could have adopted the following procedure:

SOLUTION:	*ha*	*a*	*m²*
8)12	84	36	
1	60	54	remainder 4 m².

EXPLANATION: 12 ÷ 8 = 1 with 4 ha remainder. 4 ha = 400 a, so that we now have 484 a ÷ 8 = 60 with remainder 4 a. 4 a = 400 m² giving 436 m² ÷ 8 = 54·5 m².

Practice Exercise No. 30a

1. Add the following: 50° 37′ 23″, 40° 28′ 42″, 18° 17′ 53″.
2. From 50° 45′ 32″ subtract 43° 54′ 39″.
3. Multiply 4 ha 56 a 46 m² by 10.
4. Divide 16 km² 72 ha 50 a by 6.
5. If a journey of 16 h 18 min is shortened by one-sixth how long will it take?
6. An estate of 15 km² 75 ha is to be divided into six equal lots. What is the area of one lot?
7. The mass of a car is 1·5 t. If a transporter has a mass of 3·6 t and carries six of these cars what is the total mass of the load?

Latitude, Longitude, and Time

The **latitude** of a place is its distance north or south of the equator, calculated in degrees, minutes, and seconds. (*See table of Angular or Circular*

Measure above.) Distance north of the equator is called **north latitude,** and distance south of the equator, **south latitude.**

The **longitude** of a place is its distance east or west of a given **prime meridian,** calculated in degrees, minutes, and seconds. Distance east of the prime meridian is called **east longitude,** and distance west of the prime meridian is called **west longitude.**

By international agreement the English-speaking countries and a number of others consider the prime meridian (or the meridian of 0° longitude) to be that which passes through Greenwich, a suburb of London, England.

The 180th meridian of longitude is designated as neither East nor West, since the same line on the earth's surface is both 180° east and 180° west of Greenwich.

The latitude and longitude of a place are arrived at by determining its position with relation to the sun and the time of day. Observation of the sun is performed with an optical instrument called, according to its type, a *sextant* or an *octant*. The procedures incident to calculating position from such observations constitute the science of **navigation.**

Differences of latitude or **longitude** between two places are calculated by the following rules:

1. *If the latitudes of the two given places are both north or both south, subtract the lesser latitude from the greater; but if one is north and the other south, add the two latitudes.*

2. *If the longitudes of the two given places are both east or both west, subtract the lesser longitude from the greater; but if one is east and the other west, add the two longitudes.*

3. *If the sum of two longitudes exceeds* 180°, *this sum is subtracted from* 360° *to obtain the correct difference in longitude.*

EXAMPLE 1: The longitude of New York is 74° 0′ 3″ West; of Philadelphia 75° 10′ West. What is their difference in longitude?

SOLUTION:

	75°	10′	0″
	74	0	3
	1°	9′	57″ Ans.

EXPLANATION: Since New York and Philadelphia are both west of the prime meridian of Greenwich, their difference in longitude must be the distance by which Philadelphia is farther west than New York. This distance is found by subtracting the lesser longitude of New York from the greater longitude of Philadelphia.

EXAMPLE 2: What is the difference in longitude between New York and Paris? The longitude of the former city is 74° 0′ 3″ West; that of the latter city is 2° 20′ 22″ East.

SOLUTION:

	74°	0′	3″
	2	20	22
	76°	20′	25″ Ans.

EXPLANATION: Since the distance from Paris to the prime meridian is 2° 20′ 22″, and from the prime meridian to New York is 74° 0′ 3″ farther on in the same direction, the total distance or difference in longitude between these two places must be the sum of these two distances.

The relation of time to longitude is described in the following:

The daily revolution of the earth on its axis causes the sun to appear to pass from east to west over the 360° of the earth's longitude in 24 hours. Hence in 1 hour the sun appears to pass over $\frac{1}{24}$ of 360°, or 15° of longitude; in 1 minute of time, $\frac{1}{60}$ of 15°, or 15′ of longitude; and in 1 second of time, $\frac{1}{60}$ of 15′, or 15″ of longitude. Hence 15° longitude corresponds to 1 hour of solar time; 15′ of longitude corresponds to 1 minute of solar time; and 15″ of longitude corresponds to 1 second of solar time.

The solar time (sun time) of any place depends upon the sun's relative position. It is 12 o'clock M, or noon, at a place when the sun crosses its meridian. Hence, if it is noon at our meridian, it is afternoon (p.m. or *post meridian*) at all places east of us, and forenoon (a.m. or *ante meridian*) at all places west of us.

The United States uses what is known as **Standard Time.** The country is divided into four time zones, each approximately 15° of longitude in width. Every place in each of these zones uses the same *standard time* instead of its own local sun time.

The four time zones are designated as Eastern, Central, Mountain, and Pacific, and use the solar time, respectively, of the 75th, 90th, 105th, and 120th meridians of West Longitude.

Foreign standard time. When it is 12 o'clock noon Eastern Standard Time the standard time in various other cities is as follows:

Calcutta	10.53 p.m.	Moscow	8.00 p.m.
Cape Town	6.15 p.m.	Paris	5.00 p.m.
Honolulu	6.30 a.m.	Rome	6.00 p.m.
Istanbul	7.00 p.m.	Shanghai	1.00 a.m. next day
Leningrad	8.00 p.m.	Sydney	3.00 a.m. next day
London	5.00 p.m.	Tokyo	2.00 a.m. next day

Daylight saving time originated in England during the First World War. By setting the clock an hour ahead during late spring, summer, and early autumn, an hour of daylight is taken off the beginning of the clock-day and added to the evening.

To find the difference in solar time between two places, *divide the difference in longitude in degrees, minutes, and seconds by 15, and the quotient will be the difference in solar time expressed respectively in hours, minutes, and seconds.*

EXAMPLE: What is the difference in solar time between New York and Greenwich? The longitude of New York is 74° 0′ 3″ West.

SOLUTION:

$$15 \overline{)74° \quad 0′ \quad 3″}$$
4 hr. 56 min. 0 sec. ANS.

EXPLANATION: Division is performed in the regular manner used with denominate numbers. The final remainder, $\frac{3}{15}$ or $\frac{1}{5}$ second is discarded.

To find the difference in longitude between two places when their difference in solar time is known, *multiply the difference in solar time in hours, minutes, and seconds by 15, and the product will be the difference in longitude expressed respectively in degrees, minutes, and seconds.*

EXAMPLE: The difference in solar time between Berlin and Paris is 44 min. $13\frac{3}{4}$ sec. If the longitude of Berlin is 13° 23′ 45″ East, and if Paris is west of Berlin, what is the longitude of Paris?

SOLUTION: 44 min. 13¾ sec.
                    15

                11°   3'   24"
                13 23' 45" East

                 2° 20' 21" East ANS.

EXPLANATION: Multiply the difference in time by 15 to find the difference in longitude, amounting to 11° 3' 24". Since Paris is west of Berlin, the longitude of Paris must be 11° 3' 24" nearer to the prime meridian, which is also west of Berlin. Hence, the difference in longitude is subtracted from the longitude of Berlin to find the longitude of Paris.

Since time is reckoned both east and west of Greenwich, it follows that time on the 180th meridian might be considered either 12 hours earlier or 12 hours later than that of Greenwich. Instead of taking the 180th meridian as the strict dividing line between eastern time and western time, agreement among the nations has established the International Date Line, which follows a somewhat zigzag course in the neighbourhood of the 180th meridian. This has been done so that Pacific islands in the same group may use the same time. When it is noon at Greenwich it is shortly after midnight on the morning of the same day at places slightly east of the International Date Line, and shortly before midnight on the night of the same day at places slightly west. When it is 1 p.m. at Greenwich it is about 1 a.m. of the same day at the former places and 1 a.m. of the *following day* at the latter places.

Army–Navy Time

In the military, naval, and air services time is reckoned by the 24-hour clock. Starting at midnight, or 0000, the hours are indicated by hundreds. 8 a.m. is 0800; 12 m. is 1200; 1 p.m. is 1300; 11 p.m. is 2300. Minutes are indicated by units—preceded by a 0 if less than 10, since there must always be four digits. 8.10 a.m. is 0810; 12.05 p.m. is 1205; 11.59 p.m. is 2359; 12.01 a.m. is 0001.

Practice Exercise No. 30b

1. The latitude of Washington is 38° 53' 39" North and that of Montreal is 45° 35' North. What is their difference in latitude?
2. The longitude of Calcutta is 88° 19' 2" East and that of San Francisco 122° 26' 45" West. What is their difference in longitude? (*Hint:* Subtract from 360°.)
3. The longitude of New Orleans is 90° 2' 23" West; that of San Francisco is given in Question 2. What is their difference in solar time?
4. The longitude of Cape Town is 18° 28' East and that of Adelaide is 138° 35' East. What is their difference in solar time?

Exchange Rate

The **exchange rate** on money expresses the value of the money of one country in terms of the money of another. Thus, if the **exchange rate** on English money is $U.S. 2.40, it means that one pound is considered to be worth $2.40 of U.S. money. Exchange rates vary according to the conditions under which money is wanted and also fluctuate with general conditions in the money market. Current exchange rates are quoted daily in the newspapers.

Note: It should be observed that when an individual buys foreign money on a small scale he does not secure the full advantage of the exchange rate, since the bank or money broker charges a small commission on the transaction. In the following examples and exercises such commissions are disregarded.

EXAMPLE 1: If the rate of exchange for the pound is $2.40, how much is £4·25 worth in U.S. money?

SOLUTION: £1 is worth $2.40
∴ £4·25 is worth $2.40 × 4·25 = $10.20,   ANS.

EXAMPLE 2: If the rate of exchange for the pound is $2.40, how much is $34.00 worth in English money?

SOLUTION: $2.40 is worth £1
$4.80 is worth £4·80 ÷ 2.40 = £2
∴ $34.00 is worth 34.00 ÷ 2.40 = £14·17,   ANS.

(The answer in the last example should be £14·16⅝, but in practice banks will round off the final figure.)

Fluctuations in the exchange rate can result in a profit or loss for someone exchanging currency at the appropriate time. To the ordinary man in the street going abroad for a holiday the size of the profit or loss is hardly a cause for concern, but for someone carrying out transactions involving millions of pounds the profit or loss could be considerable.

EXAMPLE: A holidaymaker visiting the U.S.A. exchanges £50 when the exchange rate is $2.40. On his return he has £10 left, which he exchanges at a rate of $2.35. Calculate his profit due to the exchange.

SOLUTION: We need only consider the profit on the £10, since he has presumably spent the other £40.
On the way out his £10 is worth $2.40 × 10 = $24.00
On the way back his $24.00 is worth £24 ÷ 2.35 = £10 21
∴ The holidaymaker has made a profit of £0·21,   ANS.

However, the fluctuation in the exchange is normally much smaller, but even so presents a large profit to the speculator in the next example.

EXAMPLE: A speculator moves £1 000 000 into the U.S.A. at a time when the exchange rate is $2.40 and withdraws it shortly afterwards when the rate is $2.39. Calculate his profit due to the exchange.

SOLUTION: On buying into the U.S.A. he receives $2.40 × 1 000 000 = $2 400 000
On buying out of the U.S.A. he receives £2 400 000 ÷ 2.39
= £1 004 184·10
∴ The speculator makes a profit of £4 184·10,   ANS.

The profit in the last example seems considerable, but would be regarded as much too small to justify the movement of such a large sum of money, especially since it disregards any commission which would have to be paid to the money broker. The revaluation of a currency provides the best opportunity for the speculator, as the following example shows.

EXAMPLE: Anticipating a revaluation of the U.S. dollar, a speculator moves £1 000 000 into the U.S.A. when the exchange rate is $2.40. A month later the revaluation takes place and the exchange rate becomes $2.00. If he now withdraws the money, what is his profit?

SOLUTION: On buying out of the U.S.A. he receives $\dfrac{£1\,000\,000 \times 2.40}{2.00} = £1\,200\,000$

∴ The speculator makes a profit of £200 000.

The size of the possible profit would make him consider it worthwhile to borrow the money at a high rate of interest, such as 20%, for the short time involved—hence the name speculator!

Practice Exercise No. 30c

1. When the exchange rate is $2.40 what is the value in United States money of £6·50?

2. How much English money can be purchased for $A1 when the rate of exchange is $A2.5? At the same rate what is £10 worth in Australian money?

3. A speculator moves £1 000 000 into the U.S.A. when the exchange rate is $2·39. The expected revaluation does not take place and he has to buy out when the rate is $2.40. Find his loss.

Surface Measurements

The *area* of a rectangular surface is the number of square units which it contains. (See also p. 160.)

In finding an *area* the unit of measure is a square, each side of which is a unit of the same denomination as the given dimensions. Hence to find the area of a rectangular surface 8 m long and 5 m wide, the measuring unit will be 1 m², since the denomination of the length and width is metres.

To find the area of a rectangular surface, *multiply the two dimensions.*

EXAMPLE: How many square metres are in a path 48 m long and 11·2 m wide?

SOLUTION: $\qquad 11·2 \times 48 = 537·6$ m², ANS.

EXPLANATION: Taking 1 m² as the unit, a path 48 m long and 1 m wide would contain 48 m². Hence a path of equal length and 11·2 m wide must contain 11·2 × 48 or 537·6 m².

To find either dimension of a rectangular surface when the other dimension and the area are given, *divide the area by the given dimension.*

EXAMPLE: If a rectangular field containing 30 ares is 40 m wide, what is its length?

SOLUTION: $\qquad$ 30 ares = 3000 m²
$\qquad\qquad$ 3000 ÷ 40 = 75 m, ANS.

EXPLANATION: A field 40 m long and 1 m wide would contain 40 m². Therefore to contain 30 ares or 3000 m², a field 40 m wide would have to be as many metres long as 40 m² are contained in 3000 m².

In estimating the cost of paving, painting, plastering, roofing, etc., the basic procedure is to multiply the area of the surface to be covered by the unit cost per square metre. There are a number of special considerations, however, and in any case the cost of such work is likely to be governed by local trade customs and the practices of individual contractors.

In the case of *painting* and *plastering*, **allowances** must be made for openings. That is to say, the area of doors and windows must be deducted from the area represented by the overall dimensions of the walls. Similarly, deductions may have to be made for baseboards, wainscotings, and the spaces concealed by tubs, ranges, and the like.

In estimating the **quantity of paint** necessary to cover a surface, the area of the surface is divided by the covering capacity of the paint as stated on the tin or in the literature of the manufacturer. *Thus*, if a litre of paint is rated to cover 40 m² with two coats a surface of 100 m² would require $2\frac{1}{2}$ l for two coats or $1\frac{1}{4}$ l for one coat.

To find the number of pieces of material necessary to cover a given surface, *divide one of the dimensions of the surface by one of the dimensions of the piece, and the other dimension of the surface by the other dimension of the piece, and multiply these two quotients.*

EXAMPLE: How many tiles 250 mm square will be necessary to floor a room measuring 3·6 m by 2·8 m?

SOLUTION:

$$3{\cdot}6 \text{ m} = 360 \text{ cm.} \qquad 360 \div 25 = 14\tfrac{2}{5}$$
$$= 15 \text{ tiles}$$
$$2{\cdot}8 \text{ m} = 280 \text{ cm.} \qquad 280 \div 25 = 11\tfrac{1}{5}$$
$$= 12 \text{ tiles}$$
$$15 \times 12 = 180 \text{ tiles,} \quad \text{Ans.}$$

EXPLANATION: Along the 3·6 m side the exact measurement equals $14\frac{2}{5}$ tiles but it will be necessary to take this as 15 tiles. Similarly, a width of 12 tiles must be taken. 15 rows of 12 tiles each come to a total of 180 tiles.

To find how many yards of material are needed to cover a given surface, *determine how many strips will be necessary and the length of each strip. The total area of all strips in square yards will be the required yardage.*

EXAMPLE: If 50 mm are to be allowed for turning under on all sides of the room how many metres of carpet 700 mm wide will be required for a room measuring 5 m by 6 m?

SOLUTION:

$$5 \text{ m} \div 0{\cdot}7 \text{ m} = 7\tfrac{1}{7}. \qquad 6 \text{ m} \div 0{\cdot}7 \text{ m} = 8\tfrac{4}{7}$$
$$6 \text{ m} + 0{\cdot}1 \text{ m} = 6{\cdot}1 \text{ m} \qquad 5 \text{ m} + 0{\cdot}1 \text{ m} = 5{\cdot}1 \text{ m}$$
$$6{\cdot}1 \times 8 = 49 \text{ m} \qquad 5{\cdot}1 \times 9 = 46 \text{ m,} \quad \text{Ans.}$$

EXPLANATION: In examples of this kind it is often necessary and always safest to determine in which direction the carpet may be laid more economically. If we divide 5 m by 0·7 m we get $7\frac{1}{7}$ which means 8 strips. If we divide 6 m by 0·7 m we get $8\frac{4}{7}$ which means 9 strips. Allowing 50 mm for turning under on both ends of each strip, we find that in one case the strip would have to be 6·1 m long while in the other case they would be 5·1 m long. 9 strips of 5·1 m give a smaller yardage than 8 strips of 6·1 m. Hence we select 46 m as the correct answer.

Practice Exercise No. 31

1. What is the difference between 8 square metres and 8 metres square?

    (A) 16 m ......   (B) 16 m² ......   (C) 56 m² ......   (D) 64 m² ......

2. A fence surrounding a 4 km track is 2 m high. How many square metres does it contain?

    (A) 6000 ......   (B) 7000 ......   (C) 8000 ......   (D) 7500 ......

3. What is the cost of flooring a room 5 m by 6 m with 700 mm wide carpet at £3 per metre? Make no allowance for turning under or matching pattern.

    (A) £250 ......   (B) £135 ......   (C) £150 ......   (D) £175 ......

4. A farm 3·6 km long and 1·6 km wide was bought for £500 per hectare. How much did the farm cost?

    (A) £300 000 ......   (B) £280 000 ......   (C) £288 000 ......   (D) £320 000 ......

5. A path 50 m long and 5 m wide is paved with bricks measuring 200 mm by 100 mm. How many bricks are in it?

    (A) 12 500 ......   (B) 12 750 ......   (C) 13 500 ......   (D) 15 000 ......

6. A roof is resurfaced at a cost of £75. If the area was 10 m by 4·5 m what was the cost per square metre?

    (A) £5 ......   (B) £4·5 ......   (C) £1·$\dot{6}$ ......   (D) £5·25 ......

Cubic Measurement

The **cubical contents** or the **volume** of a rectangular solid is the number of cubic units which it contains. Another synonymous term is **capacity**. (See also p. 168.)

The unit of *cubical measure* is a cube, each edge of which is a unit of the same denomination as the three given dimensions. Hence, to find the cubical contents of a rectangular solid which is 6 m long, 4 m wide, and 3 m deep, the measuring unit will be 1 m⁰, since the denomination of the length, width, and depth is metres.

To find the cubical contents of a rectangular solid, *multiply together the three dimensions.*

EXAMPLE: What are the cubical contents of a box 1·8 m long, 1·6 m wide, and 0·2 m in length?

SOLUTION: $1·8 \times 1·6 \times 0·2 = 0·576$ m³, ANS.

EXPLANATION: Imagining cubes measuring 1 m each way, we find that a row of 1·8 such cubes could be laid along the longest dimension of the box. The 1·6 m of width would accommodate 1·6 rows of 1·8 m³ each, and the depth of the box would have room for 0·2 layers of 1·8 × 1·6 cubes. Hence the total comes to $1·8 \times 1·6 \times 0·2$ m³.

To find a third dimension of a rectangular solid when the other two dimensions and the cubical contents are given, *divide the cubical contents by the product of the two given dimensions.*

EXAMPLE: A block of stone 0·7 m long and 0·5 m wide contains 1·4 m³. What is its height?

SOLUTION: $$140 \div (7 \times 5) = 4 \text{ m.}$$

EXPLANATION: A block 0·7 m long, 0·5 m wide, and 1 m high would contain 0·35 m³. Hence, to contain 1·4 m³, a block of this length and width would have to be as many times 1 m high as 35 is contained in 140.

Practice Exercise No. 32

1. How many cubic metres in a box 1·4 m × 1·3 m × 1·2 m?
2. How many cubic metres in a box 1·2 m × 0·6 m × 0·5 m?
3. How many cubic metres in a block 1 m × 4 m × 7 m?
4. How many cubic millimetres in 1 m³?
5. How many cubic millimetres in 1 cm³?
6. How many cubic metres in a metre cube?

THE METRIC SYSTEM

Since the metric system is based on decimal values, all ordinary arithmetical operations may be performed by simply moving the decimal point.

Consider the quantity of 4·567 metres. It is made up of the following units:

metres	decimetres	centimetres	millimetres
4	5	6	7

We may read this quantity as 4·567 metres, or as 45·67 decimetres, or as 456·7 centimetres, or as 4567 millimetres.

When it comes to dealing with square and cubic measures we must simply bear in mind that instead of moving the decimal point one place for each successive change in unit value, we move it two places in the case of square measurements and three places in the case of cubic measurements.

Thus, take 42·365 783 m². This consists of 42 m², 36 dm² 57 cm², 83 mm². It may be read as 42·365 783 m², or as 4236·5783 dm² or as 423 657·83 cm², or as 42 365 783 mm².

Now take 75·683 256 m³. It is composed of 75 m³, 683 dm³, 256 cm³. It may be read as 75·683 256 m³, or as 75 683·256 dm³, or as 75 683 256 cm³.

The metric system is a system of related masses and measures. The metre is the basis from which all other units are derived. The unit of mass, the gram, is the mass of a cubic centimetre of water (under certain conditions). The unit of capacity, the litre, is the volume of 1 kg (1000 g) of water, and thus is represented by 1000 cm³.

Practice Exercise No. 33

1. How many kilometres in 3746·23 m?
2. How many metres in 4·253 km?
3. How many square kilometres in 85·46 m²?
4. How many square millimetres in 47·386 dm²?
5. How many cubic centimetres in 3·56 m³?
6. How many cubic metres in 374 658 mm³?
7. How many kilograms in in 3426 g?

In the following tables wherever the metric equivalents of standard measures are given, metric equivalents of other denominations may be found by simply moving the decimal point to the right or the left as may be necessary.

Equivalent Values

LINEAR MEASURE

1 inch	=	2·5400	centimetres
1 foot	=	0·3048	metre
1 yard	=	0·9144	metre
1 mile	=	1·6093	kilometres
1 centimetre	=	0·3937	inch
1 decimetre	=	3·9370	inches
1 decimetre	=	0·3281	foot
1 metre	=	39·3700	inches
1 metre	=	3·2808	feet
1 metre	=	1·0936	yards
1 kilometre	=	3280·83	feet
1 kilometre	=	1093·611	yards
1 kilometre	=	0·62137	mile

SQUARE MEASURE

1 square inch	=	6·451 6	square centimetres
1 square foot	=	0·092 9	square metre
1 square yard	=	0·836 1	square metre
1 acre	=	4 046·86	square metres
1 acre	=	0·404 686	hectare
1 square mile	=	258·999	hectares
1 square mile	=	2·589 99	square kilometres
1 square centimetre	=	0·155	square inch
1 square decimetre	=	15·500	square inches
1 square metre	=	1 550·000	square inches
1 square metre	=	10·764	square feet
1 square metre	=	1·195 99	square yards
1 hectare	=	2·471 05	acres
1 square kilometre	=	247·105	acres
1 square kilometre	=	0·386 1	square mile

The hectare is the unit of land measure.

CUBIC MEASURE

1 cubic inch	= 16·387 1	cubic centimetres
1 cubic foot	= 28·316 8	cubic decimetres
1 cubic yard	= 0·764 555	cubic metre

1 cubic centimetre	= 0·061 02	cubic inch
1 cubic decimetre	= 0·035 31	cubic foot
1 cubic metre	= 1·307 95	cubic yards

CAPACITY

1 gallon U.S.	= 3·785	litres
1 gallon U.K.	= 4·546	litres

The millilitre is equivalent in volume to a cubic centimetre.

MASS

1 ounce	=	28·350 grammes
1 pound	=	0·4536 kilogramme
1 ton	=	1·0160 tonne
1 gramme	=	0·0353 ounce
1 kilogramme	=	2·2046 pounds
1 tonne	=	0·9842 ton
1 tonne	=	2204·6223 pounds

Practice Exercise No. 34

1. Reduce 385·25 hectares to acres.
2. How many miles in 153 kilometres?
3. Change 75·5 kg to pounds.
4. How many metres in 87 yd.?
5. Change 157·35 acres to hectares.
6. How many grammes in 8 ounces?

CHAPTER SIX

RATIO AND PROPORTION

RATIO

A *ratio* is the relation between two like numbers or two like values. The ratio may be written as a fraction, $\frac{3}{4}$; as a division, $3 \div 4$; or with the colon or *ratio sign* (:), 3 : 4. When the last of these forms is used, it is read, 3 *to* 4; or 3 *is to* 4. Ratios may be expressed by the word *per*, as in litres per day or revolutions per second. In arithmetic these are written litres/day, revolutions/minute, volts/ampere. Whatever the manner of writing the ratio, its value in arithmetical computations is always the same.

Since a ratio may be regarded as a fraction, you will recognize the following principle as being true:

Rule 1. *Multiplying or dividing both terms of a ratio by the same number does not change the value of the ratio.*

Thus, $2 : 4 = 4 : 8$ (multiplying both terms by 2)
or $2 : 4 = 1 : 2$ (dividing both terms by 2)

To reduce a ratio to its lowest terms, *treat the ratio as a fraction and reduce the fraction to its lowest terms.*

EXAMPLE 1: Express $\frac{2}{3}$ to $\frac{4}{9}$ in its lowest terms.

SOLUTION:

$$\frac{2}{3} \text{ to } \frac{4}{9} = \frac{2}{3} \div \frac{4}{9} = \frac{2}{3} \times \frac{9}{4} = \frac{3}{2}.$$

Hence $\frac{2}{3}$ to $\frac{4}{9}$ is the same as 3 to 2.

To separate a quantity according to a given ratio, *add the terms of the ratio to find the total number of parts. Find what fractional part each term is of the whole. Divide the total quantity into parts corresponding to the fractional parts.*

EXAMPLE 2: Three hundred tents have to be divided between two army divisions in the ratio of $1 : 2$. How many does each division get?

$1 + 2 = 3$ (adding the terms)

$\left.\begin{array}{l} \frac{1}{3} \times 300 = 100 \\ \frac{2}{3} \times 300 = 200 \end{array}\right\}$ ANS.   (Taking corresponding fractional parts of total quantity)

Check: 100 : 200 or $\frac{100}{200} = \frac{1}{2}$ or $1 : 2$.

EXAMPLE 3: 1600 kg of coffee have to be distributed to 3 wholesale dealers in the ratio of $8 : 11 : 13$. How many kg should each dealer receive?

SOLUTION: $8 + 11 + 13 = 32$

$\frac{8}{32}, \frac{11}{32}, \frac{13}{32}$ are the fractional parts

$\left.\begin{array}{l} \frac{8}{32} \times 1600 = 400 \\ \frac{11}{32} \times 1600 = 550 \\ \frac{13}{32} \times 1600 = 650 \end{array}\right\}$ ANS.

Practice Exercise No. 35

1. Reduced to its lowest terms, 24 : 32 equals what?
   (A) $\frac{1}{3}$ ......   (B) $\frac{1}{2}$ ......   (C) $\frac{6}{8}$ ......   (D) $\frac{3}{4}$ ......

2. What is the value of the ratio $7 \times 9 : 8 \times 7$?
   (A) $\frac{8}{9}$ ......   (B) $1\frac{1}{8}$ ......   (C) $8 : 9$ ......   (D) $1\frac{23}{49}$ ......

3. If 5 kg of vegetables lose 10 g in drying, what part of the original mass was water?
   (A) $\frac{1}{6}$ ......   (B) $\frac{1}{8}$ ......   (C) 12% ......   (D) $\frac{1}{500}$ ......

4. A mixture requires 2 parts of water to 3 parts of alcohol. What percentage of the mixture is water?
   (A) 40% ......   (B) 50% ......   (C) 60% ......   (D) $66\frac{2}{3}\%$ ......

5. Bronze consists of 6 parts tin to 19 parts copper. How many grammes of tin are there in a 500 g bronze statue?
   (A) 100 ......   (B) 120 ......   (C) 140 ......   (D) 200 ......

6. £2000 is to be distributed among 3 members of a family in the ratio of 5 : 14 : 21. How much greater is the largest share than the smallest share?
   (A) £900 ......   (B) £800 ......   (C) £500 ......   (D) £750 ......

PROPORTION

A **proportion** is a statement of equality between two ratios. It may be written with the double colon or **proportion sign** (::), or with the sign of equality (=).

Thus, 2 : 6 :: 3 : 9 is a proportion that is read, 2 *is to* 6 *as* 3 *is to* 9; or $\frac{2}{6}$ *equals* $\frac{3}{9}$.

In any proportion the first and last terms are called the **extremes** and the second and third terms are called the **means**. In 2 : 6 :: 3 : 9 the *extremes* are 2 and 9; and the *means* are 6 and 3.

Multiply the two extremes and the two means of the proportion 2 : 6 :: 3 : 9 and compare the products.

$$\text{Extremes: } 2 \times 9 = 18$$
$$\text{Means: } \quad 6 \times 3 = 18$$

This result illustrates **Rule 2.** *The product of the* means *is equal to the product of the* extremes.

If you write the proportion in the form of $\frac{2}{6} = \frac{3}{9}$, note that the means and extremes are diagonally opposite each other. This affords another way to pick out your equation.

No proportion is a true proportion unless the two ratios are equal. This is another way of saying that Rule 2 must be satisfied.

By means of this rule you can find the missing term of any proportion if the other 3 terms are given.

EXAMPLE 1: 2 : 6 = 8 : ? Find the value of the missing term. The letter x is traditionally used to denote a missing term or an unknown quantity. Rewriting the proportion, we get

$$2 : 6 :: 8 : x$$

(a) 2 times x = 6 times 8   (a) Product of the extremes equals product of the
    $2x = 48$ means.

(b) $\quad \dfrac{2x}{2} = \dfrac{48}{2}$ (b) Both sides of any equation may be divided by the same number without changing the equation.

$x = 24$, Ans.

The above process is the equation method of solving problems containing an unknown. This process will be treated at greater length in Chapter Eight, which deals with elementary algebra.

If you wish to use a strict arithmetic method of finding the missing term in a proportion you may employ the following two rules.

Rule 3. *The product of the means divided by either extreme gives the other extreme as the quotient.*

2 : 6 :: 8 : 24;
$6 \times 8 = 48$, $48 \div 2 = 24$, $48 \div 24 = 2$.
Thus if given   ? : 6 = 8 : 24,
multiply the two means, $6 \times 8 = 48$,
and divide this product by the known extreme; $48 \div 24 = 2$. The quotient here is the unknown term.

Rule 4. *The product of the extremes divided by either mean gives the other mean as a quotient.*

$2 : 6 = 8 : 24;$

$2 \times 24 = 48, 48 \div 6 = 8, 48 \div 8 = 6.$

Thus if given $2 : ? :: 8 : 24,$

multiply the two extremes, $2 \times 24 = 48$, and divide this product by the known mean; $48 \div 8 = 6$. The quotient again is the unknown term.

Practice Exercise No. 36

Find the missing term.

1. $2 : 3 :: 4 : ?$ 2. $20 : 10 :: ? : 6$ 3. $2 : ? :: 8 : 24$ 4. $18 : ? :: 36 : 4$
5. $12 : 4 :: ? : 7$ 6. $5 : ? :: 25 : 20$ 7. $? : 5 :: 12 : 20$ 8. $? : 25 :: 10 : 2$
9. $9 : ? :: 24 : 8$ 10. $24 : 4 :: ? : 3$

Problems in Proportion

In solving problems by the ratio and proportion method it is first necessary to recognize whether a proportion exists, and if so what kind it is.

A direct proportion is indicated when two quantities are so related that an increase in one causes a corresponding increase in the other or when a decrease in one causes a corresponding decrease in the other.

The following is a list of typical quantitative expressions in which the variables are directly related when other quantities remain unchanged:

(a) The faster the speed, the greater the distance covered.

(b) The more men working, the greater the amount of work done.

(c) The faster the speed, the greater the number of revolutions.

(d) The higher the temperature of gas, the greater the volume.

(e) The taller the object, the longer the shadow.

(f) The larger the quantity, the greater the cost.

(g) The smaller the quantity, the lower the cost.

(h) The greater the length, the greater the area.

(j) The greater the base, the larger the discount, commission, interest, and profit.

EXAMPLE 2: If 20 men assemble 8 machines in a day, how many men are needed to assemble 12 machines in a day?

SOLUTION:

8 machines need 20 men

12 machines need ? men

$8 : 12 :: 20 : x$

$8x = 240$

$x = 30,$ ANS.

EXPLANATION: Place corresponding values on the same line. Put *like numbers* together. The more machines, the more men needed. ∴ The values are a direct proportion. Solve for x.

EXAMPLE 3: If 12 drills cost £8, how much will 9 drills cost?

SOLUTION:

12 drills cost £8

9 drills cost?

$12 : 9 :: 8 = x$

$12x = 72, x = £6.$

EXPLANATION: The fewer the drills, the lower the cost. ∴ The values are in direct proportion. Solve for x.

Examples 2 and 3 are easily recognized as direct proportions, since more men can assemble more machines and fewer drills cost less money.

Cues in Solving Proportion Problems

In every proportion both ratios must be written in the same order of value, for instance in Example 2:

$$\frac{\text{Smaller no. of mach's}}{\text{Larger no. of mach's}} = \frac{\text{Smaller no. of men}}{\text{Larger no. of men}}$$

In Example 3:

$$\frac{\text{Larger no. of drills}}{\text{Smaller no. of drills}} = \frac{\text{Larger cost}}{\text{Smaller cost}}$$

An **inverse proportion** is indicated when two quantities are so related that an increase in one causes a corresponding decrease in the other, or vice versa.

The following are quantitative expressions in which the variables are inversely related.

    (*a*) The greater the speed, the less the time.
    (*b*) The slower the speed, the longer the time.
    (*c*) The greater the volume, the less the density.
    (*d*) The more men working, the shorter the time.
    (*e*) The fewer men working, the longer the time.

EXAMPLE 4: When two pulleys are belted together the revolutions per minute (rev/min) vary inversely as the size of the pulleys. A 20 cm pulley running at 180 rev/min drives an 8 cm pulley. Find the revolutions per minute of the 8 cm pulley.

SOLUTION:

20 cm makes 180 rev/min
8 cm makes ? rev/min
$$\frac{8}{20} = \frac{180}{x}$$
$$8x = 3600$$
$$x = 450 \text{ rev/min}, \quad \text{ANS.}$$

EXPLANATION: First make a table of corresponding values. Put *like numbers* together. The smaller the pulley, the greater the number of revolutions; ∴ the quantities are in inverse ratio. Inverting the first ratio, write the proportion. Solve for x.

CUE: If you write your proportion in this form, $\frac{8}{20} = \frac{180}{x}$, you may note that in the inverse proportion the corresponding numbers are arranged diagonally, i.e. 20 cm and 180 rev/min, and 8 cm and x rev/min are diagonally opposite each other.

In the direct proportion as in Example 3, $\frac{12}{9} = \frac{8}{x}$, the corresponding numbers are arranged directly on a line with each other, i.e. 12 drills and £8, 9 drills and £x.

Practice Exercise No. 37

1. If a chimney 18 m high casts a shadow 20 m long, how long a shadow would a chimney 27 m high cast?
   (A) 10 ......   (B) 25 ......   (C) 30 ......   (D) 36 ......

2. If a soldier walks 18 km in 2 hours, how long will it take him to walk 60 km?
   (A) 6 ......   (B) $6\frac{2}{3}$ ......   (C) $8\frac{1}{2}$ ......   (D) 9 ......

3. If a car runs 90 km on 5 litres of petrol, how far will it run on 20 litres?
   (A) 300 ......   (B) 360 ......   (C) 450 ......   (D) 280 ......

4. An army camp has provisions for 240 men for 28 days; but only 112 men are sent to the camp. How long will the provisions last?
   (A) 60 ......   (B) 56 ......   (C) $13\frac{3}{4}$ ......   (D) 76 ......

5. A train takes 26 hours at a speed of 35 km/h to go from Chicago to New York. How fast must the train travel to make the trip in 20 hours?
   (A) $39\frac{1}{2}$ ......   (B) 40 ......   (C) $26\frac{14}{13}$ ......   (D) $45\frac{1}{2}$ ......

6. The flywheel on an engine makes 220 revolutions in 2 seconds. How many revolutions does it make in 8 seconds?
   (A) 1760   (B) 55 ......   (C) 880 ......   (D) 800 ......

CHAPTER SEVEN

SIGNED NUMBERS AND ALGEBRAIC EXPRESSIONS

Up to the present, all the numbers used here have been *positive* numbers. That is, none was less than zero (0). In solving some problems in arithmetic it is necessary to assign a *negative* value to some numbers. This is used principally for numbers with which we wish to represent opposite quantities or qualities, and can best be illustrated by use of a diagram. For example, consider a thermometer, as in Fig. 1.

If temperatures *above* zero are taken as *positive*, then temperatures *below* zero are considered *negative*.

In measuring distances east and west, as in Fig. 2, if distance *east* of a certain point is taken as *positive*, then distance *west* of that point is considered *negative*.

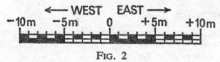

Fig. 2

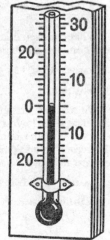

Fig. 1

Another good example may be taken from commercial book-keeping, where money in the bank and other *assets* may be considered as *positive* amounts, while money *owed* represents *negative* amounts.

Thus, in general, positive and negative numbers are used to distinguish between opposite qualities. Values above zero are considered positive and take the + sign, while values below zero are considered negative and are written with the − sign. These then become **signed numbers,** as they are called.

The + and − also continue to be used as signs of addition and subtraction. When no sign is indicated the + sign is understood.

Learning to use signed numbers is an introduction to some of the special rules for algebraic operations and also a preparation for the equation method of solving some difficult arithmetic problems in easier ways.

ADDING SIGNED NUMBERS

To add numbers of like signs, *add the numbers as in arithmetic and give to the result the common sign.*

EXAMPLE 1: − 14 added to − 8 = − 22.

EXAMPLE 2: Add + 4, + 12, and + 16,   ANS. +32.

To add numbers of unlike signs, *combine all positive and negative quantities, subtract the smaller from the larger and give the result the sign of the larger combination.*

EXAMPLE 1: Add − 4 − 8 + 2 + 6 + 10.

SOLUTION:
$$(-4) + (-8) = - 12;$$
$$2 + 6 + 10 = 18$$
$$18 - 12 = 6,   \text{ANS.}$$

EXAMPLE 2: Add 3 + 19 + 4 − 45.

SOLUTION:
$$26 - 45 = - 19,   \text{ANS.}$$

What has been done above is called finding the **algebraic sum.** Similarly, we can combine numbers that are represented by similar symbols.

EXAMPLE 3: Add $5b - 11b + 14b$.

SOLUTION:
$$19b - 11b = 8b,   \text{ANS.}$$

We cannot arithmetically add terms containing unlike symbols. For instance, if we let b stand for books and p for plates we know from arithmetic that we couldn't combine books and plates as a single quantity of either. Therefore, **to add quantities containing unlike symbols,** *collect like terms and express them separately in the answer.*

EXAMPLE 4: Add $5b + 2p + 7p + 3b$.

SOLUTION: Collecting like terms,
$$5b + 3b = 8b$$
$$2p + 7p = 9p$$

Expressing unlike terms separately, we get $8b + 9p$, which is an algebraic expression containing two terms, as the answer.

Practice Exercise No. 38
ADDITION OF SIGNED NUMBERS

1. $+ 5 + 18 =$
2. $- 5 - 17 - 14 =$
3. $+ 7 - 12 - 6 + 4 =$
4. $- 14d - 6d =$
5. $7b - 3b =$
6. $+ 22 - 14 - 17 - 12 + 18 =$
7. $5x - 7x + 14x =$
8. $3a + 4b + 2a - 2b =$
9. $6a + 3b + 9a - 5b =$
10. $6a + 3b + 9a - 5 =$

SUBTRACTING SIGNED NUMBERS

Subtraction means finding the difference between two numbers, or the difference between two values on a scale.

If you were asked what is the difference between $-4°$ and $+5°$, your answer would be $9°$. You would do this mentally. Now how did you arrive at the answer? First you counted from $-4°$ to zero, then added 5 to that. The rule for subtraction of signed numbers is therefore:

To subtract signed numbers, *change the sign of the subtrahend and apply the rules for addition.*

EXAMPLE 1: Subtract $+20$ from $+32$.

SOLUTION:
$+20$ is the subtrahend or number to be subtracted. Changing its sign and adding, we get $32 - 20 = 12$, ANS.

EXAMPLE 2: From $- 18$ subtract $- 12$.

SOLUTION:
-12 is the subtrahend. Changing its sign and adding, we get $- 18 + 12 = -6$, ANS.

Practice Exercise No. 39
SUBTRACTION OF SIGNED NUMBERS

1. $+47$ $+19$	2. -26 -17	3. -42 -18	4. $+54$ -12
5. $\quad 80$ -50	6. $-22ab$ $+18ab$	7. $(-5) - (-8)$	8. $(-7) - (-4)$

9. $(-9) - (+16)$

MULTIPLICATION AND DIVISION OF SIGNED NUMBERS

Law of signs for multiplication of signed numbers—Rule : *The product of any two numbers that have like signs is positive (+), and the product of any two numbers that have unlike signs is negative (−).*

EXAMPLE 1: Multiply -8 by -6.

SOLUTION: The signs are the same.
$$\therefore -8 \times -6 = +48, \text{ ANS.}$$

EXAMPLE 2: Multiply $+3$ by -4.

SOLUTION: The signs are unlike.

$$\therefore 3 \times -4 = -12, \text{ ANS.}$$

EXAMPLE 3: Multiply -2 by $+5$ by -3 by $+4$.

SOLUTION:
$$(-2) \times (+5) = -10,$$
$$(-10) \times (-3) = +30,$$
$$+ (+30) \times (+4) = +120, \text{ ANS.}$$

Division of signed numbers is carried out by the same process as division in arithmetic, but *the sign of the quotient is positive if the divisor and dividend have the same sign, and negative if the divisor and dividend have opposite signs.*

EXAMPLE 4: Divide -16 by -2.

SOLUTION:
$$\frac{-16}{-2} = +8, \text{ ANS.}$$

Same signs, $\therefore$ answer is plus $(+)$.

EXAMPLE 5: Divide -35 by $+5$.

SOLUTION:
$$\frac{-35}{5} = -7, \text{ ANS.}$$

Opposite signs, $\therefore$ answer is minus $(-)$.

Practice Exercise No. 40

Do the following examples:

1. $2 \times -16 =$
2. $-18 \times -12 =$
3. $-4 \times -6 \times 3 =$
4. $4 \times 3 \times -2 \times 6 =$
5. $72 \div -24 =$
6. $-68 \div -17 =$
7. $-14 \div -5 =$
8. $-24 \times 4 \div 8 =$

ALGEBRAIC EXPRESSIONS

Working with signed numbers is an introduction to using algebraic expressions. An **algebraic expression** is one in which letter symbols are used to represent numbers.

A letter symbol or other type of symbol that represents a number is called a **literal number.**

If you know the numerical values of the symbols and understand the arithmetic signs of an algebraic expression, then you can find the numerical value of any algebraic expression. *Thus:*

$a + b$ means that b is added to a.
If $a = 2$ and $b = 3$, then $a + b = 5$.
$b - a$ means that a is subtracted from b.
If $b = 6$ and $a = 4$, $b - a = 2$.
$a \times b$ means that b is multiplied by a.
If $a = 7$ and $b = 3$, $a \times b = 21$.

Multiplication can be indicated in four ways in algebra. a multiplied by b can be written $a \times b, a \cdot b, (a)(b),$ or ab. That is, multiplication can be expressed by a cross $\times$, by a dot $.$, by adjacent parentheses, and by directly joining a letter and its multiplier with no sign between them. *Thus,* $2a$ means 2 times a, and ab means a times b.

a^2 means $a \cdot a$. You read it: *a squared.*
If $a = 3$, then $a^2 = 3 \cdot 3$ or 9.
a^3 means $a \cdot a \cdot a$. You read it: *a cubed.*
If $a = 3$, then $a^3 = 3 \cdot 3 \cdot 3$ or 27.
$a^2 + b^3$ means that b^3 is to be added to a^2.
If $a = 3$ and $b = 2$, then $a^2 + b^3 = 9 + 8$, or 17.

The small 2 and 3 placed to the right and slightly above the a and the b in writing a^2 and b^3 are called **exponents.**

The number a is called the **base.** a^2 and a^3 are called **powers** of the *base a.*

$3a^2 - 2b^2$ means that $2b^2$ is subtracted from $3a^2$.

The $3a^2$ and $2b^2$ are known as **terms** in the algebraic expression.

The numbers placed before the letters are called **coefficients.** *Thus,* in $3a^2 - 2b^2$, 3 is called the *coefficient* of a^2, 2 is the *coefficient* of b^2. The coefficient so placed indicates multiplication, i.e. $3a^2$ means $3 \times a^2$.

Practice Exercise No. 41

In each of the following write the algebraic expression and find its numerical value if $x = 2$, $y = 3$, and $z = 4$.

1. x added to $y =$
2. x, y, and z added together $=$
3. Twice x added to twice $y =$
4. z subtracted from the sum of x and $y =$
5. The square of x added to the square of $y =$
6. 3 less than $y -$
7. Twice the product of x and $z =$

CHAPTER EIGHT

ALGEBRAIC FORMULAE AND EQUATIONS

In the preceding chapters rules in words were used to describe methods to be followed in solving various types of problems. For example, to find the amount of a discount the rule is to multiply the base or price by the rate of discount. By the use of symbols this rule can be expressed in a brief form known as a **formula.**

Thus a short way to express the rule in question is:

(*a*) Discount $=$ Base $\times$ Rate

A still shorter way is:

(*b*) $D = B \times R,$

in which D, B, and R mean discount, base, and rate respectively.

The shortest and algebraic way to express this is:

(*c*) $D = BR.$

DEFINITIONS

A **formula** is a shorthand method of expressing a rule by the use of symbols or letter designations (literal numbers).

At the same time it must be remembered that a formula is an equation. And what is an equation?

An **equation** is a statement that two expressions are equal.

For example, $D = BR$ states that D, the discount, is equal to B, the base, multiplied by R, the rate of discount. Before we can start working with formulae and equations there are a few things that have to be learned about them.

An equation has two equal sides or members. In the equation $D = BR$, D is the left side and BR is the right side.

Terms are made up of numbers or symbols combined by multiplication or division.

For example, $6DR$ is a term in which the factors 6, D, and R are combined by *multiplication*; $\dfrac{M}{4}$ is a term in which the quantities M and 4 are combined by *division* or in which the factors M and $\frac{1}{4}$ are combined by multiplication.

An **expression** is a collection of terms combined by addition, subtraction, or both, and frequently grouped by parentheses, as in: $(3a + 2b)$, $(2c - 4c + 3b)$, $2x - 3y$.

USING PARENTHESES

Parentheses ( ) or **brackets** [ ] mean that quantities are to be grouped together, and that quantities enclosed by them are to be considered as one quantity. The line of a fraction has the same significance in this respect as a pair of parentheses.

Thus, $18 + (9 - 6)$ is read 18 *plus the quantity* $9 - 6$.

Rule: To solve examples containing parentheses, *do the work within the parentheses first; then remove the parentheses and proceed in the usual way. Within parentheses and in examples without parentheses do multiplications from left to right before doing additions and subtractions.*

It is extremely important to observe this method of procedure, since it is otherwise impossible to solve algebraic problems.

EXAMPLE 1: $94 - (12 + 18 + 20) = ?$
$94 - 50 = 44,$ ANS.

EXAMPLE 2: $12(3 + 2) = ?$
$12 \times 5 = 60,$ ANS.

EXAMPLE 3: $\dfrac{18}{2(4 - 1)} = ?$

$\dfrac{18}{2 \times 3} = \dfrac{18}{6} = 3,$ ANS.

EXAMPLE 4: $3 \times 6 - 4$
$18 - 4 = 14,$ ANS.

Note: If in this example the 4 had been subtracted from the 6 before multiplying the answer would have been 6, but this would be wrong according to the laws of algebra. This example illustrates the absolute necessity of *doing multiplication first* in any cases similar to this.

Practice Exercise No. 42

Clear parentheses and solve.

1. $18 + (19 - 14) =$
2. $22(3 + 2) =$
3. $42 - 9 - (18 + 2) =$
4. $(6 - 4)(8 + 2) =$
5. $(18 \div 3)(9 - 7) =$
6. $(7 \times 8) - (6 \times 4) + (18 - 6) =$
7. $(6 \times 8) \div (8 \times 2) =$
8. $19 + (18 - 14 + 32) =$
9. $(7 \times 6)(6 \times 5) =$
10. $69 \div [35 - (15 - 3)] =$

TRANSLATING WORD STATEMENTS
INTO FORMULAE AND ALGEBRAIC EXPRESSIONS

To express word statements as formulae or as brief algebraic expressions, letters and symbols are substituted for words.

EXAMPLE 1: Express briefly, *What number increased by 6 gives 18 as a result?*
Substituting the letter N for the unknown *what number*, we get
$$N + 6 = 18,$$
$$N = ? \quad \text{ANS.}$$

EXAMPLE 2: Express briefly, *The product of two numbers is 85. One is 5, find the other.*
$$5N = 85, N = ? \quad \text{ANS.}$$

EXAMPLE 3: Express briefly, *Fifteen exceeds a certain number by 6. What is the number?*
$$15 - 6 = N, N = ? \quad \text{ANS.}$$

EXAMPLE 4: Express briefly, *Two thirds of a number is 20. Find the number.*
$$\tfrac{2}{3}N = 20, N = ? \quad \text{ANS.}$$

In algebra, however, the regular method of writing fractions is to place all factors, as far as may be possible, above or below a single horizontal line. The form $\tfrac{2}{3}N$, while mathematically correct, is less regular than $\dfrac{2N}{3}$. Hence to express our problem in approved form we arrive at:

$$\frac{2N}{3} = 20, N = ? \quad \text{ANS.}$$

The foregoing illustrates in simple form the general method of making algebraic statements. In engineering, scientific, industrial, and commercial practice it is common to express certain kinds of facts in algebraic *formulae*. The usual way is to state the formula with symbolic letters and to follow it immediately with an explanation (starting with the words *in which*) to make intelligible to the reader any symbols that may require definition. Examples of this method of **formula statement** follow.

EXAMPLE 1: The cost equals the selling price minus the margin of profit.

FORMULA: $C = S - M$, in which C stands for cost, S for selling price, and M for margin of profit.

EXAMPLE 2: The area of a rectangle equals the base times the height.

FORMULA: $A = bh$, in which A stands for area, b for base, and h for height.

EXAMPLE 3: To determine the resistance in ohms of an electrical circuit, divide the number of volts by the number of amperes.

FORMULA: $O = \dfrac{V}{A}$, in which O stands for ohms, V for volts, and A for amperes.

Practice Exercise No. 43

Write the following statements as equations.

Note: Most of the statements represent formulae commonly used by draftsmen, designers, carpenters, engineers, and clerks.

1. The perimeter (p) of a rectangle equals twice its length (l) added to twice its width (w).
2. The distance (d) travelled by an object that moves at a given speed (r) for a given time (t) equals the rate multiplied by the time.
3. To get the power (p) of an electric motor multiply the number of volts (v) by the number of amperes (a).
4. Interest (I) on money is calculated by multiplying the principal (P) by the rate (R) by the time (T).
5. The amperage (A) of an electrical circuit is equal to the wattage (W) divided by the voltage (V).
6. Profit (P) equals the margin (M) minus the overhead (O).
7. The distance (d) that an object will fall in any given time (t) is equal to the square of the time multiplied by 4·9.
8. The area (A) of a square figure is equal to the square of one of its sides (S).
9. Centigrade temperature (C) was equal to Fahrenheit temperature (F) minus 32°, multiplied by $\frac{5}{9}$.
10. The speed (R) of a revolving wheel is proportional to the number of revolutions (N) it makes in a given time (T).

RULE FOR SOLVING EQUATIONS

When you solve an equation you are finding the value of the unknown or literal number in terms of what has been given about the other numbers in the equation. To do this you must learn the following rules of procedure for treating equations. Primarily, *what you do to one side of an equation you must also do to the other.* This might be called the golden rule of algebra. Its observance is imperative in order to preserve equality.

Rule 1. *The same number may be added to both sides of an equation without changing its equality.*

EXAMPLE 1: If $x - 4 = 6$, what does x equal?

SOLUTION: $x - 4 + 4 = 6 + 4$. Adding 4 to both sides, $x = 10$, ANS.

To check the solution of algebraic examples, *substitute the value of the unknown quantity as determined in the answer for the corresponding symbol in the original equation. If both sides produce the same number the answer is correct.*

EXAMPLE 1: Check the correctness of 10 as the solution of $x - 4 = 6$.

METHOD: $x - 4 = 6$, original equation,
$10 - 4 = 6$, substituting answer for symbol,
$6 = 6$, proof of correctness.

Rule 2. *The same number may be subtracted from both sides of an equation.*

EXAMPLE 2: If $n + 6 = 18$, what does n equal?

SOLUTION: $n + 6 - 6 = 18 - 6$. Subtracting 6 from both sides, $n = 12$, ANS.
Check by substituting 12 for n in the original equation. *Thus*, $n + 6 = 18$ becomes
$12 + 6 = 18$ or $18 = 18$, which is correct.

Rule 3. *Both sides of an equation may be multiplied by the same number.*

EXAMPLE 3: If $\frac{1}{3}$ of a number is 10, find the number.

SOLUTION: $\frac{1}{3} n$ or $\frac{n}{3} = 10$,

$\frac{n}{3} \times 3 = 10 \times 3$, multiplying both sides by 3,

$\frac{n}{3} \times 3 = 10 \times 3$, cancelling,

$n = 30$, ANS.

Check the answer.

Rule 4. *Both sides of an equation may be divided by the same number.*

EXAMPLE 4: Two times a number is 30. What is the number?

SOLUTION: $2n = 30$,

$\frac{2n}{2} = \frac{30}{2}$, dividing both sides by 2,

$n = 15$, ANS.

Check the answer.

Transposition

Transposition is the process of moving a quantity from one side of an equation to the other side by changing its sign of operation. This is exactly what has been done in carrying out the rules in the four examples above.

Division is the operation opposite to multiplication.

Addition is the operation opposite to subtraction.

Transposition is performed in order to obtain an equation in which the unknown quantity is on one side and the known quantity on the other.

Rule: *A term may be transposed from one side of an equation to the other if its sign is changed from + to −, or from − to +.*

Rule: *A factor (multiplier) may be removed from one side of an equation by making it a divisor in the other. A divisor may be removed from one side of an equation by making it a factor (multiplier) in the other.*

Observe again the solution to Example 1.

$$x - 4 = 6,$$
$$x = 6 + 4$$
$$x = 10$$

EXPLANATION: To get x by itself on one side of the equation, the -4 was transposed from the left to the right side and made $+4$.

Observe again the solution to Example 2.

$$n + 6 = 18$$
$$n = 18 - 6$$
$$n = 12$$

EXPLANATION: To get n by itself on one side of the equation, the $+6$ was transposed from the left to the right side and made -6.

Observe again the solution to Example 3.

$$\frac{n}{3} = 10$$
$$n = 10 \times 3$$
$$n = 30$$

EXPLANATION: To get n by itself on one side of the equation, the divisor 3 on the left was changed to the multiplier $3(\frac{3}{1})$ on the right.

Observe again the solution to Example 4.

$$2n = 30$$
$$n = \frac{30}{2}$$
$$n = 15$$

EXPLANATION: To get n by itself on one side of the equation, the multiplier 2 on the left was changed to the divisor 2 on the right.

Note that transposition is essentially nothing more than a shortened method for performing like operations of addition, subtraction, multiplication, or division on both sides of the equation.

Changing $x - 4 = 6$ to $x = 6 + 4$ is the same as adding 4 to both sides:

$$
\begin{array}{rcl}
x - 4 = & & 6 \\
+ 4 = & & + 4 \\
\hline
x & = & 10
\end{array}
$$

Changing $n + 6 = 18$ to $n = 18 - 6$ is the same as subtracting 6 from both sides:

$$
\begin{array}{rcl}
n + 6 = & & 18 \\
- 6 = & & - 6 \\
\hline
n & = & 12
\end{array}
$$

Changing $\frac{n}{3} = 10$ to $n = 10 \times 3$ is the same as multiplying both sides by 3:

$\frac{n}{3} \times 3 = 10 \times 3$, in which the 3s on the left cancel.

Changing $2n = 30$ to $n = \frac{30}{2}$ is the same as dividing both sides by 2:

$\frac{2n}{2} = \frac{30}{2}$, in which the 2s on the left cancel.

When terms involving the unknown quantity occur on both sides of the equation, *perform such transpositions as may be necessary to collect all the unknown terms on one side (usually the left) and all the known terms on the other.*

EXAMPLE 5: If $3x - 6 = x + 8$, what does x equal?

SOLUTION: $3x = x + 8 + 6$ transposing -6 from left to right.

$3x - x = 14$ transposing x from right to left.

$2x = 14$ transposing 2 as a multiplier from
$x = \dfrac{14}{2}$ left to a divisor at the right.

$x = 7$,   ANS.

CHECK: $3x - 6 = x + 8$
$21 - 6 = 7 + 8$ substituting 7 for x.
$15 = 15$ proof of correctness.

When using an algebraic formula in actual practice, it may be necessary to change its form from that in which it has been originally expressed. Such changes are effected by transposition.

EXAMPLE 6: If $R = \dfrac{WC}{L}$, solve for W, C, and L.

SOLUTION:

$R = \dfrac{WC}{L}$, original formula.

$\dfrac{LR}{C} = W$. To separate W, C and L are transposed.

$\dfrac{LR}{W} = C$. To separate C, L and W are transposed.

$L = \dfrac{WC}{R}$. To separate L, L and R are transposed.

Practice Exercise No. 44

Solve by transposition:

1. $p + 3 = 8$ $p = ?$ 2. $2n = 25$ $n = ?$ 3. $\frac{1}{2}x = 14$ $x = ?$
4. $5c - 3 = 27$ $c = ?$ 5. $18 = 5y - 2$ $y = ?$ 6. $\frac{2}{3}n = 24$ $n = ?$
7. $\dfrac{a}{2} + \dfrac{a}{4} = 36$ $a = ?$ 8. $W = \dfrac{b}{c}$ $b = ?$ 9. $V = \dfrac{W}{A}$ $A = ?$

10. $H = \dfrac{P}{AW}$ $W = ?$

FUNDAMENTAL OPERATIONS

Addition is performed thus:
$$3a - 4b + 2c$$
$$-8a + 6b - 3c$$
$$6a - 4b + 8c$$
$$\overline{}$$
$$a - 2b + 7c$$

EXPLANATION: The terms are arranged in columns in such a way that all like terms are in the same column.

Subtraction is performed thus:

$$8a - 4b + 2c$$
$$5a - 6b + 8c$$
$$\overline{3a + 2b - 6c}$$

EXPLANATION: To subtract algebraically, whenever you cannot directly subtract a smaller from a larger quantity of like sign, mentally change the sign of the subtrahend and perform an addition. $8a - 5a = 3a$; $- 4b + 6b = + 2b$; $2c - 8c = -6c$.

Multiplication is performed thus:

$$a^2 - 2ab + b^2$$
$$a - b$$
$$\overline{a^3 - 2a^2b + ab^2}$$
$$\quad - a^2b + 2ab^2 - b^3$$
$$\overline{a^3 - 3a^2b + 3ab^2 - b^3}$$

EXPLANATION: Each term in the multiplicand is multiplied separately by a and then by b. Like terms are set under each other and the whole is added. $+ \times +$ gives $+$; $- \times -$ gives $+$; $+ \times -$ gives $-$.

Division is performed thus:

$$\frac{3a^2b + 3ab^2 + 3a}{3a} = ab + b^2 + 1 \text{ or } b^2 + ab + 1$$

EXPLANATION: $3a$ is a factor of each term in the dividend. Separate divisions give us $ab + b^2 + 1$. This is changed to $b^2 + ab + 1$ because it is customary to place algebraic terms in the order of their highest powers.

Practice Exercise No. 45

USE OF FORMULAE AND EQUATIONS

1. Diameters of pulleys are inversely proportional to the rev/min $\frac{D}{d} = \frac{r}{R}$. An 18 cm diameter pulley turning at 100 rev/min is driving a 6 cm diameter pulley. What is the rev/min of the smaller pulley?

(A) $36\frac{2}{3}$ ......   (B) 600 ......   (c) 300 ......   (D) 900 ......

2. What size pulley at 144 rev/min will drive a 9 cm pulley at 256 rev/min?

(A) 24 cm ......   (B) $5\frac{1}{6}$ cm ......   (c) 16 cm ......   (D) 32 cm ......

3. Three times a certain number plus twice the same number is 90. What is the number?

(A) 16 ......   (B) 18 ......   (c) 20 ......   (D) 24 ......

4. The larger of two numbers is seven times the smaller. What is the larger if their sum is 32?

(A) 28 ......   (B) 36 ......   (c) 25 ......   (D) 39 ......

5. Six hundred pairs of shoes are to be divided among three army units. The second unit is to get twice as many as the first, and the third unit is to get as many as the first and second units together. How many pairs of shoes does the second unit get?

(A) 100 ......   (B) 200 ......   (c) 300 ......   (D) 400 ......

6. Two aviators are 3000 km apart. They start towards each other, one at a rate of 200 km/h and the other at 300 km/h. How much distance does the faster one cover up to the time they meet? $R \times T = D$.
(A) 1200 ......   (B) 1400 ......   (C) 1600 ......   (D) 1800 ......

7. Two soldiers start out from camp in opposite directions. One travels twice as fast as the other. In 10 hours they are 24 km apart. What is the rate of the faster soldier?
(A) $\frac{4}{5}$ km/h ...   (B) 1 km/h ...   (C) $1\frac{3}{5}$ km/h ...   (D) $1\frac{3}{5}$ km/h ...

8. A man has 3 times as many 1p coins as 2p coins. How many coins has he if the value of both together is £2. *Hint:* Let n = No. of 2p coins and $3n$ = No. of 1p coins and multiply each by their value.
(A) 120 ......   (B) 140 ......   (C) 160 ......   (D) 180 ......

9. When two gears run together the revolutions per minute vary inversely as the number of teeth. A 48-tooth gear is driving a 72-tooth gear. Find the revolutions per minute of the larger gear if the smaller one is running at 160 rev/min.
(A) $106\frac{2}{3}$ ......   (B) 240 ......   (C) $66\frac{2}{3}$ ......   (D) 180 ......

10. A teeter board is a form of lever. It is balanced when the mass multiplied by the distance on one side equals the mass multiplied by the distance on the other side. A mass of 120 kg is placed 0·45 m from the fulcrum. What mass is needed to balance this at a distance of 0·5 m from the fulcrum on the other end?

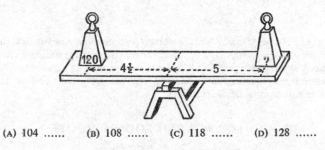

(A) 104 ......   (B) 108 ......   (C) 118 ......   (D) 128 ......

CHAPTER NINE

FACTORS AND ROOTS

A **factor** of a number is an exact divisor of that number. Thus, 2 is a factor of 6 because $6 \div 2 = 3$ exactly; 3 is the other factor of 6.

For the number 9, 3 and 3 are equal factors; and for 8, 2, 2, and 2 are equal factors. These equal factors are called **roots** of the number. Thus:

The number 3 is a *root* of 9.

The number 2 is a *root* of 8.

A root of a number is therefore one of the equal factors which, if multiplied together, produce the number.

The **square root** of a number is one of TWO equal factors which, if multiplied together, produce that number.

$3 \times 3 = 9$, hence 3 is the *square root* of 9.

The **cube root** of a number is one of THREE equal factors which if multiplied together, produce that number.

$3 \times 3 \times 3 = 27$, hence 3 is the *cube root* of 27.

A **fourth root** of a number is one of FOUR equal factors; the fifth root is one of five, and so on.

The square root is the one most frequently used in mathematics.

The sign indicating square root is $\sqrt{}$. It is placed in front of the number whose root is to be found. $\sqrt{25}$ means the square root of 25. It is called the **square root sign** or **radical sign**.

To indicate a root other than square root a small figure called the **index** of the root is placed in the radical sign. Thus: $\sqrt[3]{8}$ means the cube root of 8.

The square root of $4 = 2$, of $36 = 6$, or $49 = 7$.

To check that you have obtained the correct square root of a number, *multiply it by itself. If the product is equal to the original number the answer is correct.*

Practice Exercise No. 46

Find the roots indicated and check.

1. $\sqrt{64}$	2. $\sqrt{100}$	3. $\sqrt{81}$
4. $\sqrt[3]{27}$	5. $\sqrt[3]{125}$	6. $\sqrt{144}$
7. $\sqrt[3]{1000}$	8. $\sqrt{1}$	9. $\sqrt{0\cdot04} = \sqrt{0\cdot2 \times 0\cdot2} = 0\cdot2$
10. $\sqrt{0\cdot09}$	11. $\sqrt{1\cdot44}$	12. $\sqrt{0\cdot0025}$

Not all numbers have exact square roots. Nor can we always determine square root by *inspection* as you have done above. (Inspection means 'trial and error'.) There is an arithmetic method of extracting the square roots of numbers whereby an answer may be found that will be correct to any necessary or desired number of decimal places.

METHOD FOR FINDING SQUARE ROOTS

Find the square root of 412 164.

1. Place the square root sign over the number, and then, beginning at the right, divide it into *periods* of two figures each. Connect the digits in each period with tie-marks as shown. In the answer there will be one digit for each period.

$$\sqrt{41\ 21\ 64}$$

2. Find the largest number which when squared is contained in the first left-hand period. In this case 6 is the number. Write 6 in the answer over the first period. Square it, making 36, and subtract 36 from the first period. Bring down the next period, making the new dividend 5 21.

$$\begin{array}{r} 6 \\ \sqrt{41\ 21\ 64} \\ \underline{36} \\ 5\ 21 \end{array}$$

3. Multiply the root 6 by 2, getting 12. Place the 12 to the left of 5 21, since 12 is the new trial divisor. Allow, however, for one more digit to follow 12. The place of this missing digit may be indicated by a question mark. To find the number belonging in this place, ignore (cover over) the last number in the dividend 5 21, and see how many times 12 goes into 52. Approximately 4. Place the 4 above its period, 21, and put it in place of the ? in the divisor.

$$\begin{array}{r} 6\ \ 4 \\ \sqrt{41\ 21\ 64} \\ \underline{36} \\ 12?\ \overline{)\ 5\ 21} \end{array}$$

$$
\begin{array}{r}
6\ \ 4 \\
\sqrt{41\ 21\ 64} \\
\smile\ \smile\ \smile
\end{array}
$$

4. Multiply the divisor 124 by the new number in the root, 4. 124 × 4 = 496. Place this product under 521 and subtract. Bring down the next period, 64.

$$
\begin{array}{r}
36 \\
124\ \overline{\big)\ 5\ 21} \\
4\ 96 \\
\hline
25\ 64
\end{array}
$$

$$
\begin{array}{r}
6\ \ 4\ \ 2 \\
\sqrt{41\ 21\ 64} \\
\smile\ \smile\ \smile
\end{array}
$$

5. Multiply 64 by 2 to get 128 as the new trial divisor. 128 goes into 256 two times. Place the 2 above the next period in the root and also in the divisor. Then multiply the divisor 1,282 by the new root 2, to get 25 64. Subtracting, the remainder is zero. 642 is therefore the exact square root.

$$
\begin{array}{r}
36 \\
124\ \overline{\big)\ 5\ 21} \\
4\ 96 \\
128\tfrac{?}{2}\ \overline{\ \ 25\ 64} \\
25\ 64 \\
\hline
0
\end{array}
$$

6. CHECK: 642 × 642 = 412 164.

Finding the Square Root of Decimals

A slight variation in method is necessary when it is required to find the square root of a decimal figure.

Mark off periods beginning at the decimal point. Count to the right for the decimal quantities and to the left for the whole numbers. If the last period of the whole numbers contains one figure, leave it by itself, but remember that in such a case the first figure in the root cannot be more than 3, because the square of any number greater than 3 is a two-place number. If the last period of the decimal numbers contains only one figure you may add a zero to it. This is because two digits are necessary to make up a period, while the addition of a zero at the right of a decimal figure does not change its value.

The square root of a decimal will contain as many decimal places as there are periods, or half as many decimal places as the given number.

The operations in obtaining the square root of a decimal number are the same as for whole numbers.

Follow the steps in the example following.

EXAMPLE 1: Find the square root of 339·2964.

1. Beginning at the decimal point, mark off periods to left and right.

2. 1 is the largest whole-number square root that is contained in 3, which constitutes the first period.

3. Place decimal point in root after the 8 because the root of the next period has a decimal value.

4. Bring down 29 next to the 15, making 1529 the new dividend. Multiply the root 18 by 2, making 36 the new divisor.

$$
\begin{array}{r}
1\ \ 8\cdot\ 4\ \ 2 \\
\sqrt{3\ 39\cdot29\ 64} \\
\smile\ \smile\ \smile\ \smile \\
1 \\
2\tfrac{?}{8}\big)\ 2\ 39 \\
2\ 24 \\
36\tfrac{?}{4}\big)\ \ 15\ 29 \\
14\ 56 \\
368\tfrac{?}{2}\big)\ \ 73\ 64 \\
73\ 64 \\
\hline
0
\end{array}
$$

5. Covering the 9 of 1529, 36 seems to be contained about 4 times in this number. Place a 4 in the root above 29, and multiply 364 by 4 to get 1456. Subtract this from 1529.

6. Bring down the 64 and repeat the previous process. Since the number is a perfect square, the remainder is zero.

When the given number is not a perfect square, *add zeros after the decimal point, or after the last figure if the original number is already in decimal form, and carry out the answer to the required or desired number of decimal places. Usually two places are sufficient.*

Note: In working a square root example, when a divisor is larger than the corresponding dividend, write zero in the trial divisor and bring down the next period. This is illustrated in the next example.

EXAMPLE 2: Find the square root of 25·63 to two decimal places.

$$
\begin{array}{r}
5 \cdot \ 0 \ \ 6 \ \ 2 + , \\
\sqrt{25 \cdot 63\ \ 00\ \ 00} \\
25 \quad\quad\quad\quad\quad \\
\overline{100_6^?\,\lfloor\ 0\ 63\ \ 00} \\
60\ 36 \quad\quad \\
\overline{1012_2^?\,\lfloor\ 2\ 64\ \ 00} \\
2\ 02\ 44 \quad \\
\overline{61\ 56}\ \text{remainder}\quad\ 5\cdot06, \text{Ans.}
\end{array}
$$

To find the square of a fraction, determine separately the square roots of the numerator and of the denominator, and reduce to lowest terms or to a decimal.

EXAMPLE 3:

$$\sqrt{\frac{33}{67}}.$$

$$\sqrt{\frac{33}{67}} = \frac{5\cdot745}{8\cdot185} = 0\cdot70, \quad \text{Ans. (two decimal places)}$$

USE OF SQUARE ROOTS

Although in many test situations the student may be required to work out square roots as above, in actual practice it is inconvenient to stop work for such calculations. Most mathematics books therefore contain tables giving powers and roots of numbers. Table V is a simple form of such a table.

Any formula or problem containing the square of a number or factor requires a knowledge of square roots for its solution. You will find many such problems and formulae in the material contained in Chapter Fourteen on geometry.

EXAMPLE 1: Find the length of one side of a square whose area is $225\,\text{m}^2$

SOLUTION: Let x = length of one side.
Area = base × height. ∴ Area = x^2,
  $x^2 = 225$,
  $x = 15$, extracting the square root of both sides of the equation.

Table V. Squares and Square Roots, Cubes, and Cube Roots of Numbers from 1 *to* 100

No. n	Sq. n²	Cube n³	Sq. Root √n	Cube Root ∛n	n	n²	n³	√n	∛n
1	1	1	1·000	1·000	51	2 601	132 651	7·141	3·708
2	4	8	1·414	1·259	52	2 704	140 608	7·211	3·733
3	9	27	1·732	1·442	53	2 809	148 877	7·280	3·756
4	16	64	2·000	1·587	54	2 916	157 464	7·348	3·780
5	25	125	2·236	1·710	55	3 025	166 375	7·416	3·803
6	36	216	2·449	1·817	56	3 136	175 616	7·483	3·826
7	49	343	2·646	1·913	57	3 249	185 193	7·550	3·849
8	64	512	2·828	2·000	58	3 364	195 112	7·616	3·871
9	81	729	3·000	2·080	59	3 481	205 379	7·681	3·893
10	100	1 000	3·162	2·154	60	3 600	216 000	7·746	3·915
11	121	1 331	3·317	2·224	61	3 721	226 981	7·810	3·936
12	144	1 728	3·464	2·289	62	3 844	238 328	7·874	3·958
13	169	2 197	3·606	2·351	63	3 969	250 047	7·937	3·979
14	196	2 744	3·742	2·410	64	4 096	262 144	8·000	4·000
15	225	3 375	3·873	2·466	65	4 225	274 625	8·062	4·021
16	256	4 096	4·000	2·520	66	4 356	287 496	8·124	4·041
17	289	4 913	4·123	2·571	67	4 489	300 763	8·185	4·062
18	324	5 832	4·243	2·621	68	4 624	314 432	8·246	4·082
19	361	6 859	4·359	2·668	69	4 761	328 509	8·307	4·102
20	400	8 000	4·472	2·714	70	4 900	343 000	8·367	4·121
21	441	9 261	4·583	2·759	71	5 041	357 911	8·426	4·141
22	484	10 648	4·690	2·802	72	5 184	373 248	8·485	4·160
23	529	12 167	4·796	2·844	73	5 329	389 017	8·544	4·179
24	576	13 824	4·899	2·884	74	5 476	405 224	8·602	4·198
25	625	15 625	5·000	2·924	75	5 625	421 875	8·660	4·217
26	676	17 576	5·099	2·962	76	5 776	438 976	8·718	4·236
27	729	19 683	5·196	3·000	77	5 929	456 533	8·775	4·254
28	784	21 952	5·292	3·037	78	6 084	474 552	8·832	4·273
29	841	24 389	5·385	3·072	79	6 241	493 039	8·888	4·291
30	900	27 000	5·477	3·107	80	6 400	512 000	8·944	4·309
31	961	29 791	5·568	3·141	81	6 561	531 441	9·000	4·327
32	1 024	32 768	5·657	3·175	82	6 724	551 368	9·055	4·344
33	1 089	35 937	5·745	3·208	83	6 889	571 787	9·110	4·362
34	1 156	39 304	5·831	3·240	84	7 056	592 704	9·165	4·380
35	1 225	42 875	5·916	3·271	85	7 225	614 125	9·220	4·397
36	1 296	46 656	6·000	3·302	86	7 396	636 056	9·274	4·414
37	1 369	50 653	6·083	3·332	87	7 569	658 503	9·327	4·431
38	1 444	54 872	6·164	3·362	88	7 744	681 472	9·381	4·448
39	1 521	59 319	6·245	3·391	89	7 921	704 969	9·434	4·465
40	1 600	64 000	6·325	3·420	90	8 100	729 000	9·487	4·481
41	1 681	68 921	6·403	3·448	91	8 281	753 571	9·539	4·498
42	1 764	74 088	6·481	3·476	92	8 464	778 688	9·592	4·514
43	1 849	79 507	6·557	3·503	93	8 649	804 357	9·644	4·531
44	1 936	85 184	6·633	3·530	94	8 836	830 584	9·695	4·547
45	2 025	91 125	6·708	3·557	95	9 025	857 375	9·747	4·563
46	2 116	97 336	6·782	3·583	96	9 216	884 736	9·798	4·579
47	2 209	103 823	6·856	3·609	97	9 409	912 673	9·849	4·595
48	2 304	110 592	6·928	3·634	98	9 604	941 192	9·899	4·610
49	2 401	117 649	7·000	3·659	99	9 801	970 299	9·950	4·626
50	2 500	125 000	7·071	3·684	100	10 000	1 000 000	10·000	4·641

EXAMPLE 2: $d = 16t^2$. If $d = 10\,000$ find t.

SOLUTION: $10\,000 = 16t^2$,

$\dfrac{10\,000}{16} = t^2$,   dividing both sides by 16

$625 = t^2$,

$25 = t$,   extracting square root of both sides of the equation.

Practice Exercise No. 47

Work out examples 1–5. Find answers in Table V for 6–10.

1. $\sqrt{5329}$ 2. $\sqrt{1225}$ 3. $\sqrt{2937 \cdot 64}$ 4. $\sqrt{312 \cdot 649}$
5. $\sqrt{428}$ to 2 places   6. $\sqrt{676}$ 7. $\sqrt{1849}$ 8. $\sqrt{3136}$
9. $\sqrt{7225}$ 10. $\sqrt{9409}$

CHAPTER TEN

POWERS

To square a number is to use that number as a factor twice. Thus, $4 \times 4 = 16$, and 16 is said to be the **square** of 4. This is also called raising a number to its **second power**. Using a number as a factor three times (for instance, $4 \times 4 \times 4 = 64$) is called raising it to the **third power**. The given case would be written 4^3, and be read *four cubed*. 4^4 is read *four to the fourth power*, 4^5 is read *four to the fifth power*, etc.

A **power** of a number is the product obtained by multiplying the number by itself a given number of times. Raising a number to a given power is the opposite process of finding the corresponding root of a number.

To raise a given number to its indicated power, multiply the number by itself as many times as the power indicated. Thus, $3^5 = 3 \times 3 \times 3 \times 3 \times 3 = 243$. The small five used in writing 3^5 is called an *exponent*, while the number 3 is called the **base**.

An **exponent** indicates the power to which a number is to be raised. *Thus, x^3 means that x is to be raised to the third power.*

$$\text{If } x^3 = 125,$$
$$x = \,?$$

To raise a fraction to a given power, *raise both the numerator and the denominator to the given power.*

$$(\tfrac{1}{3})^2 = \tfrac{1}{3} \times \tfrac{1}{3} = \tfrac{1}{9}$$
$$(\tfrac{3}{5})^2 = \tfrac{3}{5} \times \tfrac{3}{5} = \tfrac{9}{25}$$

Any power or root of 1 is 1, because 1 multiplied or divided by 1 any number of times is 1.

Any number **without an exponent** is considered to be the first power or first root of itself. Neither the exponent nor the index 1 is written. *Thus*, x means x^1.

Any number raised to the **zero power**, such as 5^0, is equal to 1. The reason for this will appear when we consider the multiplication of powers of numbers.

When a number has a **negative exponent**, i.e. when the exponent is preceded by the minus sign, as in 3^{-3}, it indicates the reciprocal of the indicated power of the number. Since $3^3 = 27$, $3^{-3} = \frac{1}{27}$, the reciprocal of 27. 12^{-2} means the reciprocal of 12^2, or $\frac{1}{144}$.

When a number has a **fractional exponent** with a numerator of 1, as has $x^{\frac{1}{2}}$, it signifies that the corresponding *root* is to be taken of the number. In other words, $16^{\frac{1}{2}} = \sqrt{16} = 4$.

When a fractional exponent has a numerator greater than one, as has $x^{\frac{3}{2}}$, the numerator indicates the power to which the number is to be raised, while the denominator indicates the root that is to be taken. Accordingly, $4^{\frac{3}{2}} = \sqrt{4^3} = \sqrt{64} = 8$. To reverse this example, $8^{\frac{2}{3}} = \sqrt[3]{8^2} = \sqrt[3]{64} = 4$.

Powers of 10

$$10^1 = 10 \qquad 10^{-1} = \frac{1}{10} \text{ or } 0 \cdot 1$$

$$10^2 = 100 \qquad 10^{-2} = \frac{1}{100} \text{ or } 0 \cdot 01$$

$$10^3 = 1000 \qquad 10^{-3} = \frac{1}{1000} \text{ or } 0 \cdot 001$$

$$10^4 = 10\,000 \qquad 10^{-4} = \frac{1}{10\,000} \text{ or } 0 \cdot 0001$$

$$10^5 = 100\,000 \qquad 10^{-5} = \frac{1}{100\,000} \text{ or } 0 \cdot 000\,01$$

From this it is apparent that 10 raised to any positive power is equal to a multiple of 10 bearing as many zeros as are represented by the quantity of the exponent.

Also, 10 raised to any negative power is equal to a multiple of 10 containing as many decimal places as the quantity of the negative exponent.

The above forms are used for writing very large and very small numbers in an abbreviated way. *Thus*,

$$32\,000 \text{ may be written as } 32 \times 10^3,$$
$$6\,900\,000 \text{ may be written as } 6 \cdot 9 \times 10^6,$$
$$0 \cdot 000\,008 \text{ may be written as } 8 \times 10^{-6},$$
$$0 \cdot 000\,000\,023\,5 \text{ may be written as } 2 \cdot 35 \times 10^{-8}.$$

A positive exponent moves the decimal point a corresponding number of places to the right. *Thus*, $8 \cdot 2 \times 10^7 = 82\,000\,000$.

A negative exponent moves the decimal point a corresponding number of places to the left. *Thus*, $6 \cdot 3 \times 10^{-5} = 0 \cdot 000\,063$.

Laws of Exponents

To multiply powers of the same base, *add their exponents.*

$$Thus, \quad 2^2 \text{ times } 2^3 = 2^5.$$
$$\text{PROOF: } 2^2 = 4; 2^3 = 8; 2^5 = 32;$$
$$4 \times 8 = 32.$$

To divide powers of the same base, *subtract the exponent of the divisor from the exponent of the dividend.*

$$Thus, \quad 3^5 \div 3^3 = 3^2.$$
$$\text{PROOF: } 3^5 = 243; 3^3 = 27; 3^2 = 9;$$
$$243 \div 27 = 9.$$

It will now become apparent why *any* number with an exponent of zero is equal to 1. According to the laws just stated—

$$x^3 \times x^0 = x^{3+0} = x^3$$

because if equals are multiplied by equals the products are equal;

but
$$x^3 \times 1 = x^3$$
$$\therefore x^0 = 1$$

To generalize this fact, let *n* denote any positive exponent whatever. Then $x^n \times x^0 = x^n$ and x^0 necessarily equals 1. The same conclusion will be reached if the process is division and the exponents are subtracted.
Thus,

$$x^n \div x^0 = x^{n-0} = x^n, \therefore x^0 = 1.$$

Practice Exercise No. 48

Perform the indicated operations.

1. $6^2 =$ 2. $9^3 =$ 3. $25^{\frac{1}{2}} =$ 4. $4^{-3} =$
5. $432^2 =$ 6. $8^5 =$ 7. $\sqrt{81} =$ 8. $\sqrt[3]{125} =$
9. $43 \times 10^6 =$ 10. $6 \cdot 2 \times 10^5 =$ 11. $(\frac{2}{3})^2 =$ 12. $6^{\frac{1}{2}} =$
13. $2 \cdot 8 \times 10^{-7} =$ 14. $25 \times 10^{-4} =$ 15. $12 \cdot 2 \times 10^7 =$

METHOD FOR FINDING CUBE ROOTS

In studying the following example, read step by step the rule that follows it and note how the example illustrates the rule.

EXAMPLE: What is the cube root of 264 609 288 ?

$$\begin{array}{r} 6 \quad 4 \quad 2 \\ \sqrt[3]{264 \ 609 \ 288} \end{array}$$

$$6^3 = 216$$

1st Part. Div. 3×60^2 =	10 800	48 609
$3 \times 60 \times 4$ =	720	
4^2 =	16	
1st Comp. Div.	11 536	46 144
		2 465 288
2nd Part. Div. 3×640^2 =	1 228 800	
$3 \times 640 \times 2$ =	3 840	
2^2 =	4	
2nd Comp. Div.	1 232 644	2 465 288

The following rule is more readily understood if we bear in mind the formula for the cube of the sum of two numbers:

$$(a + b)^3 = a^3 + 3a^2b + 3ab^2 + b^3$$

Rule: 1. *Separate the given number into periods of three figures each, beginning at the right, and place over it the radical sign with the proper index.*

The extreme left-hand period may contain one, two, or three figures.

2. *Determine the greatest cube that is smaller than the first left-hand period, and write its cube root, in the position shown, as the first figure of the required root.*

This root corresponds to a in the formula.

3. *Subtract the cube of this root from the first period and annex the next period to the remainder.*

4. *Multiply this root mentally by ten and write three times the square of this as a partial divisor.*

5. *Make a trial division to determine what the next figure in the root will be and write it in its proper place.*

6. *Add to the partial divisor* (1) *the product of 3 times the first part of the root considered as tens multiplied by the second part of the root; and* (2) *the square of the second part of the root. The sum of these numbers is the complete divisor.*

7. *Multiply the complete divisor by the second part of the root and subtract the product from the new dividend.*

Note in the example that at this point $a = 60$ and $b = 4$. When we subtracted 216 we took 216 000 or a^3 out of the given figure. When we multiply the first complete divisor by 4 this is equivalent to multiplying $3a^2$ (10 800) by b, producing $3a^2b$; $3ab$ (720) by b, producing $3ab^2$; and b^2 (16) by b, producing b^3. Hence when we write 46 144 previous to performing the subtraction we have fulfilled up to this point all the requirements of the formula $(a + b)^3 = a^3 + 3a^2b + 3ab^2 + b^3$.

8. *Bring down the next period and continue the same process until all the figures of the root have been determined.*

When the third figure of the root is found in the example a becomes 640 and b becomes 2. The student should check the manner in which multiplication of the second complete divisor by 2 fulfils the requirements of the formula. The correctness of the complete extraction may, of course, be checked by multiplying the determined root to its third power.

APPROXIMATE ROOTS OF FRACTIONS

We have seen (p. 92) that the square root of a fraction is the square root of its numerator placed over the square root of the denominator, subject to further reduction or to conversion to a decimal.

When the terms of a fraction are not perfect squares it is often desirable to approximate a square root without going to the trouble of making an exact calculation. This is done by multiplying the terms of the fraction by any number that will make the denominator a perfect square, as in the following example.

EXAMPLE: What is the approximate square root of $\frac{19}{8}$?

$\frac{19}{8} = \frac{38}{16}$, of which the approximate square root, $\frac{6}{4}$, is correct to within $\frac{1}{4}$; or $\frac{19}{8} \times \frac{32}{32} = \frac{608}{256}$, of which the approximate square root, $\frac{25}{16}$, is correct to within $\frac{1}{16}$.

EXPLANATION: We select a factor that will make the denominator a perfect square. We then extract the square root of the denominator and the square root of the perfect square that is nearest to the numerator. If we write the fraction as $\frac{38}{16}$ the square root of the denominator is 4 and the square root of the nearest perfect square to 38 is 6. The resulting approximate square root, $\frac{6}{4}$, reducible to $\frac{3}{2}$, is correct to within $\frac{1}{4}$.

If we want a closer approximation than this we multiply by a larger factor. Using 32 as a factor, we get $\frac{608}{256}$. The square root of the denominator is 16. The nearest perfect square to 608 is 625, the square root of which is 25. The resulting approximate square root, $\frac{25}{16}$, is correct to within $\frac{1}{16}$.

It will be noted that the larger the factor, the more closely will the result approximate the correct value.

The approximate cube root of a fraction may be found by a similar process.

EXAMPLE: Find the approximate cube root of $\frac{173}{32}$.

$\frac{173}{32} = \frac{346}{64}$, of which the approximate cube root, $\frac{7}{4}$, is correct to within $\frac{1}{4}$; or $\frac{173}{32} = \frac{2768}{512}$, of which the approximate cube root, $\frac{14}{8}$, is correct to within $\frac{1}{8}$.

EXPLANATION: The denominator has been multiplied by two different factors in order to demonstrate again that the higher factor produces the more nearly accurate answer. It will be noted that the final result in both cases has the same ultimate value, since $\frac{14}{8} = \frac{7}{4}$. If, however, we had not worked out the second solution we would not know that $\frac{7}{4}$ is actually correct to within $\frac{1}{8}$.

HIGHER ROOTS

If the index of a higher root contains no other prime factors than 2 and 3 we can find the required root by repeated extraction of square or cube roots, according to the nature of the problem.

EXAMPLE 1: What is the fourth root of 923 521?

SOLUTION:
$$\sqrt{923\ 521} = 961,$$
$$\sqrt{961} = 31, \quad \text{ANS.}$$

EXPLANATION: Since the fourth power of a number is its square multiplied by its square, we find the fourth root of a given number representing such a power by extracting the square root of the square root.

EXAMPLE 2: What is the sixth root of 191 102 976?

SOLUTION:
$$\sqrt{191\ 102\ 976} = 13\ 824,$$
$$\sqrt[3]{13\ 824} = 24, \quad \text{ANS.}$$

EXPLANATION: The sixth root is found by taking the cube root of the square root. The order of making the extractions is, of course, immaterial.

Higher roots with indexes that are prime to 2 and 3 are found by methods based on the same general theory as that underlying the methods for extracting square and cube roots. Thus, if it be required to find the fifth root of a number, we consider that $(a + b)^5 = a^5 + 5a^4b + 10a^3b^2 + 10a^2b^3 + 5ab^4 + b^5$. After subtracting a^5 from the first period we must construct a complete divisor which when multiplied by b will satisfy the whole formula. Dividing what follows a^5 in the formula by b, we get as the requirement of our complete divisor, $5a^4 + 10a^3b + 10a^2b^2 + 5ab^3 + b^4$. We use the first term of this, $5a^4$, as a trial divisor, but where the complete divisor is so complex several estimates may have to be tried before finding the correct value for b. In actual practice, however, higher roots are more commonly found by the use of logarithms and the slide rule, as in Chapter Twelve.

HANDY ALGEBRAIC FORMULAE

The following formulae should be memorized:

$$(a + b)^2 = a^2 + 2ab + b^2$$
$$(a - b)^2 = a^2 - 2ab + b^2$$
$$(a + b)(a - b) = a^2 - b^2$$
$$(a + b)^3 = a^3 + 3a^2b + 3ab^2 + b^3$$

These formulae have many applications, and they are particularly applicable to doing arithmetic by short-cut methods. Compare what is said below with the methods of multiplication presented on p. 8.

Translating Numbers into Algebra

In the following consider that a represents a number of the tens order, like 10, 20, 30, etc., while b represents a number of the units order.

Squaring a number:

EXAMPLE 1: Multiply 63 by 63.

60×60 combined with $3 \times 3 = 3609$,
6×60, or 360, added to $3609 = 3969$, Ans.

EXPLANATION: 60×60 represents the a^2 of the formula, to which we at once add 9 as the b^2. The $2ab$ is most quickly worked out as $2 \times 3 \times 60$.

EXAMPLE 2: What is the square of 65?

60×70 combined with $25 = 4225$, Ans.

EXPLANATION: Doing this example by the previous method we would get $3625 + (2 \times 5 \times 60)$. But $(2 \times 5) \times 60 = 10 \times 60$. Hence we at once multiply 60 by 10 more than we otherwise would, or 70.

EXAMPLE 3: What is the square of 89?

8100 combined with $1 = 8101$,
$8101 - 180 = 7921$, Ans.

EXPLANATION: Since the digits are large and 89 is near 90, it is preferable here to use the square of $a - b$, taking a as 90 and b as 1. $a^2 + b^2 = 8101$; $2ab = 2 \times 1 \times 90$ or 180, which in accordance with the formula is subtracted from 8101.

Multiplying a sum by a difference:

EXAMPLE 4: How much is 53×47?

$2500 - 9 = 2491$, Ans.

EXPLANATION: $a = 50$, $b = 3$. $53 = a + b$; $47 = a - b$. $(a + b)(a - b) = a^2 - b^2 = 2500 - 9$.

Note that this method is applicable whenever the units add up to 10 and the tens differ by 10.

Cubing a number:

EXAMPLE: What is the cube of 23?

$$\begin{array}{r} 8027 \\ (69 \times 60) \quad 4140 \\ \hline 12\ 167, \quad \text{ANS.} \end{array}$$

EXPLANATION: $(a + b)^3 = a^3 + 3a^2b + 3ab^2 + b^3$. $a^3 + b^3$ may be quickly written down as 8027. $3a^2b + 3ab^2 = 3ab(a + b)$. $a + b$ is the given number, in this case 23. We therefore want 3×23 or 69 multiplied by ab or 3×20 or 60. In other words, to the cubes of the digits properly placed add three times the number multiplied by the product of its digits with an added 0. A little practice makes all this quite simple. For small numbers the method is very much quicker than performing separate multiplications.

In solving examples like the preceding there is, of course, no reason why a may not represent a number of the hundreds plus the tens order instead of one of the tens order. Consider a few examples:

EXAMPLE 1: Square 116.
$$12\ 136 + 1320 = 13\ 456, \quad \text{ANS.}$$

EXAMPLE 2: Square 125.
$$130 \times 120 \text{ combined with } 25 = 15\ 625, \quad \text{ANS.}$$

EXAMPLE: 3: Multiply 127 by 113.
$$14\ 400 - 49 = 14\ 351, \quad \text{ANS.} \quad \text{from } (a^2 - b^2).$$

Practice Exercise No. 49

Find the required roots (approximate in the case of fractions).

1. $\sqrt[3]{2460375}$ 2. $\sqrt[3]{11089567}$ 3. $\sqrt[3]{40353607}$ 4. $\sqrt[3]{403583419}$
5. $\sqrt[3]{115501303}$ 6. $\sqrt{\frac{2}{3}} (\times \frac{48}{48})$ 7. $\sqrt{\frac{3.8}{5}} (\times \frac{5}{5})$ 8. $\sqrt{\frac{4.5}{7}} (\times \frac{343}{343})$
9. $\sqrt{10\frac{1}{2}} (\times \frac{200}{200})$ 10. $\sqrt{7\frac{1}{8}} (\times \frac{20000}{20000})$ 11. $\sqrt[3]{\frac{3}{2}} (\times \frac{72}{72})$ 12. $\sqrt[3]{\frac{2}{3}} (\times \frac{15552}{15552})$
13. $\sqrt[3]{\frac{2}{3}} (\times \frac{124416}{124416})$ 14. $\sqrt[3]{5\frac{13}{32}} (\times \frac{2}{2})$ 15. $\sqrt[3]{\frac{125}{256}} (\times \frac{2}{2})$ 16. $\sqrt[4]{6561}$
17. $\sqrt[6]{117649}$ 18. $\sqrt[4]{29\frac{52}{81}}$ 19. $\sqrt[4]{104\frac{536}{625}}$ 20. $\sqrt[6]{11\frac{25}{64}}$

Do the following mentally by algebraic methods.

21. 21^2 22. 23^3 23. 33^2 24. 37^2
25. 39^2 26. 35^2 27. 65^2 28. 95^2
29. 105^2 30. 205^2 31. $29^2 (a - b)^2$ 32. 39^2
33. 99^2 34. 28^2 35. 38^3 36. 19×21
37. 28×32 38. 37×43 39. 46×54 40. 48×72

CHAPTER ELEVEN

ALGEBRAIC PROCESSES

DEFINITIONS

A **monomial** is an algebraic expression of one term. *Thus*, $8a$ and $16a^2b$ and $\sqrt{3}ax$ are monomials.

A **polynomial** is an algebraic expression of more than one term. *Thus*, $a + b$ and $a^2 + 2ab + b^2$ and $a^3 + 3a^2b + 3ab^2 + b^3$ are three different *polynomials*.

A **binomial** is a polynomial that contains *two* terms. *Thus*, $a + b$ and $a + 1$ and $\sqrt{2} + \sqrt{3}$ are *binomials*. A **trinomial** contains *three* terms.

FACTORING

Factoring is the process of separating, or resolving, a quantity into factors. No general rule can be given for factoring. In most cases the operation is performed by inspection and trial. The methods are best explained by examples.

Principle: *If every term of a polynomial contains the same monomial factor, then that monomial is one factor of the polynomial, and the other factor is equal to the quotient of the polynomial divided by the monomial factor.*

EXAMPLE: Factor the binomial $8a^2x^2 + 4a^3x$.

SOLUTION: $$8a^2x^2 + 4a^3x = 4a^2x(2x + a).$$

EXPLANATION: We see by inspection that $4a^2x$ is a factor common to both terms. Dividing by $4a^2x$, we arrive at the other factor.

Principle: *If a trinomial contains three terms two of which are squares, and if the third term is equal to plus or minus twice the product of the square roots of the other two, the expression may be recognized as the square of a binomial.*

Thus, $a^2x^2 + 2acx + c^2 = (ax + c)^2$ and $9a^2b^2 - 24a^3bc + 16a^2c^2 = (3ab - 4ac)^2$.

Principle: *If an expression represents the difference between two squares it can be factored as the product of the sum of the roots by the difference between them.*

Thus, $4x^2 - 9y^2 = (2x + 3y)(2x - 3y)$ and $25a^4b^4x^4 - 4z^2 = (5a^2b^2x^2 + 2z)(5a^2b^2x^2 - 2z)$.

Principle: *If the factors of an expression contain like terms these should be collected so as to present the result in the simplest form.*

EXAMPLE: Factor $(5a + 3b)^2 - (3a - 2b)^2$.

SOLUTION:
$$
\begin{aligned}
&(5a + 3b)^2 - (3a - 2b)^2 \\
&= [(5a + 3b) + (3a - 2b)][(5a + 3b) - (3a - 2b)] \\
&= (5a + 3b + 3a - 2b)(5a + 3b - 3a + 2b) \\
&= (8a + b)(2a + 5b), \quad \text{ANS.}
\end{aligned}
$$

101

Principle: *A trinomial in the form of $a^4 + a^2b^2 + b^4$ can be written in the form of the difference between two squares.*

EXAMPLE: Resolve $9x^4 + 26x^2y^2 + 25y^4$ into factors.

SOLUTION:

$$9x^4 + 26x^2y^2 + 25y^4$$
$$+ \ 4x^2y^2 \qquad\qquad - 4x^2y^2$$

$$\overline{}$$

$$(9x^4 + 30x^2y^2 + 25y^4) - 4x^2y^2$$
$$= (3x^2 + 5y^2)^2 - 4x^2y^2$$
$$= (3x^2 + 5y^2 + 2xy)(3x^2 + 5y^2 - 2xy)$$
$$= (3x^2 + 2xy + 5y^2)(3x^2 - 2xy + 5y^2)$$

EXPLANATION: We note that the given expression is nearly a perfect square. We therefore add $4x^2y^2$ to it to make it a square and also subtract from it the same quantity. We then write it in the form of a difference between two squares. We resolve this into factors and rewrite the result so as to make the terms follow in the order of the powers of x.

Principle: *If a trinomial has the form $x^2 + ax + b$ and is factorable into two binomial factors the first term of each factor will be x; the second term of the binomials will be two numbers whose product is b and whose sum is equal to a, which is the coefficient of the middle term of the trinomial.*

EXAMPLE 1: Factor $x^2 + 10x + 24$.

SOLUTION: $\qquad\qquad x^2 + 10x + 24 = (x + 6)(x + 4)$.

EXPLANATION: We are required to find two numbers whose product is 24 and whose sum is 10. The following pairs of factors will produce 24: 1 and 24, 2 and 12, 3 and 8, 4 and 6. From among these we select the pair whose sum is 10.

EXAMPLE 2: Factor $x^2 - 16x + 28$.

SOLUTION: $\qquad\qquad x^2 - 16x + 28 = (x - 14)(x - 2)$.

EXPLANATION: We are required to find two numbers whose product is 28 and whose algebraic sum is -16. Since their product is positive, they must both have the same sign, and since their sum is negative, they must both be negative. The negative factors that will produce 28 are -1 and -28, -2 and -14, -4 and -7. We select the pair whose algebraic sum is -16.

EXAMPLE 3: Factor $x^2 + 5x - 24$.

SOLUTION: $\qquad\qquad x^2 + 5x - 24 = (x + 8)(x - 3)$.

EXPLANATION: We are required to find two numbers whose product is -24 and whose algebraic sum is 5. Since their product is negative, the numbers must have unlike signs, and since their sum is $+5$, the larger number must be positive. The pairs of numbers that will produce 24, without considering signs, are 1 and 24, 2 and 12, 3 and 8, 4 and 6. From these we select the pair whose difference is 5. This is 3 and 8. We give the plus sign to the 8 and the minus sign to the 3.

EXAMPLE 4: Factor $x^2 - 7x - 18$.

SOLUTION: $\qquad\qquad x^2 - 7x - 18 = (x - 9)(x + 2)$.

EXPLANATION: We are required to find two numbers whose product is -18 and whose algebraic sum is -7. Since their product is negative, the signs of the two numbers are unlike, and since their sum is negative, the larger number must be negative. The pairs of numbers that will produce 18, without considering signs, are 1 and 18, 2 and 9, 3 and 6. We select the pair whose difference is 7, giving the minus sign to the 9 and the plus sign to the 2.

EXAMPLE 5: Factor $x^2 - 7xy + 12y^2$.

SOLUTION: $$x^2 - 7xy + 12y^2 = (x - 4y)(x - 3y).$$

EXPLANATION: We are required to find two terms whose product is $12y^2$ and whose algebraic sum is $-7y$. Since their product is positive and their sum negative, they must both be negative terms. From the pairs of negative terms that will produce $+12y^2$ we select $-4y$ and $-3y$ as fulfilling the requirements.

When a trinomial factorable into two binomials has the form $ax^2 \pm bx \pm c$ it is resolved into factors by a process of trial and error which is continued until values are found that satisfy the requirements.

EXAMPLE 1: Factor $4x^2 + 26x + 22$

$$
\begin{array}{l}
4x + 11 \\
\qquad \diagdown\!\!\!\diagup \quad + 11x + 8x = 19x \ (reject) \\
x + 2
\end{array}
$$

$$
\begin{array}{l}
x + 11 \\
\qquad \diagdown\!\!\!\diagup \quad + 44x + 2x = 46x \ (reject) \\
4x + 2
\end{array}
$$

$$
\begin{array}{l}
2x + 11 \\
\qquad \diagdown\!\!\!\diagup \quad 22x + 4x = 26x \ (correct) \\
2x + 2
\end{array}
$$

$$\therefore 4x^2 + 26x + 22 = (2x + 11)(2x + 2), \quad \text{ANS.}$$

EXPLANATION: We use what is called the *cross multiplication* method to find the required binomials. We consider the pairs of terms that will produce the first and last terms of the trinomials. We write down the various forms of examples that can be worked out with these, and we reject one trial result after another until we find the arrangement that will give us the correct value for the middle term of the given trinomial.

Instead of making a separate example out of each of the possibilities, the process is shortened by simply listing the possible factors involved, in the following manner:

$$
\begin{array}{ccc|c}
1 & 4 & 2 & 11 \\
4 & 1 & 2 & 2
\end{array}
$$

Factors to the left of the vertical line represent possible coefficients of x; those to the right of the line represent possible numerical values of second terms. Each pair of x coefficients is written in two positions (1 over 4, 4 over 1, etc.). Accordingly, it is not necessary to write second-term values in more than one position in order to exhaust the possibilities. We proceed with cross multiplication of the numbers on both sides of the vertical line. $(1 \times 2) + (4 \times 11) = 46$ (*too large—reject*); $(4 \times 2) + (1 \times 11) = 19$ (*too small—reject*); $(2 \times 2) + (2 \times 11) = 26$ (*correct*).

EXAMPLE 2: Factor $24x^2 - 2x - 15$.

SOLUTION:

24	1	2	12	3	8	4	6	1	3
1	24	12	2	8	3	6	4	15	5

We select $\frac{4}{6}$ and $\frac{3}{5}$ as the combination of numbers that will give us the required middle term.

$\therefore 24x^2 - 2x - 15 = (4x + 3)(6x - 5)$, ANS.

EXPLANATION: We write down the possible numerical values in the manner previously described. Inasmuch as the third term of the trinomial is negative, the two second terms of its binomial factors must have unlike signs. Considering that the given middle term has a very small value, we conclude that we are more likely to find the answer quickly if we start our cross multiplication at the right of the numerical arrangement rather than at the left. In carrying out this cross multiplication we give the negative sign in each case to the larger of the two products involved. *Thus:* $6(-5) + 4(+3) = -18$ *(reject)*; $4(-5) + 6(+3) = -2$ *(correct)*. We have been fortunate in finding the correct values so soon. Otherwise we should have had to continue the process of trial and error with the numerical listing—though this is not as lengthy a process as may appear, since many of the wrong results are recognized at a glance without taking the trouble to calculate them. Having selected the correct combination of numbers, we write the factors as $4x + 3$ and $6x - 5$.

Practice Exercise No. 50

Resolve the following into factors

1. $7a^2bc^3 - 28abc$
2. $15a^2cd + 20ac^2d - 15acd^2$
3. $4x^2 + 12xy + 9y^2$
4. $9a^2b^2 - 24a^2bc + 16a^2c^2$
5. $9a^2x^2 - 16a^2y^2$
6. $49x^4 - 16y^2$
7. $(2x + y + z)^2 - (x - 2y + z)^2$
8. $a^4 + a^2 + 1$
   (*Hint:* Add and subtract a^2.)
9. $x^2 + 10x + 21$
10. $x^2 - 18x + 45$
11. $x^2 + 5x - 36$
12. $x^2 - 13x - 48$
13. $x^2 - 14xy + 33y^2$
14. $6x^2 + 21x + 9$
15. $15x^2 - 6x - 21$
16. $12x^2 + 27x - 39$

SIMULTANEOUS EQUATIONS

Simultaneous equations are equations involving the same unknown quantities. *Thus,* $a + 2b = 11$ and $2a + b = 10$ are *simultaneous equations,* since they both involve the same unknowns, namely, a and b.

Simultaneous equations involving two unknown quantities are solved as follows:

Rule 1. *Eliminate one of the unknowns.*

Rule 2. *Solve for the other unknown.*

Rule 3. *Find the value of the unknown previously eliminated.*

Elimination may be performed by any one of three different methods:

(1) *by addition or subtraction;*
(2) *by substitution;*
(3) *by comparison.*

Elimination by Addition or Subtraction

Rule 1. *Multiply one or both of the equations by such a number or numbers as will give one of the unknowns the same coefficient in both equations.*

Rule 2. *Add or subtract the equal coefficients according to the nature of their signs.*

EXAMPLE: $5x + 2y = 32$, $2x - y = 2$. Find x and y.

SOLUTION:

$$5x + 2y = 32$$
$$4x - 2y = 4, \text{ multiplying } 2x - y \text{ by } 2$$
$$\overline{}$$
$$9x = 36$$
$$x = 4.$$
$$20 + 2y = 32, \text{ substituting 4 for } x \text{ in first equation}$$
$$2y = 32 - 20, \text{ transposing}$$
$$y = 6$$

Elimination by Substitution

Rule 1. *From one of the equations find the value of one of the unknowns in terms of the other.*

Rule 2. *Substitute the value thus found for the unknown in the other of the given equations.*

EXAMPLE: $2x + 4y = 50$, $3x + 5y = 66$. Find x and y.

SOLUTION:

$$2x + 4y = 50$$
$$2x = 50 - 4y, \text{ transposing}$$
$$x = 25 - 2y.$$
$$3(25 - 2y) + 5y = 66, \text{ substituting for } x \text{ in other equation}$$
$$75 - 6y + 5y = 66$$
$$-y = 66 - 75 = -9, y = 9$$
$$2x + 36 = 50, \text{ substituting 9 for } y \text{ in first equation}$$
$$2x = 50 - 36 = 14, x = 7$$

Elimination by Comparison

Rule 1. *From each equation find the value of one of the unknowns in terms of the other.*

Rule 2. *Form an equation from these equal values.*

EXAMPLE: $3x + 2y = 27$, $2x - 3y = 5$. Find x and y.

SOLUTION: $3x + 2y = 27$, $3x = 27 - 2y$, $x = \dfrac{27 - 2y}{3}$

$$2x - 3y = 5, 2x = 5 + 3y, x = \frac{5 + 3y}{2}$$

$$\frac{27 - 2y}{3} = \frac{5 + 3y}{2}, \text{ both being equal to } x$$

$$27 - 2y = \frac{3(5 + 3y)}{2}, \text{ multiplying both sides by 3}$$

$$2(27 - 2y) = 3(5 + 3y), \text{ multiplying both sides by 2}$$

$$54 - 4y = 15 + 9y, \text{ carrying out multiplication}$$

$$-4y - 9y = 15 - 54 = -39, y = 3$$

$$3x + 6 = 27, 3x = 21, x = 7$$

Of the foregoing methods, select the one which appears most likely to make the solution simple and direct.

Practice Exercise No. 51

1. The hands of a clock are together at 12 o'clock. When do they next meet? (x = minute spaces passed over by minute hand; y = number passed over by hour hand.)

2. A man has £22 000 invested, and on it he earns £1220. Part of the money is out at 5% interest and part at 6%. How much is in each part?

3. Jack is twice as old as Joe. Twenty years ago Jack was four times as old as Joe. What are their ages?

4. There are two numbers: the first added to half the second gives 35; the second added to half the first equals 40. What are the numbers?

5. The inventory of one department of a store increased by one-third of that of a second department amounts to £1700; the inventory of the second increased by one-fourth of that of the first amounts to £1800. What are the inventories?

6. Find two numbers such that $\frac{1}{2}$ of the first plus $\frac{1}{3}$ of the second shall equal 45, and $\frac{1}{2}$ of the second plus $\frac{1}{3}$ of the first shall equal 40.

7. A and B invest £918 in a partnership venture and clear £153. A's share of the profit is £45 more than B's. What was the contribution of each one to the capital?

8. Two girls receive $153 for baby-sitting. Ann is paid for 14 days and Mary for 15. Ann's pay for 6 days' work is $3 more than Mary gets for 4. How much does each earn per day?

9. In 80 kg of an alloy of copper and tin there are 7 kg of copper to 3 of tin. How much copper must be added so that there may be 11 kg of copper to 4 of tin?

10. Brown owes £1200 and Jones £2500, but neither has enough money to pay his debts. Brown says to Jones, 'Lend me one-eighth of your bank account and I'll pay my creditors.' Jones says to Brown, 'Lend me one-ninth of yours and I'll pay mine.' How much money has each?

FRACTIONS

To reduce a fraction to its lowest terms, *resolve the numerator and the denominator into their prime factors and cancel all the common factors, or divide the numerator and the denominator by their highest common factor.*

EXAMPLE 1: Reduce $\dfrac{12a^2b^3c^4}{9a^3bc^2}$.

SOLUTION:
$$\frac{12a^2b^3c^4}{9a^3bc^2} = \frac{2 \times 2 \times 3a^2bc^2(b^2c^2)}{3 \times 3a^2bc^2(a)}$$

$$= \frac{4b^2c^2}{3a}, \text{ ANS.}$$

EXPLANATION: The numerical parts of the fraction are separated into their prime factors, and the algebraic parts are divided by their highest common factor. The terms that cancel out are then eliminated. As a guide for determining the highest common factor of monomial terms note that such a factor is made up of the lower (or lowest) of the given powers of each letter involved.

EXAMPLE 2: Reduce $\dfrac{12x^2 + 15x - 63}{4x^2 - 31x + 42}$.

SOLUTION:
$$\frac{12x^2 + 15x - 63}{4x^2 - 31x + 42} = \frac{(3x + 9)(4x - 7)}{(x - 6)(4x - 7)}$$
$$= \frac{3x + 9}{x - 6}, \quad \text{Ans.}$$

EXPLANATION: Numerator and denominator are factored, and the common factor is then cancelled.

A fraction may be reduced to an integral or mixed expression if the degree (power) of its numerator equals or exceeds that of its denominator.

To reduce a fraction to an integral or mixed expression, *divide the numerator by the denominator.*

EXAMPLE 1: Reduce $\dfrac{x^2 - y^2}{x - y}$ to an integral expression.

SOLUTION:
$$\frac{x^2 - y^2}{x - y} = \frac{(x - y)(x + y)}{x - y} = x + y.$$

EXAMPLE 2: Reduce $\dfrac{x^2 + y^2}{x + y}$ to a mixed expression.

SOLUTION:
$$\frac{x^2 + y^2}{x + y} = \frac{(x^2 - y^2) + 2y^2}{x + y}$$
$$= \frac{(x + y)(x - y) + 2y^2}{x + y}$$
$$= x - y + \frac{2y^2}{x + y}, \quad \text{Ans.}$$

EXPLANATION: While $x^2 + y^2$ is not evenly divisible by $x + y$, we recognize that it would be so divisible if it were $x^2 - y^2$. Hence we subtract $2y^2$ to convert it to $x^2 - y^2$ and also add to it the same amount. We divide $x^2 - y^2$ by $x + y$ and write the remainder as a fraction that has $x + y$ for its denominator.

To reduce a mixed expression to a fraction, *multiply the integral expression by the denominator of the fraction; add to this product the numerator of the fraction and write under this result the given denominator.*

EXAMPLE: Reduce $x + 1 + \dfrac{x + 1}{x - 1}$ to a fraction.

SOLUTION:
$$\left(x + 1 + \frac{x + 1}{x - 1}\right)\left(\frac{x - 1}{x - 1}\right)$$
$$= \frac{x^2 - 1 + x + 1}{x - 1} = \frac{x^2 + x}{x - 1}, \quad \text{Ans.}$$

To reduce fractions to their lowest common denominator, *find the lowest common multiple of the denominators and proceed on the same principles that govern arithmetical fractions.*

EXAMPLE: Reduce $\dfrac{1}{x^2 + 3x + 2}$, $\dfrac{2}{x^2 + 5x + 6}$ and $\dfrac{3}{x^2 + 4x + 3}$ to fractions having the lowest common denominator.

SOLUTION:
$$\frac{1}{x^2 + 3x + 2}, \frac{2}{x^2 + 5x + 6}, \frac{3}{x^2 + 4x + 3}$$

$$= \frac{1}{(x + 1)(x + 2)}, \frac{2}{(x + 2)(x + 3)}, \frac{3}{(x + 1)(x + 3)}$$

The L.C.D. is $(x + 1)(x + 2)(x + 3)$.

Dividing this by each of the denominators and multiplying each numerator by the resulting quotient we obtain

$$\frac{x + 3}{(x + 1)(x + 2)(x + 3)}, \frac{2x + 2}{(x + 1)(x + 2)(x + 3)},$$

$$\frac{3x + 6}{(x + 1)(x + 2)(x + 3)}, \quad \text{ANS.}$$

Addition and Subtraction of Fractions

EXAMPLE 1: Simplify $\dfrac{2a - 4b}{4} - \dfrac{a - b + c}{3} + \dfrac{a - b - 2c}{12}$

SOLUTION:
$$\frac{2a - 4b}{4} - \frac{a - b + c}{3} + \frac{a - b - 2c}{12}$$

$$= \frac{6a - 12b - 4a + 4b - 4c + a - b - 2c}{12}$$

$$= \frac{3a - 9b - 6c}{12} = \frac{a - 3b - 2c}{4}, \quad \text{ANS.}$$

EXAMPLE 2: Simplify $\dfrac{u + 2x}{a - 2x} - \dfrac{u - 2x}{a + 2x}$.

SOLUTION:
$$\frac{a + 2x}{a - 2x} - \frac{a - 2x}{a + 2x}$$

$$= \frac{(a + 2x)^2 - (a - 2x)^2}{a^2 - 4x^2}$$

$$= \frac{a^2 + 4ax + 4x^2 - a^2 + 4ax - 4x^2}{a^2 - 4x^2}$$

$$= \frac{8ax}{a^2 - 4x^2}, \quad \text{ANS.}$$

Multiplication and Division of Fractions

Principle: *The product of two or more fractions is equal to the product of the numerators multiplied together, divided by the product of the denominators multiplied together.*

EXAMPLE 1: Multiply $\dfrac{7x}{5y}$ by $\dfrac{3a}{4c}$.

SOLUTION:
$$\frac{7x}{5y} \cdot \frac{3a}{4c} = \frac{21ax}{20cy}, \quad \text{ANS.}$$

EXAMPLE 2: Multiply $\dfrac{2x}{x-y}$ by $\dfrac{x^2-y^2}{3}$.

SOLUTION:
$$\left(\frac{2x}{x-y}\right)\left(\frac{x^2-y^2}{3}\right)=\frac{2x(x+y)(x-y)}{3(x-y)}$$

$$=\frac{2x(x+y)}{3}, \quad \text{Ans.}$$

EXAMPLE 3: Multiply $\dfrac{2(x+y)}{x-y}$ by $\dfrac{x^2-y^2}{x^2+2xy+y^2}$.

SOLUTION:
$$\left[\frac{2(x+y)}{x-y}\right]\left[\frac{x^2-y^2}{x^2+2xy+y^2}\right]$$

$$=\frac{2(x+y)(x+y)(x-y)}{(x-y)(x+y)^2}=2, \quad \text{Ans.}$$

Principle: *Division by a fraction is equivalent to multiplication by the reciprocal of the fraction, i.e. the fraction inverted.*

EXAMPLE: Divide $\dfrac{3a^2}{a^2-b^2}$ by $\dfrac{a}{a+b}$.

SOLUTION:
$$\frac{3a^2}{a^2-b^2}\div\frac{a}{a+b}=\frac{3a^2}{a^2-b^2}\cdot\frac{a+b}{a}$$

$$=\frac{3a^2(a+b)}{a(a+b)(a-b)}=\frac{3a^2}{a(a-b)}=\frac{3a}{a-b}, \quad \text{Ans.}$$

Practice Exercise No. 52

1. Reduce $\dfrac{45x^3y^3z}{36abx^2y^2z}$ to its lowest terms.

2. Reduce $\dfrac{x^2+2ax+a^2}{3(x^2-a^2)}$ to its lowest terms.

3. Reduce $\dfrac{x^2+a^2+3-2ax}{x-a}$ to a mixed quantity.

4. Reduce $a+\dfrac{ax}{a-x}$ to a fraction.

5. Reduce $1+\dfrac{c}{x-y}$ to a fraction.

6. Reduce $\dfrac{x+a}{b}$, $\dfrac{a}{b}$ and $\dfrac{a-x}{a}$ to fractions with the L.C.D.

7. Reduce $\dfrac{x}{1-x}$, $\dfrac{x^2}{(1-x)^2}$ and $\dfrac{x^3}{(1-x)^3}$ to fractions with the L.C.D.

8. Add $\dfrac{x+y}{2}$ and $\dfrac{x-y}{2}$.

9. Add $\dfrac{2}{(x-1)^2}$, $\dfrac{3}{(x-1)^3}$ and $\dfrac{4}{x-1}$.

10. Subtract $2a - \dfrac{a - 3b}{c}$ from $4a + \dfrac{2a}{c}$.

11. Subtract $\dfrac{x}{a + x}$ from $\dfrac{a}{a - x}$.

12. Multiply $\dfrac{2}{x - y}$ by $\dfrac{x^2 - y^2}{a}$.

13. Multiply $\dfrac{x^2 - 4}{3}$ by $\dfrac{4x}{x + 2}$.

14. Divide $\dfrac{3x}{2x - 2}$ by $\dfrac{2x}{x - 1}$.

15. Divide $\dfrac{(x + y)^2}{x - y}$ by $\dfrac{x + y}{(x - y)^2}$.

CHAPTER TWELVE

LOGARITHMS

Logarithms are a means of simplifying the manipulation of numbers containing many digits or decimal places. The system of common logarithms, which is the one in most common use, is based on powers of 10.

By this system **the logarithm of a given number** *is the exponent to which* 10 *must be raised to obtain that number. Thus:*

> $10^1 = 10$; ∴ the logarithm of 10 is 1.
> $10^2 = 100$; ∴ the logarithm of 100 is 2.
> $10^3 = 1000$; ∴ the logarithm of 1000 is 3.
> $10^4 = 10\,000$; ∴ the logarithm of 10 000 is 4.
> and so on up.

The logarithm of a number between 10 and 99 is therefore an exponent greater than 1 and less than 2.

The logarithm of a number between 100 and 200 is an exponent greater than 2 and less than 3.

The logarithm of any number other than a multiple of 10 is therefore a whole number plus a decimal.

FINDING THE LOGARITHM OF A NUMBER

The logarithm of 45 should be between 1 and 2. That is, it must be 1 plus something. To find out what this something is, we refer to what is known as a **table of logarithms,** and then we find the logarithm of 45 to be equal to $1 \cdot 6532$. This is written:

$$\log 45 = 1 \cdot 6532$$

The method of finding a logarithm from the table will be explained in detail later.

The **characteristic** is the whole-number part of the logarithm. In the above case the *characteristic* is 1.

The **mantissa** is the decimal part of the logarithm, and is the part found in the table of logarithms. In the above case the *mantissa* is 0·6532.

Finding the characteristic. *The characteristic is not found in the table but is determined by rule. It is positive for numbers equal to 1 or greater, and negative for numbers less than 1.*

By definition,

For numbers between these limits	the characteristic is
10 000 and 100 000 *minus*	4
1000 and 10 000 *minus*	3
100 and 1000 *minus*	2
10 and 100 *minus*	1
1 and 10 *minus*	0
0·1 and 1 *minus*	−1
0·01 and 0·1 *minus*	−2
0·001 and 0·01 *minus*	−3
0·0001 and 0·001 *minus*	−4

Note: The characteristic 4 would apply to numbers from 10 000 to 99 999·999 999 + carried to any number of places; characteristic 3, from 1000 to 9999·999 999 . . . etc. For the sake of simplicity the latter numbers in these groups are expressed as 100 000 minus, 10 000 minus, 1000 minus, etc.

Rule 1. *For whole numbers the characteristic is one less than the number of figures to the left of the decimal point.*

EXAMPLE 1: What is the characteristic of 82 459·23?

SOLUTION: There are 5 figures to the left of the decimal. $5 - 1 = 4$. ∴ the characteristic is 4.

Rule 2. *The characteristic of decimal numbers is equal to minus the number of places to the right from the decimal point to the first significant figure* (number other than zero).

EXAMPLE 2: What is the characteristic of 0·001 326?

SOLUTION: From the decimal point to 1, the first significant figure, there are 3 places. ∴ the characteristic is −3.

EXAMPLE 3: What is the characteristic of 0·443?

SOLUTION: There is but one place from the decimal point to the first significant figure. ∴ the characteristic is −1.

Note: If the characteristic of a number (0·023) is −2, and the mantissa is 3617, the whole logarithm is written $\bar{2}$·3617. The mantissa is always considered positive, and therefore negative characteristics are denoted by the placing of the minus sign *above* the characteristic.

Table VI. Logarithms

N	0	1	2	3	4	5	6	7	8	9	1	2	3	4	5	6	7	8	9
10	·0000	0043	0086	0128	0170	0212	0253	0294	0334	0374	4	8	12	17	21	25	29	33	37
11	·0414	0453	0492	0531	0569	0607	0645	0682	0719	0755	4	8	11	15	19	23	26	30	34
12	·0792	0828	0864	0899	0934	0969	1004	1038	1072	1106	3	7	10	14	17	21	24	28	31
13	·1139	1173	1206	1239	1271	1303	1335	1367	1399	1430	3	6	10	13	16	19	23	26	29
14	·1461	1492	1523	1553	1584	1614	1644	1673	1703	1732	3	6	9	12	15	18	21	24	27
15	·1761	1790	1818	1847	1875	1903	1931	1959	1987	2014	3	6	8	11	14	17	20	22	25
16	·2041	2068	2095	2122	2148	2175	2201	2227	2253	2279	3	5	8	11	13	16	18	21	24
17	·2304	2330	2355	2380	2405	2430	2455	2480	2504	2529	2	5	7	10	12	15	17	20	22
18	·2553	2577	2601	2625	2648	2672	2695	2718	2742	2765	2	5	7	9	12	14	16	19	21
19	·2788	2810	2833	2856	2878	2900	2923	2945	2967	2989	2	4	7	9	11	13	16	18	20
20	·3010	3032	3054	3075	3096	3118	3139	3160	3181	3201	2	4	6	8	11	13	15	17	19
21	·3222	3243	3263	3284	3304	3324	3345	3365	3385	3404	2	4	6	8	10	12	14	16	18
22	·3424	3444	3464	3483	3502	3522	3541	3560	3579	3598	2	4	6	8	10	12	14	15	17
23	·3617	3636	3655	3674	3692	3711	3729	3747	3766	3784	2	4	6	7	9	11	13	15	17
24	·3802	3820	3838	3856	3874	3892	3909	3927	3945	3962	2	4	5	7	9	11	12	14	16
25	·3979	3997	4014	4031	4048	4065	4082	4099	4116	4133	2	3	5	7	9	10	12	14	15
26	·4150	4166	4183	4200	4216	4232	4249	4265	4281	4298	2	3	5	7	8	10	11	13	15
27	·4314	4330	4346	4362	4378	4393	4409	4425	4440	4456	2	3	5	6	8	9	11	13	14
28	·4472	4487	4502	4518	4533	4548	4564	4579	4594	4609	2	3	5	6	8	9	11	12	14
29	·4624	4639	4654	4669	4683	4698	4713	4728	4742	4757	1	3	4	6	7	9	10	12	13
30	·4771	4786	4800	4814	4829	4843	4857	4871	4886	4900	1	3	4	6	7	9	10	11	13
31	·4914	4928	4942	4955	4969	4983	4997	5011	5024	5038	1	3	4	6	7	8	10	11	12
32	·5051	5065	5079	5092	5105	5119	5132	5145	5159	5172	1	3	4	5	7	8	9	11	12
33	·5185	5198	5211	5224	5237	5250	5263	5276	5289	5302	1	3	4	5	6	8	9	10	12
34	·5315	5328	5340	5353	5366	5378	5391	5403	5416	5428	1	3	4	5	6	8	9	10	11
35	·5441	5453	5465	5478	5490	5502	5514	5527	5539	5551	1	2	4	5	6	7	9	10	11
36	·5563	5575	5587	5599	5611	5623	5635	5647	5658	5670	1	2	4	5	6	7	8	10	11
37	·5682	5694	5705	5717	5729	5740	5752	5763	5775	5786	1	2	3	5	6	7	8	9	10
38	·5798	5809	5821	5832	5843	5855	5866	5877	5888	5899	1	2	3	5	6	7	8	9	10
39	·5911	5922	5933	5944	5955	5966	5977	5988	5999	6010	1	2	3	4	5	7	8	9	10
40	·6021	6031	6042	6053	6064	6075	6085	6096	6107	6117	1	2	3	4	5	6	8	9	10
41	·6128	6138	6149	6160	6170	6180	6191	6201	6212	6222	1	2	3	4	5	6	7	8	9
42	·6232	6243	6253	6263	6274	6284	6294	6304	6314	6325	1	2	3	4	5	6	7	8	9
43	·6335	6345	6355	6365	6375	6385	6395	6405	6415	6425	1	2	3	4	5	6	7	8	9
44	·6435	6444	6454	6464	6474	6484	6493	6503	6513	6522	1	2	3	4	5	6	7	8	9
45	·6532	6542	6551	6561	6571	6580	6590	6599	6609	6618	1	2	3	4	5	6	7	8	9
46	·6628	6637	6646	6656	6665	6675	6684	6693	6702	6712	1	2	3	4	5	6	7	7	8
47	·6721	6730	6739	6749	6758	6767	6776	6785	6794	6803	1	2	3	4	5	5	6	7	8
48	·6812	6821	6830	6839	6848	6857	6866	6875	6884	6893	1	2	3	4	4	5	6	7	8
49	·6902	6911	6920	6928	6937	6946	6955	6964	6972	6981	1	2	3	4	4	5	6	7	8
50	·6990	6998	7007	7016	7024	7033	7042	7050	7059	7067	1	2	3	3	4	5	6	7	8
51	·7076	7084	7093	7101	7110	7118	7126	7135	7143	7152	1	2	3	3	4	5	6	7	8
52	·7160	7168	7177	7185	7193	7202	7210	7218	7226	7235	1	2	2	3	4	5	6	7	7
53	·7243	7251	7259	7267	7275	7284	7292	7300	7308	7316	1	2	2	3	4	5	6	6	7
54	·7324	7332	7340	7348	7356	7364	7372	7380	7388	7396	1	2	2	3	4	5	6	6	7
	0	1	2	3	4	5	6	7	8	9	1	2	3	4	5	6	7	8	9

Table VI. Logarithms—contd.

N	0	1	2	3	4	5	6	7	8	9	1	2	3	4	5	6	7	8	9
55	·7404	7412	7419	7427	7435	7443	7451	7459	7466	7474	1	2	2	3	4	5	5	6	7
56	·7482	7490	7497	7505	7513	7520	7528	7536	7543	7551	1	2	2	3	4	5	5	6	7
57	·7559	7566	7574	7582	7589	7597	7604	7612	7619	7627	1	2	2	3	4	5	5	6	7
58	·7634	7642	7649	7657	7664	7672	7679	7686	7694	7701	1	1	2	3	4	4	5	6	7
59	·7709	7716	7723	7731	7738	7745	7752	7760	7767	7774	1	1	2	3	4	4	5	6	7
60	·7782	7789	7796	7803	7810	7818	7825	7832	7839	7846	1	1	2	3	4	4	5	6	6
61	·7853	7860	7868	7875	7882	7889	7896	7903	7910	7917	1	1	2	3	4	4	5	6	6
62	·7924	7931	7938	7945	7952	7959	7966	7973	7980	7987	1	1	2	3	3	4	5	6	6
63	·7993	8000	8007	8014	8021	8028	8035	8041	8048	8055	1	1	2	3	3	4	5	5	6
64	·8062	8069	8075	8082	8089	8096	8102	8109	8116	8122	1	1	2	3	3	4	5	5	6
65	·8129	8136	8142	8149	8156	8162	8169	8176	8182	8189	1	1	2	3	3	4	5	5	6
66	·8195	8202	8209	8215	8222	8228	8235	8241	8248	8254	1	1	2	3	3	4	5	5	6
67	·8261	8267	8274	8280	8287	8293	8299	8306	8312	8319	1	1	2	3	3	4	5	5	6
68	·8325	8331	8338	8344	8351	8357	8363	8370	8376	8382	1	1	2	3	3	4	4	5	6
69	·8388	8395	8401	8407	8414	8420	8426	8432	8439	8445	1	1	2	2	3	4	4	5	6
70	·8451	8457	8463	8470	8476	8482	8488	8494	8500	8506	1	1	2	2	3	4	4	5	6
71	·8513	8519	8525	8531	8537	8543	8549	8555	8561	8567	1	1	2	2	3	4	4	5	5
72	·8573	8579	8585	8591	8597	8603	8609	8615	8621	8627	1	1	2	2	3	4	4	5	5
73	·8633	8639	8645	8651	8657	8663	8669	8675	8681	8686	1	1	2	2	3	4	4	5	5
74	·8692	8698	8704	8710	8716	8722	8727	8733	8739	8745	1	1	2	2	3	4	4	5	5
75	·8751	8756	8762	8768	8774	8779	8785	8791	8797	8802	1	1	2	2	3	3	4	5	5
76	·8808	8814	8820	8825	8831	8837	8842	8848	8854	8859	1	1	2	2	3	3	4	5	5
77	·8865	8871	8876	8882	8887	8893	8899	8904	8910	8915	1	1	2	2	3	3	4	4	5
78	·8921	8927	8932	8938	8943	8949	8954	8960	8965	8971	1	1	2	2	3	3	4	4	5
79	·8976	8982	8987	8993	8998	9004	9009	9015	9020	9025	1	1	2	2	3	3	4	4	5
80	·9031	9036	9042	9047	9053	9058	9063	9069	9074	9079	1	1	2	2	3	3	4	4	5
81	·9085	9090	9096	9101	9106	9112	9117	9122	9128	9133	1	1	2	2	3	3	4	4	5
82	·9138	9143	9149	9154	9159	9165	9170	9175	9180	9186	1	1	2	2	3	3	4	4	5
83	·9191	9196	9201	9206	9212	9217	9222	9227	9232	9238	1	1	2	2	3	3	4	4	5
84	·9243	9248	9253	9258	9263	9269	9274	9279	9284	9289	1	1	2	2	3	3	4	4	5
85	·9294	9299	9304	9309	9315	9320	9325	9330	9335	9340	1	1	2	2	3	3	4	4	5
86	·9345	9350	9355	9360	9365	9370	9375	9380	9385	9390	1	1	1	2	3	3	4	4	5
87	·9395	9400	9405	9410	9415	9420	9425	9430	9435	9440	0	1	1	2	2	3	3	4	4
88	·9445	9450	9455	9460	9465	9469	9474	9479	9484	9489	0	1	1	2	2	3	3	4	4
89	·9494	9499	9504	9509	9513	9518	9523	9528	9533	9538	0	1	1	2	2	3	3	4	4
90	·9542	9547	9552	9557	9562	9566	9571	9576	9581	9586	0	1	1	2	2	3	3	4	4
91	·9590	9595	9600	9605	9609	9614	9619	9624	9628	9633	0	1	1	2	2	3	3	4	4
92	·9638	9643	9647	9652	9657	9661	9666	9671	9675	9680	0	1	1	2	2	3	3	4	4
93	·9685	9689	9694	9699	9703	9708	9713	9717	9722	9727	0	1	1	2	2	3	3	4	4
94	·9731	9736	9741	9745	9750	9754	9759	9763	9768	9773	0	1	1	2	2	3	3	4	4
95	·9777	9782	9786	9791	9795	9800	9805	9809	9814	9818	0	1	1	2	2	3	3	4	4
96	·9823	9827	9832	9836	9841	9845	9850	9854	9859	9863	0	1	1	2	2	3	3	4	4
97	·9868	9872	9877	9881	9886	9890	9894	9899	9903	9908	0	1	1	2	2	3	3	4	4
98	·9912	9917	9921	9926	9930	9934	9939	9943	9948	9952	0	1	1	2	2	3	3	4	4
99	·9956	9961	9965	9969	9974	9978	9983	9987	9991	9996	0	1	1	2	2	3	3	3	4
N	0	1	2	3	4	5	6	7	8	9	1	2	3	4	5	6	7	8	9

Table VIa. Antilogarithms

N	0	1	2	3	4	5	6	7	8	9	1	2	3	4	5	6	7	8	9
·00	1000	1002	1005	1007	1009	1012	1014	1016	1019	1021	0	0	1	1	1	1	2	2	2
·01	1023	1026	1028	1030	1033	1035	1038	1040	1042	1045	0	0	1	1	1	1	2	2	2
·02	1047	1050	1052	1054	1057	1059	1062	1064	1067	1069	0	0	1	1	1	1	2	2	2
·03	1072	1074	1076	1079	1081	1084	1086	1089	1091	1094	0	0	1	1	1	1	2	2	2
·04	1096	1099	1102	1104	1107	1109	1112	1114	1117	1119	0	1	1	1	1	2	2	2	2
·05	1122	1125	1127	1130	1132	1135	1138	1140	1143	1146	0	1	1	1	1	2	2	2	2
·06	1148	1151	1153	1156	1159	1161	1164	1167	1169	1172	0	1	1	1	1	2	2	2	2
·07	1175	1178	1180	1183	1186	1189	1191	1194	1197	1199	0	1	1	1	1	2	2	2	2
·08	1202	1205	1208	1211	1213	1216	1219	1222	1225	1227	0	1	1	1	1	2	2	2	3
·09	1230	1233	1236	1239	1242	1245	1247	1250	1253	1256	0	1	1	1	1	2	2	2	3
·10	1259	1262	1265	1268	1271	1274	1276	1279	1282	1285	0	1	1	1	1	2	2	2	3
·11	1288	1291	1294	1297	1300	1303	1306	1309	1312	1315	0	1	1	1	2	2	2	2	3
·12	1318	1321	1324	1327	1330	1334	1337	1340	1343	1346	0	1	1	1	2	2	2	3	3
·13	1349	1352	1355	1358	1361	1365	1368	1371	1374	1377	0	1	1	1	2	2	2	3	3
·14	1380	1384	1387	1390	1393	1396	1400	1403	1406	1409	0	1	1	1	2	2	2	3	3
·15	1413	1416	1419	1422	1426	1429	1432	1435	1439	1442	0	1	1	1	2	2	2	3	3
·16	1445	1449	1452	1455	1459	1462	1466	1469	1472	1476	0	1	1	1	2	2	2	3	3
·17	1479	1483	1486	1489	1493	1496	1500	1503	1507	1510	0	1	1	1	2	2	2	3	3
·18	1514	1517	1521	1524	1528	1531	1535	1538	1542	1545	0	1	1	1	2	2	2	3	3
·19	1549	1552	1556	1560	1563	1567	1570	1574	1578	1581	0	1	1	1	2	2	3	3	3
·20	1585	1589	1592	1596	1600	1603	1607	1611	1614	1618	0	1	1	1	2	2	3	3	3
·21	1622	1626	1629	1633	1637	1641	1644	1648	1652	1656	0	1	1	2	2	2	3	3	3
·22	1660	1663	1667	1671	1675	1679	1683	1687	1690	1694	0	1	1	2	2	2	3	3	3
·23	1698	1702	1706	1710	1714	1718	1722	1726	1730	1734	0	1	1	2	2	2	3	3	4
·24	1738	1742	1746	1750	1754	1758	1762	1766	1770	1774	0	1	1	2	2	2	3	3	4
·25	1778	1782	1786	1791	1795	1799	1803	1807	1811	1816	0	1	1	2	2	2	3	3	4
·26	1820	1824	1828	1832	1837	1841	1845	1849	1854	1858	0	1	1	2	2	3	3	3	4
·27	1862	1866	1871	1875	1879	1884	1888	1892	1897	1901	0	1	1	2	2	3	3	3	4
·28	1905	1910	1914	1919	1923	1928	1932	1936	1941	1945	0	1	1	2	2	3	3	4	4
·29	1950	1954	1959	1963	1968	1972	1977	1982	1986	1991	0	1	1	2	2	3	3	4	4
·30	1995	2000	2004	2009	2014	2018	2023	2028	2032	2037	0	1	1	2	2	3	3	4	4
·31	2042	2046	2051	2056	2061	2065	2070	2075	2080	2084	0	1	1	2	2	3	3	4	4
·32	2089	2094	2099	2104	2109	2113	2118	2123	2128	2133	0	1	1	2	2	3	3	4	4
·33	2138	2143	2148	2153	2158	2163	2168	2173	2178	2183	0	1	1	2	2	3	3	4	4
·34	2188	2193	2198	2203	2208	2213	2218	2223	2228	2234	1	1	2	2	3	3	4	4	5
·35	2239	2244	2249	2254	2259	2265	2270	2275	2280	2286	1	1	2	2	3	3	4	4	5
·36	2291	2296	2301	2307	2312	2317	2323	2328	2333	2339	1	1	2	2	3	3	4	4	5
·37	2344	2350	2355	2360	2366	2371	2377	2382	2388	2393	1	1	2	2	3	3	4	4	5
·38	2399	2404	2410	2415	2421	2427	2432	2438	2443	2449	1	1	2	2	3	3	4	4	5
·39	2455	2460	2466	2472	2477	2483	2489	2495	2500	2506	1	1	2	2	3	3	4	5	5
·40	2512	2518	2523	2529	2535	2541	2547	2553	2559	2564	1	1	2	2	3	4	4	5	5
·41	2570	2576	2582	2588	2594	2600	2606	2612	2618	2624	1	1	2	2	3	4	4	5	5
·42	2630	2636	2642	2649	2655	2661	2667	2673	2679	2685	1	1	2	2	3	4	4	5	6
·43	2692	2698	2704	2710	2716	2723	2729	2735	2742	2748	1	1	2	3	3	4	4	5	6
·44	2754	2761	2767	2773	2780	2786	2793	2799	2805	2812	1	1	2	3	3	4	4	5	6
·45	2818	2825	2831	2838	2844	2851	2858	2864	2871	2877	1	1	2	3	3	4	5	5	6
·46	2884	2891	2897	2904	2911	2917	2924	2931	2938	2944	1	1	2	3	3	4	5	5	6
·47	2951	2958	2965	2972	2979	2985	2992	2999	3006	3013	1	1	2	3	3	4	5	5	6
·48	3020	3027	3034	3041	3048	3055	3062	3069	3076	3083	1	1	2	3	4	4	5	6	6
·49	3090	3097	3105	3112	3119	3126	3133	3141	3148	3155	1	1	2	3	4	4	5	6	6
N	0	1	2	3	4	5	6	7	8	9	1	2	3	4	5	6	7	8	9

N	0	1	2	3	4	5	6	7	8	9	1	2	3	4	5	6	7	8	9
·50	3162	3170	3177	3184	3192	3199	3206	3214	3221	3228	1	1	2	3	4	4	5	6	7
·51	3236	3243	3251	3258	3266	3273	3281	3289	3296	3304	1	2	2	3	4	5	5	6	7
·52	3311	3319	3327	3334	3342	3350	3357	3365	3373	3381	1	2	2	3	4	5	5	6	7
·53	3388	3396	3404	3412	3420	3428	3436	3443	3451	3459	1	2	2	3	4	5	6	6	7
·54	3467	3475	3483	3491	3499	3508	3516	3524	3532	3540	1	2	2	3	4	5	6	6	7
·55	3548	3556	3565	3573	3581	3589	3597	3606	3614	3622	1	2	2	3	4	5	6	7	7
·56	3631	3639	3648	3656	3664	3673	3681	3690	3698	3707	1	2	3	3	4	5	6	7	8
·57	3715	3724	3733	3741	3750	3758	3767	3776	3784	3793	1	2	3	3	4	5	6	7	8
·58	3802	3811	3819	3828	3837	3846	3855	3864	3873	3882	1	2	3	4	4	5	6	7	8
·59	3890	3899	3908	3917	3926	3936	3945	3954	3963	3972	1	2	3	4	5	5	6	7	8
60	3981	3990	3999	4009	4018	4027	4036	4046	4055	4064	1	2	3	4	5	6	6	7	8
·61	4074	4083	4093	4102	4111	4121	4130	4140	4150	4159	1	2	3	4	5	6	7	8	9
·62	4169	4178	4188	4198	4207	4217	4227	4236	4246	4256	1	2	3	4	5	6	7	8	9
·63	4266	4276	4285	4295	4305	4315	4325	4335	4345	4355	1	2	3	4	5	6	7	8	9
·64	4365	4375	4385	4395	4406	4416	4426	4436	4446	4457	1	2	3	4	5	6	7	8	9
·65	4467	4477	4487	4498	4508	4519	4529	4539	4550	4560	1	2	3	4	5	6	7	8	9
·66	4571	4581	4592	4603	4613	4624	4634	4645	4656	4667	1	2	3	4	5	6	7	9	10
·67	4677	4688	4699	4710	4721	4732	4742	4753	4764	4775	1	2	3	4	5	7	8	9	10
·68	4786	4797	4808	4819	4831	4842	4853	4864	4875	4887	1	2	3	4	6	7	8	9	10
·69	4898	4909	4920	4932	4943	4955	4966	4977	4989	5000	1	2	3	5	6	7	8	9	10
·70	5012	5023	5035	5047	5058	5070	5082	5093	5105	5117	1	2	4	5	6	7	8	9	11
·71	5129	5140	5152	5164	5176	5188	5200	5212	5224	5236	1	2	4	5	6	7	8	10	11
·72	5248	5260	5272	5284	5297	5309	5321	5333	5346	5358	1	2	4	5	6	7	9	10	11
·73	5370	5383	5395	5408	5420	5433	5445	5458	5470	5483	1	3	4	5	6	8	9	10	11
·74	5495	5508	5521	5534	5546	5559	5572	5585	5598	5610	1	3	4	5	6	8	9	10	12
·75	5623	5636	5649	5662	5675	5689	5702	5715	5728	5741	1	3	4	5	7	8	9	10	12
·76	5754	5768	5781	5794	5808	5821	5834	5848	5861	5875	1	3	4	5	7	8	9	11	12
·77	5888	5902	5916	5929	5943	5957	5970	5984	5998	6012	1	3	4	5	7	8	10	11	12
·78	6026	6039	6053	6067	6081	6095	6109	6124	6138	6152	1	3	4	6	7	8	10	11	13
·79	6166	6180	6194	6209	6223	6237	6252	6266	6281	6295	1	3	4	6	7	9	10	11	13
·80	6310	6324	6339	6353	6368	6383	6397	6412	6427	6442	1	3	4	6	7	9	10	12	13
·81	6457	6471	6486	6501	6516	6531	6546	6561	6577	6592	2	3	5	6	8	9	11	12	14
·82	6607	6622	6637	6653	6668	6683	6699	6714	6730	6745	2	3	5	6	8	9	11	12	14
·83	6761	6776	6792	6808	6823	6839	6855	6871	6887	6902	2	3	5	6	8	9	11	13	14
·84	6918	6934	6950	6966	6982	6998	7015	7031	7047	7063	2	3	5	6	8	10	11	13	15
·85	7079	7096	7112	7129	7145	7161	7178	7194	7211	7228	2	3	5	7	8	10	12	13	15
·86	7244	7261	7278	7295	7311	7328	7345	7362	7379	7396	2	3	5	7	8	10	12	13	15
·87	7413	7430	7447	7464	7482	7499	7516	7534	7551	7568	2	3	5	7	9	10	12	14	16
·88	7586	7603	7621	7638	7656	7674	7691	7709	7727	7745	2	4	5	7	9	11	12	14	16
·89	7762	7780	7798	7816	7834	7852	7870	7889	7907	7925	2	4	5	7	9	11	13	14	16
·90	7943	7962	7980	7998	8017	8035	8054	8072	8091	8110	2	4	6	7	9	11	13	15	17
·91	8128	8147	8166	8185	8204	8222	8241	8260	8279	8299	2	4	6	8	9	11	13	15	17
·92	8318	8337	8356	8375	8395	8414	8433	8453	8472	8492	2	4	6	8	10	12	14	15	17
·93	8511	8531	8551	8570	8590	8610	8630	8650	8670	8690	2	4	6	8	10	12	14	16	18
·94	8710	8730	8750	8770	8790	8810	8831	8851	8872	8892	2	4	6	8	10	12	14	16	18
·95	8913	8933	8954	8974	8995	9016	9036	9057	9078	9099	2	4	6	8	10	12	15	17	19
·96	9120	9141	9162	9183	9204	9226	9247	9268	9290	9311	2	4	6	8	11	13	15	17	19
·97	9333	9354	9376	9397	9419	9441	9462	9484	9506	9528	2	4	7	9	11	13	15	17	20
·98	9550	9572	9594	9616	9638	9661	9683	9705	9727	9750	2	4	7	9	11	13	16	18	20
·99	9772	9795	9817	9840	9863	9886	9908	9931	9954	9977	2	5	7	9	11	14	16	18	20
N	0	1	2	3	4	5	6	7	8	9	1	2	3	4	5	6	7	8	9

Practice Exercise No. 53

Write the characteristics of the following:

1. 17	2. 342	3. 78 943	4. 4320	5. 0·42
6. 67·48	7. 7·4	8. 0·000 571	9. 0·021	10. 1

Finding the mantissa. The mantissa is found in the table of logarithms (Table VI on pp. 112–113). The mantissa is not related to the position of the decimal point in any number. For example, the mantissa of 34 562 is the same as the mantissa of 3456·2 or 345·62. But the logarithm of these numbers differs with respect to the *characteristic*, which you have learned to find by inspection of the number.

Note: The reason why the mantissa for a given set of digits does not change, no matter how they may be pointed off decimally, will appear from the following. Let us assume that m is any number and the logarithm of this number is $n + p$, in which n is the characteristic and p the mantissa. By definition $m = 10^{n+p}$. If we multiply or divide 10^{n+p} by 10, 100, 1000, etc., we make corresponding changes in the decimal pointing of m. But by the laws of algebra, multiplication or division of 10^{n+p} by 10, 100, 1000, etc., would be performed by adding or subtracting the exponents of 10^1, 10^2, 10^3, etc. Hence to arrive at any desired decimal pointing of the number m, only the whole-number part of the exponent of 10^{n+p} is modified. This part is n, the characteristic. The mantissa, p, always remains unchanged. Similar considerations will also make it clear why the mantissa still remains positive even when the characteristic is negative.

Let us now use the table of logarithms on pp. 112–113 to find the mantissa of the number 345. Find 34 in the left-hand column headed by N. Then move across to the column headed 5. The mantissa is 5378. The characteristic is 2; therefore log 345 = 2·5378.

By using the same mantissa and simply changing the characteristic we arrive at the following logarithms for various decimal pointings of the digits 345:

$$\log 34·5 \quad = 1·5378$$
$$\log 3·45 \quad = 0·5378$$
$$\log 0·345 \quad = \bar{1}·5378$$
$$\log 0·0345 = \bar{2}·5378$$

To find the mantissa of 3457 we repeat the above for 345 then move across to the second column headed 7. The number found in this final column is **added** to the mantissa of 345.

Thus:

$$\log 3457 \quad = 3·5387$$
$$\log 345·7 \quad = 2·5387$$
$$\log 0·3457 = \bar{1}·5387$$

EXAMPLE: Find the log of 0·8374.

SOLUTION: Find 83 in the column N, move across to column headed 7. The mantissa so far is 0·9227. Move farther across to the second column headed 4 and we see that 2 has to be **added**. *Notice that this is* 0·0002, *space not permitting the decimal point to be inserted in the tables.*

$$\therefore \log 0·8374 = \bar{1}·9229$$

The reader should now check the following results:

1. log 54·63 = 1·7374
2. log 37·06 = 1·5689
3. log 23·91 = 1·3786
4. log 1·472 = 0·1679
5. log 0·03265 = 2̄·5139

Practice Exercise No. 54

Find the logarithms of the following:

1. 354   2. 76   3. 8   4. 6346   5. 3·657
6. 0·234   7. 0·003 52 8. 6·04   9. 0·000 532 4   10. 672·8

Finding the antilogarithm. The number which corresponds to a given logarithm is called its **antilogarithm.**

The antilogarithm of a logarithm is found by obtaining the number corresponding to the **mantissa** and determining the position of the decimal point from the characteristic.

Antilogarithm tables are read in exactly the same way as logarithm tables.

EXAMPLE: Find the antilogarithm of 1·8536.

SOLUTION: Find 0·85 in the first column then move across to the column headed 3. The number which corresponds to the mantissa 0·853 is 7129. Now move farther across to the second column headed 6, which indicates that 10 should be added to the number so far.

∴ Mantissa 0·8536 corresponds to number 7139.

The characteristic being 1, it follows that the antilog of 1·8536 is 71·39.

The reader should now check the following results:

1. Antilog 1·9722 =   93·80
2. Antilog 2·4619 = 289·7
3. Antilog 0·0143 =   1·034
4. Antilog 1̄·6732 =   0·4712
5. Antilog 3̄·4708 =   0·002956
6. Antilog 2̄·5319 =   0·03403

How to Use Logarithms

To multiply by the use of logarithms, *add the logarithms of the numbers to be multiplied and find the antilogarithm corresponding to this sum.*

EXAMPLE: Multiply 25·31 by 42·18.

SOLUTION: log 25·31 = 1·4033,
                                  log 42·18 = 1·6251.
                                  Sum = 3·0284,
            Product = antilog of 3·0284 = 1068·0,   ANS.

To divide by the use of logarithms, *subtract the logarithm of the divisor from the logarithm of the dividend; the difference is the logarithm of the quotient.*

EXAMPLE 1: Divide 5280 by 67·82.

SOLUTION: log 5280 = 3·7226,
                                  log 67·82 = 1·8313,
                                  difference = 1·8913,
            Quotient = antilog 1·8913 = 77·85,   ANS.

EXAMPLE 2: Divide 5280 by 0·06782.

SOLUTION:

$$\log 5280 = 3·7226$$
$$\log 0·06782 = \bar{2}·8313$$

$$\text{difference} = 4·8913$$
$$\text{antilog} = 77850·0, \quad \text{ANS.}$$

EXPLANATION: $\log 0·06782 = \bar{2}·8313$.

This is a convenient way of writing
$$\log 0·06782 = -2 + 0·8313.$$
The subtraction in EXAMPLE 2 may be written
$$3·7226 - (-2 + 0·8313)$$
$$= 3·7226 + 2 - 0·8313$$
$$= 5·7226 - 0·8313$$
$$= 4·8913$$

Alternatively, we may perform the subtraction by the method of equal addition (p. 7).

$$3·7226$$
$$\bar{2}·8313$$
$$\bar{1}·9$$

$$4·8913$$

Subtracting
    3 from 6 leaves 3
    1 from 2 leaves 1
    3 from 12 leaves 9
    9 from 17 leaves 8
at this stage we are adding   1 to $\bar{2}$ to make $\bar{1}$
    $\bar{1}$ from 3 = 4

A further illustration is

$$\bar{2}·3704$$
$$\bar{1}·8425$$
$$0 \quad 53$$

$$\bar{2}·5279$$

Subtracting
    5 from 14 leaves 9
    3 from 10 leaves 7
    5 from 7 leaves 2
    8 from 13 leaves 5
at this stage we are adding 1 to $\bar{1}$ to make 0.
    0 from $\bar{2}$ = $\bar{2}$

EXAMPLE 3: Divide 52·80 by 6782.

$$\log 52·80 = 1·7226$$
$$\log 6782 = \overset{4\ 9}{3·8313}$$

$$= \bar{3}·8913$$
$$\text{antilog} = 0·007785, \quad \text{ANS.}$$

To raise to a given power by the use of logarithms, *multiply the logarithm of the number by the given exponent of the number and find the antilogarithm.*

The reason for this may be explained as follows. Let m be a number and n its logarithm. Then—

$$m = 10^n,$$
$$m^2 = 10^n \times 10^n = 10^{n+n} = 10^{2n}$$
$$m^3 = 10^{3n}, \text{ etc.}$$

EXAMPLE 1: Find 46^4.

SOLUTION:

$$\log 46 = 1 \cdot 6628$$
$$\times 4$$
$$\overline{}$$
$$\log 46^4 = 6 \cdot 6512$$
$$46^4 = \text{antilog } 6 \cdot 6512 = 4\,479\,000, \quad \text{ANS.}$$

To find a given root by the use of logarithms, *divide the logarithm of the number by the index of the root and find the antilogarithm.*

This may be demonstrated thus:

Let $\qquad\qquad m = 10^n$

Then $\qquad\qquad \sqrt{m} = \sqrt{10^n} = 10^{\frac{n}{2}}$

$\qquad\qquad\qquad \sqrt[3]{m} = 10^{\frac{n}{3}}, \text{ etc.}$

EXAMPLE 1: Find $\sqrt[3]{75}$.

SOLUTION:

$$\log 75 = 1 \cdot 8751,$$
$$\frac{1 \cdot 8751}{3} = 0 \cdot 6250$$

$$\text{Root} = \text{antilog } 0 \cdot 6250 = 4 \cdot 217, \quad \text{ANS.}$$

EXAMPLE 2: Find $\sqrt{0 \cdot 251}$.

SOLUTION:

$$\log 0 \cdot 251 = \bar{1} \cdot 3997$$

$$\frac{\bar{1} \cdot 3997}{2} = \frac{\bar{2} + 1 \cdot 3997}{2} = \bar{1} \cdot 6999 \text{ (4 decimal places)}$$

$$\text{Root} = \text{antilog } \bar{1} \cdot 6999 = 0 \cdot 501, \quad \text{ANS.}$$

EXAMPLE 3: Find $\sqrt[3]{0 \cdot 75}$.

SOLUTION:

$$\log. 0 \cdot 75 = \bar{1} \cdot 8751$$

$$\frac{\bar{1} \cdot 8751}{3} = \frac{\bar{3} + 2 \cdot 8751}{3}$$

$$= \bar{1} \cdot 9584 \text{ (4 decimal places)}$$
$$\text{antilog} = 0 \cdot 9086, \quad \text{ANS.}$$

EXPLANATION: Starting in this case with a negative characteristic, we cannot make a direct division by 3 because dividing $\bar{1}$ by 3 would result in a fractional characteristic, which is impossible. We therefore increase and decrease the logarithm by 2 in order to make the division possible.

A further example is $\dfrac{\bar{2} \cdot 4936}{3} = \dfrac{\bar{3} + 1 \cdot 4936}{3}$

$$= \bar{1} \cdot 4979 \text{ (4 decimal places)}$$

Practice Exercise No. 55

Solve by logarithms:

1. $3984 \times 5 \cdot 6$
2. $25 \cdot 31 \times 42 \cdot 18$
3. $220 \cdot 2 \times 2209$
4. $5280 \div 33 \cdot 81$
5. $7256 \div 879 \cdot 2$
6. $9783 \div 0 \cdot 1234$
7. 77^3
8. $\sqrt[3]{85}$
9. $\sqrt[5]{356 \cdot 0}$
10. $2 \cdot 43^5$
11. $\dfrac{5}{-7}$
12. $\dfrac{-17}{32}$
13. $\dfrac{6+3}{4}$
14. $\dfrac{8+7}{7}$
15. $\dfrac{13-9}{3}$
16. $\dfrac{11}{16-7}$
17. $\dfrac{8}{3 \times 5}$
18. $\dfrac{4 \times 6}{11}$
19. $\dfrac{7 \div 3}{4}$
20. $\dfrac{16}{18 \div 5}$

Give answers 11 to 20 correct to 2 decimal places.

Clearly we can only find the logarithm of a positive number, i.e. log n will exist only when n is positive. Thus log $(-2 \cdot 675)$ is meaningless. However, in examples such as $10 \cdot 98 \times (-2 \cdot 675)$ we can, of course, work out the result for $10 \cdot 98 \times 2 \cdot 675$ by logarithms and then restore the negative sign in the final answer.

EXAMPLE 1: Multiply $10 \cdot 98$ by $(-2 \cdot 675)$.

SOLUTION:
$$\begin{aligned} \log 10 \cdot 98 &= 1 \cdot 0407 \\ \log \ 2 \cdot 675 &= 0 \cdot 4273 \\ \text{Sum} &= 1 \cdot 4680 \end{aligned}$$

Product = antilog of $1 \cdot 4680 = 29 \cdot 38$

$$\therefore 10 \cdot 98 \times (-2 \cdot 675) = -29 \cdot 38, \text{ Ans.}$$

EXAMPLE 2: Divide 5280 by $(-67 \cdot 82)$.

SOLUTION: From the example on page 117 we see that $5280 \div 67 \cdot 82 = 77 \cdot 85$

$$\therefore 5280 \div (-67 \cdot 82) = -77 \cdot 85, \text{ Ans.}$$

Another fact about logarithms that may have been noticed is that as the number n increases so does log n. That is, if N is greater than n, then log N is greater than log n. For example, log 30 is greater than log 29, which in turn is greater than log $28 \cdot 6$ and so on down.

Practice Exercise No. 56

Use logarithms to calculate the following:

1. $79 \cdot 34 \times (-6 \cdot 258)$
2. $(-79 \cdot 34) \times 6 \cdot 258$
3. $(-8 \cdot 419) \times (-27 \cdot 23)$
4. $(-28)^{\ddagger}$
5. $(-36)^{\ddagger}$
6. $(-69)^{\ddagger}$

ERRORS IN THE USE OF LOGARITHMS

The use of logarithms, although convenient, may only give an approximate answer, depending on the number of figures provided by the tables we use. In general, we must not expect complete accuracy for our results. The following examples will illustrate this remark.

EXAMPLE 1: Find the error in using logarithms to calculate 16 ÷ 4.

SOLUTION: log 16 = 1·2041
    log 4 = 0·6021
   difference = 0·6020

Quotient = antilog of 0·6020 = 3·999
Since the answer should have been 4, it follows that the error is 0·001, ANS.

EXAMPLE 2: Use logarithms to calculate (i) $\dfrac{12 \times 6}{9}$, (ii) $\dfrac{72}{9}$.

SOLUTION: For the calculation (i) we have:

  log 12 = 1·0792
   log 6 = 0·7782
∴ log (12 × 6) = 1·8574
   log 9 = 0·9542
∴ log $\dfrac{(12 \times 6)}{9}$ = 0·9032

∴ $\dfrac{12 \times 6}{9}$ = antilog of 0·9032 = 8·002, ANS.

For the calculation (ii) we have:

  log 72 = 1·8573
   log 9 = 0·9542
∴ log (72 ÷ 9) = 0·9031
∴ $\dfrac{72}{9}$ = antilog of 0·9031 = 8·000, ANS.

Simple arithmetic tells us that the exact answer is 8, as given by (ii). It is clearly possible for the fourth figure to be inaccurate when using four-figure tables as supplied on page 112. This error is quite harmless provided we are on our guard, but consider the results of multiplying the answer to (i) by £1 000 000 and assuming the final answer to be correct!

Practice Exercise No. 57

Find the error in using logarithms to calculate the following:

1. $\dfrac{(8 \times 27)}{9}$ 2. $\dfrac{(11 \times 12)}{4}$ 3. $\dfrac{(16 \times 4)}{8}$

4. The correct four-figure logarithm of 7 is 0·8451. The correct logarithm of 7 according to a pocket calculator is 0·84509804. Using the tables in this book, show that we get the same result for 7^{10} no matter which logarithm we use.

5. An order is placed for 1000 articles which are quoted at £33 per dozen. Find the difference between the costs when calculated by logarithms and simple arithmetic.

Naturally, the more fractions we have to multiply in a problem, the more convenient does the use of logarithms become, since we merely increase the list of additions. Consider the convenience of this when dealing with the following two examples:

EXAMPLE 1: Calculate $\dfrac{9·837 \times 0·5672 \times 48·13}{12·64 \times 3·511}$.

SOLUTION: We calculate the numerator and the denominator separately.

$$\log 9 \cdot 8370 = 0 \cdot 9929$$
$$\log 0 \cdot 5672 = \bar{1} \cdot 7538$$
$$\log 48 \cdot 1300 = 1 \cdot 6824$$
$$\therefore \log \text{(numerator)} = 2 \cdot 4291$$
$$\log 12 \cdot 640 = 1 \cdot 1018$$
$$\log 3 \cdot 511 = 0 \cdot 5454$$
$$\therefore \log \text{(denominator)} = 1 \cdot 6472$$
$$\text{difference} = 0 \cdot 7819$$

$\therefore$ Quotient $=$ antilog of $0 \cdot 7819 = 6 \cdot 052$,   ANS.

CHECKING BY SIMPLE APPROXIMATIONS

It is surprisingly easy to make mistakes and place the final decimal point in the wrong position. It is always a good idea to check the results obtained by logarithms by using simple approximations to the numbers involved. We emphasise **simple** approximations, since anything else may not justify the work which arises. For instance, the last worked example may be dealt with as follows (we shall use the symbol $\approx$ to mean 'approximately equal to'):

$9 \cdot 837 \approx 10$, $0 \cdot 5672 \approx 0 \cdot 5$, $48 \cdot 13 \approx 50$

$\left. \begin{array}{l} 12 \cdot 64 \approx 10 \\ 3 \cdot 511 \approx 4 \end{array} \right\}$ Here we really think of them both together as being ≈ 40.

Naturally we suggest approximations which we think will be reasonably easy to work with. That is, we chose 10 as an approximation to $12 \cdot 64$ even though we know that 13 would be a better approximation. However, in this problem 13 would not be easier to deal with.

$$\therefore \frac{9 \cdot 837 \times 0 \cdot 5672 \times 48 \cdot 13}{12 \cdot 64 \times 3 \cdot 511} \approx \frac{\overset{1}{\cancel{10}} \times 0 \cdot 5 \times 50}{\underset{1}{\cancel{10}} \times 4} = \frac{25}{4} = 6 \cdot 25$$

This is a good check on the result which was obtained in the example above.

EXAMPLE 2: Calculate x from the following product after obtaining an approximation to the result:

$$x = \frac{13 \cdot 78 \times 67 \cdot 25 \times 0 \cdot 0193}{139 \cdot 2 \times 0 \cdot 431 \times 0 \cdot 0338}$$

SOLUTION: In this layout we do the approximations as we go.

$$x \approx \frac{\overset{1}{\cancel{13 \cdot 78}} \times 67 \cdot 25 \times \overset{1}{\cancel{0 \cdot 0193}}}{\underset{10}{\cancel{139 \cdot 2}} \times 0 \cdot 431 \times \underset{2}{\cancel{0 \cdot 0338}}} \approx \frac{70}{8} \approx 9$$

At this stage we have a crude idea of the size of the answer to be expected. This is quite good enough to warn us that if we obtain an answer like 100 or $0 \cdot 1$ when using logarithms then we have certainly made an error.

Using logarithms our result is as follows:

$$\log 13 \cdot 78 = 1 \cdot 1393$$
$$\log 67 \cdot 25 = 1 \cdot 8277$$
$$\log 0 \cdot 0193 = \bar{2} \cdot 2856$$
$$\log \text{(numerator)} = 1 \cdot 2526$$
$$\log 139 \cdot 2 = 2 \cdot 1436$$
$$\log 0 \cdot 431 = \bar{1} \cdot 6345$$
$$\log 0 \cdot 0338 = \bar{2} \cdot 5289$$
$$\log \text{(denominator)} = 0 \cdot 3070$$
$$\text{difference} = 0 \cdot 9456$$

∴ Quotient = antilog of $0 \cdot 9456 = 8 \cdot 822$,   ANS.

Practice Exercise No. 58

Use logarithms to calculate the following quotients after finding approximations to the results using simple approximations:

1. $\dfrac{2 \cdot 9 \times 18 \cdot 6 \times 351 \cdot 9}{710 \cdot 4 \times 39 \cdot 83}$

2. $\dfrac{67 \cdot 05 \times 0 \cdot 1219 \times 4 \cdot 83}{143 \cdot 1 \times 0 \cdot 4012}$

3. $\dfrac{9 \cdot 837 \times 0 \cdot 82 \times 50 \cdot 3}{10 \cdot 4 \times 0 \cdot 16 \times 4 \cdot 92}$

4. $\dfrac{113 \cdot 4 \times 92 \cdot 38 \times 84 \cdot 7}{0 \cdot 16 \times 75 \cdot 48 \times 11 \cdot 46}$

CHAPTER THIRTEEN

SEQUENCES AND SERIES

SEQUENCES

In this chapter we shall discuss sequences and series of numbers. A sequence of numbers is a set of numbers which can be arranged in an order which enables the numbers to be counted or listed. Each number then becomes a particular term of the sequence.

Consider the sequence whose terms are listed as follows:

$$2, 4, 6, 8, 10, 12, 14, 16, 18, 20, 22, 24.$$

The terms of the sequence are separated from each other by commas and we show that 24 is the last term of the sequence by ending with one full stop. The sequence has been listed in a counting order in the sense that the first term is 2, the second term is 4, the third term is 6, and so on up to the twelfth and last term, which is 24.

A convenient notation to represent this description is to write

$$t_1 = 2, t_2 = 4, t_3 = 6, t_4 = 8, \ldots, t_{12} = 24.$$

Now suppose that we rearrange this sequence and list a new sequence:

$$4, 2, 8, 6, 12, 10, 16, 14, 20, 18, 24, 22.$$

In this new sequence we have

$t_1 = 4, t_2 = 2, t_3 = 8, t_4 = 6, \ldots, t_{12} = 22$, and again the sequence is in a counting order.

Comparing the two sequences above, we see that the terms in the first sequence are arranged in **ascending** order of magnitude—that is, each term is greater than the previous term. In the second sequence, however, the terms are neither ascending nor descending. When we can count the total number of terms in a sequence we say that the sequence is **finite**. If we cannot count the number of terms in a sequence, then we say it is an **infinite** sequence, i.e. it has an infinite number of terms.

Since we cannot list all the terms of an infinite sequence, we do the next best thing and offer enough evidence to indicate how the terms may be worked out. Consider the infinite sequence whose first five terms are listed as follows:

$$1, 3, 5, 7, 9, \ldots$$

The run of full stops after the 9 is the way in which we say that 'this is an infinite sequence' and it is impossible to write down all the terms in the sequence. However, it is assumed that all the terms after the 9 are to be taken as the 'obvious' ones. Thus we shall assume that since all the terms so far shown increase by 2, then they will continue to do so. Therefore, $t_6 = 11$, $t_7 = 13$, $t_8 = 15$, $t_9 = 17$, and so on.

EXAMPLE 1: Find the missing terms indicated by ? in the following sequence, 2, 5, 8, ?, 14, 17, 20, ?, ... and find the fifteenth term.

SOLUTION: From the evidence given by the terms which have been listed we assume that it is intended that the terms 'increase by 3'. Therefore the sequence intended is

$$2, 5, 8, 11, 14, 17, 20, 23, \ldots$$

To find any term in general we need to discover the rule or formula of the sequence. We reason as follows:

$$t_2 = 2 + 3 \qquad = 5$$
$$t_3 = 2 + (3 \times 2) = 8$$
$$t_4 = 2 + (3 \times 3) = 11$$
$$t_5 = 2 + (3 \times 4) = 14$$

We now see that the pattern on the right-hand side suggests that

$$t_{15} = 2 + (3 \times 14) = 44, \quad \text{ANS.}$$

ALTERNATIVELY: There are other ways to find the rule of the sequence. For example, we could have reasoned along the following lines:

$$t_2 = (3 \times 2) - 1 = 5$$
$$t_3 = (3 \times 3) - 1 = 8$$
$$t_4 = (3 \times 4) - 1 = 11$$
$$t_5 = (3 \times 5) - 1 = 14$$

We now see that the pattern on the right-hand side suggests that

$$t_{15} = (3 \times 15) - 1 = 44 \text{ as before}, \quad \text{ANS.}$$

EXAMPLE 2: Find the tenth and twentieth terms of the sequence whose first three terms are 1000, 992, 984, respectively.

SOLUTION: We must assume that these three terms are supposed to suggest the 'obvious' rule for finding the other terms of the sequence. The first three terms suggest that the terms 'decrease by 8'. Therefore,

$$t_4 = 976, t_5 = 968, t_6 = 960, \text{ and so on.}$$

To find the terms which are further ahead without writing them all down beforehand, we try to find the rule of the sequence by looking for a pattern in the terms.

$$t_1 = 1000$$
$$t_2 = 992 = 1000 - 8$$
$$t_3 = 984 = 1000 - (8 \times 2)$$
$$t_4 = 976 = 1000 - (8 \times 3)$$
$$t_5 = 968 = 1000 - (8 \times 4)$$

We now see that the pattern on the right-hand side suggests that

$$t_{10} = 1000 - (8 \times 9) \quad = 1000 - 72 \quad = 928$$
$$t_{20} = 1000 - (8 \times 19) = 1000 - 152 = 848, \quad \text{Ans.}$$

Practice Exercise No. 59

Find the missing terms indicated by ? and t_{21} in each of the following sequences:

1. 5, 8, 11, ?, 17, ?, 23, 26, . . .
2. 952, 944, 936, ?, 920, ?, 904, . . .
3. 1, 3, 5, ?, ?, 11, 13, . . .
4. 26, 22, ?, 14, 10, ?, . . .
5. 1, 6, 11, ?, ?, . . .
6. 8, 2, ?, ?, −16, −22, . . .

Finding the *n*th term

In mathematics wherever possible we should try to generalize our ideas and results, and this is particularly important in sequences and series. For this reason instead of referring to the tenth term, the hundredth term, or the seventy-ninth term of a sequence separately, we refer to the *n*th term, t_n. Thus, in the example 1 above we have

When $n = 1$, $t_n = t_1 - 2 - (3 \times 1)$ 1
When $n = 2$, $t_n = t_2 = 5 = (3 \times 2) - 1$
When $n = 3$, $t_n = t_3 = 8 = (3 \times 3) - 1$, but most important we see that

$$t_n = (3 \times n) - 1, \text{ i.e. } t_n = 3n - 1,$$ and this is the formula for any term of the sequence.

For example, $t_8 = (3 \times 8) - 1 = 23$

$$t_{50} = (3 \times 50) - 1 = 149,$$ and we are now able to produce immediately any term we please.

Now if a sequence is to be completely given by listing only a few of its terms then we must assume that the 'obvious' results are intended, i.e. 7, 10, 13, . . . is meant to be continued 16, 19, 22, . . . But suppose the sequence had a more varied pattern for its terms such as 7, 10, 13, 23, 26, 29, 39, . . ., i.e. 'add 3 twice then add 10'. More terms must be listed to make this rule 'obvious', and mathematically this is not very satisfactory. The only way to make our intentions perfectly clear is to quote the *n*th term.

EXAMPLE 1: The *n*th term of a sequence is given by $t_n = 100 - 5n$. Write down the first four terms and the twenty-first term of this sequence.

SOLUTION: Since $t_n = 100 - 5n$ it follows that,

$$t_1 = 100 - 5 = 95; \qquad t_2 = 100 - (5 \times 2) = 90$$
$$t_3 = 100 - (5 \times 3) = 85; t_4 = 100 - 20 = 80$$

Also, $\qquad t_{21} = 100 - (5 \times 21) = -5$, Ans.

∴ The sequence may be listed as 95, 90, 85, 80, 75, . . ., Ans.

(Without ambiguity!)

EXAMPLE 2: Find the first four terms and the thirtieth term of the sequence whose *n*th term is given by $t_n = 4n + 3$. Find the first term which is greater than 100.

SOLUTION: Since $t_n = 4n + 3$ it follows that,
$$t_1 = 4 + 3 = 7; \quad t_2 = 8 + 3 = 11$$
$$t_3 = 12 + 3 = 15; \quad t_4 = 16 + 3 = 19$$
Also, $\quad t_{30} = (4 \times 30) + 3 = 123,$ ANS.

The first term greater than 100 will require $4n + 3$ greater than 100. (The symbol for 'is greater than' is $>$ and we shall use this from here on.) This is the equivalent to saying that $4n + 3 > 100$
or $\qquad 4n > 97,$
which is satisfied first, by $n = 25$.
∴ $t_{25} = (4 \times 25) + 3 = 103$, is the first term of the sequence to be greater than 100, ANS.
The sequence may be listed as 7, 11, 15, 19, 23, . . ., ANS.

Practice Exercise No. 60

Write down the first five terms of the sequences whose nth terms are given by the following:
1. $t_n = 2n + 5$, and find the term which is equal to 65.
2. $t_n = 3n - 3$, and find the term which is equal to 42.
3. $t_n = 105 - 5n$, and find the first term which is less than 50.
4. $t_n = 12 - n$, and find the term which is equal to 0.
5. $t_n = 48 - 4n$, and find the term which is equal to 12.
6. $t_n = 4n - 48$, and find the term which is equal to 12.

The sequences we have discussed so far have all been **arithmetic** sequences, i.e. their successive terms are formed as a result of adding or subtracting the same number. Sequences whose successive terms are formed by multiplying or dividing by the same number are called **geometric** sequences. The four sequences below are all geometric sequences.

1, 2, 4, 8, 16, 32, 64, . . . (multiply each term by 2 to get the next term)
0·1, 0·3, 0·9, 2·7, 8·1, . . . (multiply each term by 3 to get the next term)
100, 10, 1, 1/10, 1/100, . . . (divide each term by 10 to get the next term)
32, 16, 8, 4, 2, 1, $\frac{1}{2}$, . . . (divide each term by 2 to get the next term)

In the case of the geometric sequence we can have terms which alternate in sign. For example, the next sequence is produced by starting with 3 and multiplying by -2 to get,

$$3, -6, 12, -24, 48, -96, \ldots$$

Similarly, we may produce an alternating geometric sequence by dividing by -2. Starting with 128 we would get,

$$128, -64, 32, -16, 8, -4, 2, -1, \tfrac{1}{2}, -\tfrac{1}{4}, \ldots$$

EXAMPLE 1: The first term of a geometric sequence is 5, and each successive term is obtained by multiplying by 2. List the first five terms of the sequence and find the first term which is greater than 1000. Also find t_{11}.

SOLUTION: Since $t_1 = 5$, then $t_2 = 10 = 5 \times 2$
$$t_3 = 20 = 5 \times 2 \times 2 \qquad\qquad = 5 \times 2^2$$
$$t_4 = 40 = 5 \times 2 \times 2 \times 2 \qquad = 5 \times 2^3$$
$$t_5 = 80 = 5 \times 2 \times 2 \times 2 \times 2 = 5 \times 2^4$$

We write down the terms just as we did before in order to find a pattern in the results which will give a clue to the rule of the sequence.

The pattern of the right-hand side suggests that
$$t_{11} = 5 \times 2^{10} = 5 \times 1024 = 5120, \quad \text{ANS.}$$

Returning to the list of terms we continue with

$$t_8 = 5 \times 2^7 = 5 \times 128 = 640$$
$$t_9 = 5 \times 2^8 = 5 \times 256 = 1280$$

∴ The ninth term is the first term greater than 1000,   ANS.

EXAMPLE 2: The fourth and fifth terms of a geometric sequence are 54 and 18 respectively. Find the first three terms and the first term which is less than $\frac{1}{8}$.

SOLUTION: It is clear that we get from t_4 to t_5 as a result of dividing t_4 by 3. If we continue the listing of the sequence we shall get

$$?, ?, ?, 54, 18, 6, 2, \tfrac{2}{3}, \tfrac{2}{9}, \tfrac{2}{27}, \ldots$$

Now we reason that t_3 was divided by 3 to get $t_4 = 54$,

$$\therefore \tfrac{1}{3}t_3 = 54$$
$$t_3 = 54 \times 3 = 162$$

Similarly, $\qquad \tfrac{1}{3}t_2 = t_3 = 162$

$$\therefore \; t_2 = 162 \times 3 = \; 486$$

Finally, $\qquad\qquad t_1 = 486 \times 3 = 1458$

The sequence can now be listed as, 1458, 486, 162, 54, 18, 6, 2, $\tfrac{2}{3}$, $\tfrac{2}{9}$, $\tfrac{2}{27}$, ...
Examining this list we see at once that $t_{10} = \tfrac{2}{27}$ is the first term less than $\frac{1}{8}$, ANS.

EXAMPLE 3: List the first four terms of the sequence whose nth term is given by $t_n = 2 \times 5^n$. Find the first term which is greater than 2 000 000.

SOLUTION: Since $t_n = 2 \times 5^n$ it follows that

$$t_1 = 2 \times 5^1 = 10; \quad t_2 = 2 \times 5^2 = 50$$
$$t_3 = 2 \times 5^3 = 250; \quad t_4 = 5t_3 = 1250$$

To find the first term which is greater than 2 000 000 we require the first value of n which gives $t_n = 2 \times 5^n > 2\,000\,000$
This is equivalent to finding when

$$5^n > 1\,000\,000$$

Using logarithms we see this is equivalent to finding when,

$$\log 5^n > \log 1\,000\,000$$
i.e. $n \log 5 > 6$

i.e. $\qquad n > \dfrac{6}{\log 5} = \dfrac{6}{0 \cdot 699} = 8 \cdot 6 \text{ (1 decimal place)}$

The first value of $n > 8 \cdot 6$ is $n = 9$.
∴The ninth term is the first term which is greater than 2 000 000,   ANS.
(In fact, $t_8 = 781\,250$ and $t_9 = 3\,906\,250$)

Practice Exercise No. 61

Write down the first five terms of the sequences whose nth terms are given by the following:

1. $t_n = 3 \times 2^n$, and find the term which is equal to 192.
2. $t_n = 2 \times 3^n$, and find the term which is equal to 4374.
3. $t_n = \dfrac{1458}{3^n}$, and find the first term which is less than $\frac{1}{8}$.
4. $t_n = \dfrac{128}{2^n}$, and find the first term which is less than 1.
5. $t_n = \dfrac{34}{10^n}$, and find the first term which is less than 0·0001.
6. $t_n = \dfrac{2^n}{128}$, and find the first term which is greater than 5.

SERIES

A series of numbers is a sequence whose terms are added together. Thus a series is written like

$$2 + 4 + 6 + 8 + 10 + 12 + \ldots + 24.$$

As before we shall refer to 2 as the first term of the series, 4 as the second term of the series, and 24 as the twelfth and last term of the series.

An **arithmetic series** is one in which the successive terms are formed as a result of adding or subtracting the same number. For example, starting with a first term of 91 and subtracting 6 each time we get the series

$$91 + 85 + 79 + 73 + 67 + 61 + \ldots$$

Arithmetic series are also called **arithmetical progressions**, and it is usual to abbreviate this reference to A.P.

A **geometric series** or **progression (G.P.)** is one in which the successive terms are formed as a result of dividing or multiplying by the same number. For example, starting with a first term of $\frac{1}{9}$ and multiplying by 3 we get the G.P.

$$\frac{1}{9} + \frac{1}{3} + 1 + 3 + 9 + 27 + 81 + 243 + 729 + \ldots$$

Solving Series problems

In series problems you are generally required to add more terms to the series, prefix numbers to a given term to find earlier terms, or to find the sum of the terms.

The secret of solving any kind of a series is to analyse the pattern—determine 'how it goes'—by inspection. There are, however, rules of procedure that can be followed.

ARITHMETIC PROGRESSIONS

To find a formula for the *n*th term of an A.P.

We have already done this for the arithmetic sequences, so there is nothing new to learn. However, consider the series

$$2 + 11 + 20 + 29 + 38 + 47 + \ldots$$

The common difference (which we shall always call d) between the terms is found by subtracting any term from the next term. In this case we get the common difference 9.

As soon as we know the first term and the common difference d we can write down as much of the series as we please, because we can find the *n*th term.

Here we get,
$$
\begin{aligned}
t_1 &= 2 &&= 2 \\
t_2 &= 2 + 9 &&= 11 \\
t_3 &= 2 + 2 \times 9 &&= 20 \\
t_4 &= 2 + 3 \times 9 &&= 29 \\
t_5 &= 2 + 4 \times 9 &&= 38
\end{aligned}
$$

and consequently,

$$t_n = 2 + (n - 1) \times 9$$

In words, this result may be described as: The nth term (t_n) is equal to the first term (2) plus $n - 1$ times the common difference (9).

The result that we have just found can be simplified to give

$$t_n = 2 + 9n - 9 = 9n - 7.$$

EXAMPLE 1: Find the common difference and the nth term in the following A.P.

$$100 + 94 + 88 + 82 + \ldots$$

SOLUTION: Subtracting the first term from the second term we get

$$d = 94 - 100 = -6$$

(a negative result is to be expected since the series is decreasing). Writing the terms down we get

$$t_1 = 100$$
$$t_2 = 100 + \quad (-6) = 94$$
$$t_3 = 100 + 2(-6) = 88$$
$$t_4 = 100 + 3(-6) = 82$$
$$t_5 = 100 + 4(-6) = 76$$

and consequently,

$$t_n = 100 + (n - 1)(-6)$$

This result checks with the written description above, that the nth term (t_n) is equal to the first term (100) plus $n - 1$ times the common difference (-6).

Again the result may be simplified to give

$$t_n = 100 - 6n + 6 = 106 - 6n, \quad \text{ANS.}$$

EXAMPLE 2: The third and fifth terms of an A.P. are 13 and 23 respectively. Find the common difference and the nth term of the series.

SOLUTION: We can get an idea of what to do by writing down

$$t_1 + t_2 + 13 + t_4 + 23 + t_6 + t_7 + \ldots$$

Now it is clear that to get from 13 to 23 we must add the common difference twice, i.e. add $2d$.

$$\therefore 23 = 13 + 2d, \text{ and we find that } d = 5.$$
$$\therefore t_2 = 13 - 5 = 8, t_1 = 8 - 5 = 3,$$
$$\therefore t_4 = 13 + 5 = 18, t_6 = 23 + 5 = 28, \text{ and so on.}$$

The given series was, $3 + 8 + 13 + 18 + 23 + 28 + 33 + \ldots$
Now the nth term is equal to the first term (3) plus $n - 1$ times the common difference (5).
$$\therefore t_n = 3 + (n - 1)(5) \text{ which simplifies to } t_n = 5n - 2, \quad \text{ANS.}$$

To find the sum of the first n terms of an A.P.

This means that if the series is

$$1 + 4 + 7 + 10 + 13 + 16 + 19 + 22 + 25 + 28 + \ldots$$

then the sum of the first 6 terms is

$$1 + 4 + 7 + 10 + 13 + 16 = 51.$$

A suitable notation for this is to write $S_6 = 51$, where S stands for sum, and S_6 stands for the sum of the **first six terms**.

Using the same series we have

$$S_1 = 1, S_2 = 5, S_3 = 12, S_4 = 22, S_5 = 35, \ldots$$

There is an interesting shortcut for getting this result. We discover that the result is found by adding the first term to the last term and multiplying the sum by half the number of terms.

For S_6: the first term is 1 and the last term is 16. Their sum is 17. Half the number of terms is 3. We note that $S_6 = 17 \times 3 = 51$.

For S_9: the first term is 1 and the last term is 25. Their sum is 26. Half the number of terms is $\frac{9}{2}$. We note that $S_9 = 26 \times \frac{9}{2} = 117$, which is confirmed by addition.

There is a **standard formula** for the sum of n terms, S_n. In this formula we represent the first term by a, the last term by l and the number of terms by n.

$$\therefore S_n = \frac{n}{2}(a + l)$$

EXAMPLE 1: Find the sum of the natural numbers from 1 to 10 inclusive.

SOLUTION: There are ten terms here. Substituting for n, a, and l in the formula we arrive at

$$S_{10} = \frac{10}{2}(1 + 10) = 55, \quad \text{ANS.}$$

Check: $1 + 2 + 3 + 4 + 5 + 6 + 7 + 8 + 9 + 10 = 55$

The sum of the first 100 terms is just as easy to find since,

$$S_{100} = \frac{100}{2}(1 + 100) = 5050, \text{ but not so easy to check!}$$

Practice Exercise No. 62

1. In an A.P. $a = 5$, $d = 6$. Find t_{10} and S_{10}.
2. In an A.P. $a = 50$, $d = -2$. Find t_{21} and S_{21}.
3. In an A.P. $t_3 = 6$ and $t_5 = 10$. Find the common difference d, the first term and the sum of the first twelve terms.
4. Find the sum of the first 100 even numbers starting with 2.
5. Find the sum of the first 100 odd numbers starting with 1.
6. A parent puts money in the bank for a daughter every time she has a birthday. The rule is, £1 on the first birthday, £2 on the second birthday, £3 on the third birthday and so on. How much has been paid in when the daughter is 21?

GEOMETRIC PROGRESSIONS

An example of a G.P. is $3 + 6 + 12 + 24 + 48 + 96 + 192 + \ldots$, where the first term is 3 and each term is multiplied by 2 to get the next term. The multiplier 2 is called the **common ratio** of the G.P., and we can always find this by dividing one term into the next one. In this series the fifth term is 48, and if we divide this into the next term, which is 96, we get the common ratio which $96 \div 48 = 2$. In general, we represent the common ratio by the letter r.

EXAMPLE 1: Find the common ratio and the next two terms of the following G.P.

$$2 + 6 + 18 + ? + ? + \ldots$$

SOLUTION: The common ratio is $r = 6 \div 2 = 3$ or
$$r = 18 \div 6 = 3.$$

The next two terms are therefore $18 \times 3 = 54$ and $54 \times 3 = 162$, ANS.

EXAMPLE 2: A G.P. has a first term $a = 16$ and a common ratio $r = \frac{1}{2}$. Find the first five terms.

SOLUTION: With a first term of 16 and $r = \frac{1}{2}$, the second term must be $16 \times \frac{1}{2} = 8$, and from this the third term must be $8 \times \frac{1}{2} = 4$.
We can now list the series as

$$16 + 8 + 4 + 2 + 1 + \tfrac{1}{2} + \tfrac{1}{4} + \ldots, \quad \text{ANS.}$$

To find a formula for the nth term of a G.P.

As before we rely upon being able to discover the rule of the series by looking at a pattern of results from a particular example. Consider, therefore, the G.P.

$$9 + 18 + 36 + 72 + 144 + 288 + 576 + \ldots$$

Now the first term $t_1 = a = 9$, and dividing this 9 into the next term will give $r = 2$, and so we have

$$
\begin{aligned}
t_2 &= 9 \times 2 = 9 \times 2^1 & &= 18 \\
t_3 &= 9 \times 2 \times 2 = 9 \times 2^2 & &= 36 \\
t_4 &= 9 \times 2 \times 2 \times 2 = 9 \times 2^3 & &= 72 \\
t_5 &= 9 \times 2 \times 2 \times 2 \times 2 = 9 \times 2^4 &= 144
\end{aligned}
$$

and consequently $t_n = 9 \times 2^{n-1}$,
the power of the common ratio being always one less than the term number.

Thus, $t_{31} = 9 \times 2^{30}$, $t_{41} = 9 \times 2^{40}$, $t_{58} = 9 \times 2^{57}$ and so on.

As a general formula we have $t_n = a \times r^{n-1}$, where a is the first term and r is the common ratio.

EXAMPLE 1: Find the ninth term of the G.P. whose first term is 10 and whose common ratio is 3.

SOLUTION: Substituting in the formula $t_n = a \times r^{n-1}$ with $a = 10$ and $r = 3$ we get $t_9 = 10 \times 3^8 = 65610$, ANS.

EXAMPLE 2: The third term of a G.P. is 28 and the fifth term is 112. Find the first and the tenth terms.

SOLUTION: Just for something to look at we can list the series as

$$t_1 + t_2 + 28 + t_4 + 112 + t_6 + \ldots$$

Now we see that we need to multiply 28 by r to get t_4 and then multiply by r again to get 112, that is

$$28 \times r \times r = 112$$
$$\therefore \quad r \times r = \frac{112}{28} = 4$$
$$\therefore r = 2$$

Looking back at the list it is now clear that

$$t_2 = 28 \div 2 = 14 \text{ and } t_1 = 14 \div 2 = 7, \quad \text{ANS.}$$

With $a = 7$ and $r = 2$ we now see that $t_{10} = 7 \times 2^9 = 3584$, ANS.

Practice Exercise No. 63

Find the *n*th term of each of the following geometric progressions:

1. $6 + 18 + 54 + \ldots$
2. $8 + 4 + 2 + \ldots$
3. $18 + 36 + 72 + \ldots$
4. $1 + 2 + 4 + \ldots$
5. $3 + 6 + 12 + \ldots$
6. $2 + 10 + 50 + \ldots$

To find the sum of *n* terms of a G.P.

As for A.P.s we use the symbol S_n to represent the sum of the first *n* terms of our series. The formula for the sum is

$$S_n = \frac{rl - a}{r - 1}$$

where *r* is the common ratio, *a* is the first term and *l* is the last term.

EXAMPLE 1: Find the sum of the G.P. whose first term is 1 and whose last term is 1024, listed as $1 + 4 + 16 + \ldots + 1024$.

SOLUTION: From the list we have $a = 1$, $r = 4$, $l = 1024$.
Using the formula we get

$$S_n = \frac{4 \times 1024 - 1}{4 - 1} = \frac{4096 - 1}{3} = 1365.$$

By filling in the missing terms we note that there were six terms in all.
Thus $S_6 = 1 + 4 + 16 + 64 + 256 + 1024 = 1365$,   ANS.

EXAMPLE 2: Find the sum of the first eight terms of the G.P. whose first term is 5 and whose common ratio is 3.

SOLUTION: We need to find the last term of the series, t_8.
With $a = 5$ and $r = 3$ we have

$$t_8 = 5 \times 3^7,$$
$$\therefore t_8 = 5 \times 2187$$
$$= 10935$$

Using the formula for S_8 we have

$$S_8 = \frac{3 \times 10935 - 5}{3 - 1} = \frac{32805 - 5}{2} = 16400,   \text{ANS.}$$

Practice Exercise No. 64

1. In a G.P. $a = 4$, $r = 3$. List the first five terms and find S_5.
2. In a G.P. $a = 288$, $r = \frac{1}{2}$. List the first five terms and find S_6.
3. The third and fourth terms of a G.P. are 120 and 240 respectively. Find the common ratio, the fifth term, and S_5.
4. A G.P. is listed as 26411, 3773, 539, . . . Find the first term which is less than 2 and the sum of the first five terms.
5. The common ratio of a G.P. is -2 and the first term is 5. Find S_4.
6. The common ratio of a G.P. is -3 and the first term is 1. List the first five terms and find S_5.

CHAPTER FOURTEEN

GEOMETRY

DEFINITIONS AND TERMS

Elementary geometry is the branch of mathematics that deals with space relationships.

Application of the principles of geometry requires an ability to use arithmetic and elementary algebra as taught in the previous sections of this book. A knowledge of geometry in addition to simple algebra and arithmetic is basic to so many occupations (carpentry, stone-masonry, dress design, hat design, display design, sheet-metal work, machine-shop work, tool-making, architecture, drafting, engineering, etc.) that no serious student should be without it.

A **geometric figure** is a point, line, surface, solid, or any combination of these.

A **point** is the *position* of the intersection of two lines. It is *not* considered to have length, breadth, or thickness.

A **line** is the intersection of two surfaces. It has *length* but neither breadth nor thickness. It may be *straight*, *curved*, or *broken*.

A **surface** has *two* dimensions: *length* and *breadth*. A *flat* surface may be called a **plane**, i.e. the line joining any two points in the plane lies wholly in the plane.

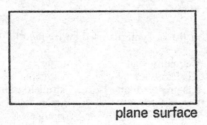

plane surface

A **solid** has *three* dimensions: *length*, *breadth*, and *thickness*.

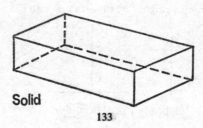

Solid

In solving geometric problems we apply certain general principles called **theorems**. These are systematically demonstrated by means of more basic principles called *axioms* and *postulates*.

Different writers use these last two terms somewhat differently. We may think of the **axioms** used in geometry, however, as *basic mathematical principles* which are so elementary that they cannot be demonstrated by means of still simpler principles. They were once widely called 'self-evident truths'. Note that the first seven 'axioms' listed below are the principles with which you have already become familiar in performing operations upon algebraic equations (Chapter Eight).

The **postulates** used in geometry are of two different, but closely related, kinds. Some are merely restatements of more general mathematical axioms in specific geometric terms. Others are axiom-like statements which apply only to geometry. For instance, the last three 'axioms' below may also be thought of as *geometric postulates*.

Axioms

1. Things equal to the same thing are equal to each other.
2. If equals are added to equals, the sums are equal.
3. If equals are subtracted from equals, the remainders are equal.
4. If equals are multiplied by equals, the products are equal.
5. If equals are divided by equals, the quotients are equal.
6. The whole is greater than any of its parts, and is equal to the sum of all its parts.
7. A quantity may be substituted for an equal one in an equation or in an inequality.
8. Only one straight line can be drawn through two points.
9. A straight line is the shortest distance between two points.
10. A straight line may be produced to any required length.

Symbols

The following is a list of symbols used so frequently that they should be memorized.

$=$	equality sign	$\angle$	angle
$<$	is less than	$°$	degree
$>$	is greater than	$\square$	parallelogram
$\therefore$	therefore	$\odot$	circle
$\parallel$	parallel	$\triangle$	triangle
$\perp$	perpendicular	$\neq$	unequal

Lines

A **horizontal** line is a straight line that is level with the horizon.

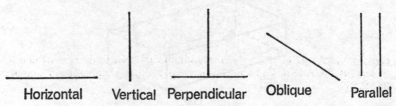

Horizontal   Vertical Perpendicular   Oblique   Parallel

A **vertical** line is a straight line that is perpendicular to the horizon.

Two lines are **perpendicular** to each other when the angles at which they intersect are all equal. Such lines are said to be at right angles to each other.

An **oblique** line is neither horizontal nor vertical.

Parallel lines are two or more straight lines which are equally distant from each other at all points and would never meet no matter how far they might be extended.

ANGLES

An **angle** is the figure formed by two lines proceeding from a common point called the **vertex**. The lines that form an angle are called its **sides**. If three letters are used to designate an angle the *vertex* is read between the others. Thus, Fig. 3 is written ∠*ABC*, and is read *angle ABC*; the sides are *AB* and *BC*.

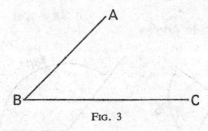

FIG. 3

In measuring an angle remember that you can think of it as composed of the spokes or radii emanating from a point (the vertex) which is at the centre of a circle. As shown, there are 360 degrees around a point. The unit of measure for angles is the *degree* (°). One degree is $\frac{1}{360}$th part of the circumfer-

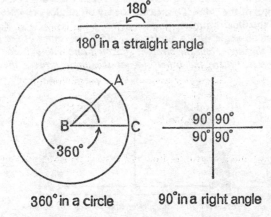

180° in a straight angle

360° in a circle   90° in a right angle

ence of a circle. It is divided into 60 minutes ('). The minute is divided into 60 seconds ("). An angle of 85 degrees, 15 minutes, 3 seconds would be written 85° 15' 3".

A **straight angle** is one of 180°. Its two sides lie in the same straight line.

A **right angle** is one of 90°. Hence it is half a straight angle.

An **acute angle** is any angle that is less than (<) a right angle. Thus it must be less than 90°.

An **obtuse angle** is greater than (>) a right angle but less than (<) a straight angle. Hence, it must be *between 90° and 180°*.

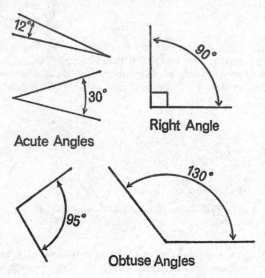

Right Angle

Acute Angles

Obtuse Angles

Measuring Angles

Angles are measured by determining the part of a circle that the sides intersect. Therefore one measures the *opening between* the sides of an angle rather than the length of the sides. To measure or lay off angles one uses a protractor as shown in the illustration.

To measure an angle with a protractor, *place the centre of the protractor at the vertex of the angle, and the straight side on a line with one side of the angle. Read the degrees where the other side of the angle crosses the scale of the protractor.*

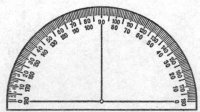

Protractor for Measuring and Laying off Angles

To draw an angle with a protractor, *draw a straight line for one side of the angle. Place the centre of the protractor at the point of the line that is to be the vertex of the angle, and make the straight side of the protractor coincide with* *the line. Place a dot on your paper at the point on the scale of the protractor that corresponds to the size of the angle to be drawn. Connect this dot and the vertex to obtain the desired angle.*

Practice Exercise No. 65

1. Draw a straight angle.
2. Draw a right angle.
3. Draw an acute angle of 30°.
4. Draw an obtuse angle of 120°

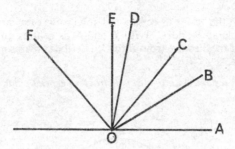

Use the diagram for the following problems.

5. Measure angle *AOB*.
6. Measure angle *AOC*.
7. Measure angle *AOD*.
8. Measure ∠*AOE*.
9. Measure ∠*AOF*.
10. Measure ∠*BOF*.
11. Measure ∠*BOD*.

GEOMETRICAL CONSTRUCTIONS

Geometrical constructions, in the strict sense, involve only the use of a straight-edge (unscaled ruler) and a pair of compasses. These are the only instruments needed to carry out the following constructions. Of course, in actual mechanical drawing the draughtsman is not thus limited.

Problem 1: *To bisect a straight line.* (Bisect means to divide in half.)

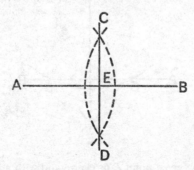

Method: With *A* and *B* as centres and with a radius greater than half the line *AB*, draw arcs intersecting at points *C* and *D*. Draw *CD*, which bisects *AB* at *E*. (It should be noted that *CD* is perpendicular to *AB*.)

Problem 2: *To bisect any angle.*

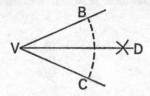

Method: With the vertex as centre and any radius draw an arc cutting the sides of the angle at *B* and *C*. With *B* and *C* as centres and with a radius greater than half the distance from *B* to *C*, describe two arcs intersecting at *D*. The line *DV* bisects ∠ *CVB*.

Problem 3: *At a point on a line to construct a perpendicular to the line.*

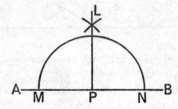

Method: From point *P* as centre with any radius describe an arc which cuts the line *AB* at *M* and *N*. From *M* and *N* as centres and with a radius greater than *MP*, describe arcs which intersect at *L*. Draw the line *PL*, which is the required perpendicular.

Problem 4: *From a given point away from a straight line to drop a perpendicular to the line.*

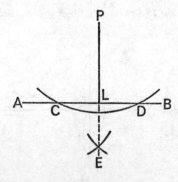

Method: From the given point *P* as centre and with a large enough radius describe an arc which cuts line *AB* at *C* and *D*. From *C* and *D* as centres and with a radius greater than half *CD*, describe two arcs that intersect at *E*. Connect *PE*. The line *PL* is the required perpendicular to the line *AB*.

Note: For some of the previous constructions and some that are to follow, more than one method is available. To avoid confusion in learning, only one method is here presented.

Problem 5: *To duplicate a given angle.*

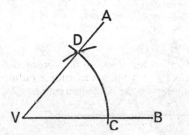 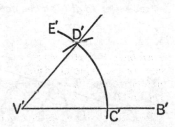

Method: Let the given angle be ∠*AVB*. Then from the vertex *V* as centre and with a convenient radius, draw an arc that intersects the sides at *C* and *D*. Draw any straight line equal to or greater in length than *VB* and call it *V'B'*. (Read *V prime B prime*.) With *V'* as centre and with the same radius, describe an arc *C'E* that cuts the line at *C'*. From *C'* as centre and with a radius equal to *DC*, describe an arc intersecting arc *C'E* at *D'*. Draw *D'V'*. ∠*D'V'C'* is the required angle.

Problem 6: *To duplicate a given triangle.*

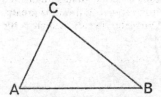 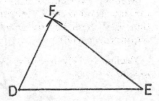

Method: Draw any straight line from any point *D* as centre, and with a radius equal to *AB* lay off *DE* equal to *AB*. With *E* as centre and *BC* as radius, draw an arc. With *D* as centre and *AC* as radius, draw an arc which intersects the other arc at *F*. Draw *FE* and *FD*. *DEF* is the required triangle.

Problem 7: *To construct a line parallel to a given line at a given distance.*

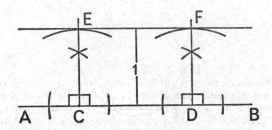

Method: If the given line is *AD* and the given distance is 1 cm, then at any two points *C* and *D* on the given line *AB* erect perpendiculars to *AB* (see Problem 3). With *C* and *D* as centres and with a radius equal to 1 cm, describe arcs cutting the perpendiculars at *E* and *F*. Draw the line *EF*, which is the required parallel line at a distance of 1 cm from *AB*.

Problem 8: *To divide a line into a given number of equal parts.*

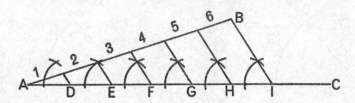

Method: If *AB* is the given line, and if it is to be divided into six parts, then draw line *AC* making an angle (most conveniently an acute angle) with *AB*. Starting at *A*, mark off on *AC* with a compass six equal divisions of any convenient length. Connect the last point *I* with *B*. Through points *D*, *E*, *F*, *G*, and *H* draw lines parallel to *IB* by making equal angles. The parallel lines divide *AB* into six equal parts.

Problem 9: *To find the centre of a circle or arc of a circle.*

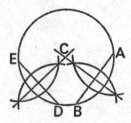

Method: Draw any two chords *AB* and *DE*. Draw the perpendicular bisectors of these chords (see Problem 1). The point *C* where they intersect is the centre of the circle or arc.

Problem 10: *To inscribe a regular hexagon in a circle.*

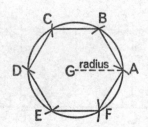

Note: A regular hexagon is a polygon with six equal sides and six equal angles. The length of a side of a hexagon is equal to the radius of a circle circumscribing it.

Method: The radius of the circle is equal to *AG*. Starting at any point on the circle and using the length of the radius as the distance, lay off successive points *B*, *C*, *D*, *E*, *F* on the circumference of the circle. Connect the points with straight lines to obtain the required hexagon.

LINE AND ANGLE RELATIONSHIPS

Having learned some basic geometric definitions, axioms, and constructions, you are now prepared to understand some important relationships between lines and angles.

In demonstrating these relationships it is necessary to introduce additional *definitions*, *postulates*, *propositions*, *theorems*, and *corollaries*.

For example, the following are important *postulates*.

Postulate 1. *A geometric figure may be moved from one place to another without changing its size or shape.*

Postulate 2. *Two angles are equal if they can be made to coincide.*

Postulate 3. *A circle can be drawn with any point as centre.*

Postulate 4. *Two straight lines can intersect in only one point.*

Postulate 5. *All straight angles are equal.*

A **corollary** is a geometric truth that follows from one previously given and needs little or no proof.

For example, from Postulate 3 above we derive the *corollary*:

Corollary 1. *An arc of a circle can be drawn with any point as centre.*

Adjacent angles are angles that have a common vertex and a common side between them.

For example, ∠*CPB* is *adjacent* to ∠*BPA*, but not to ∠*DRC*.

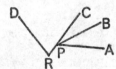

Adding Angles

Postulate 6. *Adjacent angles can be added. Thus:*

∠*AOB* + ∠*BOC* = ∠*AOC*.
∠*DOC* + ∠*COB* + ∠*BOA* = ∠*DOA*.
∠*EOD* + ∠*DOC* + ∠*COB* = ∠*EOB*.

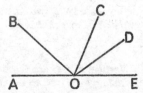

Postulate 7. *The sum of all the adjacent angles about a point on one side of a straight line is equal to one straight angle. Thus:*

If you measure ∠*AOB* + ∠*BOC* + ∠*COD* + ∠*DOE*, it should total 180°. Does it?

Complements and Supplements

Two angles whose sum is 90°, or one right angle, are called **complementary**. Each of the angles is called the **complement** of the other. *Thus:*

∠*AOB* is the *complement of* ∠*BOC*,
or 35° *is* *complementary to* 55°,
or 55° *is* *complementary to* 35°.

Two angles whose sum is 180° or a straight angle are said to be **supplementary** to each other. *Thus:*

∠*AOC* is the *supplement of* ∠*COB*,
or 150° *is* *supplementary to* 30°,
or 30° *is* *supplementary to* 150°.

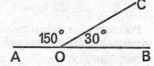

The postulates that follow concerning complementary and supplementary angles are mostly corollaries of axioms and postulates already stated. Hence, the references in parentheses are to axioms and postulates on pp. 134, 141, and this page.

Postulate 8. *All right angles are equal.* Since all straight angles are equal (POST. 5) and halves of equals are equal (Ax. 5).

Postulate 9. *When one straight line meets another two supplementary angles are formed.*

∠1 + ∠2 = ∠*AOB*, which is a straight angle (Ax. 6).

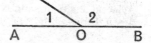

Postulate 10. *Complements of the same angle or of equal angles are equal* (Ax. 3).

Postulate 11. *Supplements of the same angle or of equal angles are equal* (Ax. 3).

Postulate 12. *If two adjacent angles have their exterior sides in a straight line they are supplementary.*

Postulate 13. *If two adjacent angles are supplementary their exterior sides are in the same straight line.*

Vertical angles are the pairs of opposite angles formed by the intersection of straight lines. *Thus:*

∠1 and ∠2 are *vertical angles.* ∠5 and ∠6 are *vertical angles.* What other pairs are vertical angles?

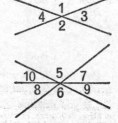

THE METHOD OF DEMONSTRATION IN GEOMETRY

A **proposition** is a statement of either a *theorem* or a *problem*.

A **theorem** is a relationship to be demonstrated.

A **problem** is a construction to be made.

In proving theorems or the correctness of constructions the procedure is as follows.

If the proposition is a *theorem* requiring proof you break it up into its two parts: the *hypothesis* and the *conclusion*. In the *hypothesis* certain facts are assumed. You use these given facts in conjunction with other previously accepted geometric propositions to prove the conclusion.

If the proposition is a *problem* you make the construction and then proceed to prove that it is correct. You do this by listing the given elements and bringing forward previously established geometric facts to build up the necessary proof of correctness.*

For example, let us take the statement, *vertical angles are equal*. This theorem is given as Proposition No. 1 in many geometry textbooks, and is presented as follows.

Given:　　Vertical angles 1 and 2 as in the diagram next to the definition of vertical angles.

To prove:　$\angle 1 = \angle 2$.

Steps	*Reasons*
1. $\angle 2$ is the supplement of $\angle 3$.	1. Two angles are supplementary if their sum is a straight $\angle$.
2. $\angle 1$ is the supplement of $\angle 3$.	2. Same as Reason 1.
3. $\angle 1 = \angle 2$.	3. Supplements of the same $\angle$ are equal (Post. 4)

Abbreviations

The following abbreviations are used:

adj.	adjacent	def.	definition
alt.	alternate	ext.	exterior
	altitude	hyp.	hypotenuse
ax.	axiom	iden.	identity
comp.	complementary	int.	interior
cong.	congruent	rt.	right
const.	construction	st.	straight
cor.	corollary	supp.	supplementary
corr.	corresponding	vert.	vertical

It should also be noted that the plurals of a number of the symbols listed on p. 134 are formed by adding an *s* after the symbol. Thus, $\angle s$ means angles; $\triangle s$, triangles; $\| s$, parallels; $\bigcirc s$, circles; $\square s$, parallelograms, etc.

Practice Exercise No. 66

1. $\angle 1$ coincides with $\angle 2$. $\angle 1 = 30°$. Find $\angle 2$.

---

* This is the method of procedure followed in most geometry textbooks for demonstrating the truth of established geometric principles. For the purposes of this book, however, it will not be necessary to give formal demonstrations of theorems and problems. It is our purpose to give you a working knowledge of the essential geometric principles, facts, and skills that can be put to practical application in office, in industry, in military pursuits, in indulging a hobby, or in studying higher mathematics as presented in this book and in other more advanced textbooks.

2. *BD* is the bisector of ∠*ABC*, which is 45°.
Find ∠*ABD*.

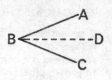

3. ∠1 = ∠5, ∠2 = ∠1, and ∠3 = ∠5. What
is the relationship between:

  (*a*) ∠1 and ∠3; (*b*) ∠2 and ∠5;
  (*c*) ∠4 and ∠7?

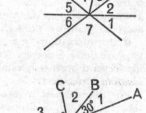

4. In the same figure list the pairs of adj. ∠*s*.
5. In the same figure list the pairs of vert. ∠*s*.

6. In the accompanying figure the opposite ∠*s*
are vert. ∠*s*; ∠1 = 30° and ∠3 = 100°. Find the
remaining four angles.

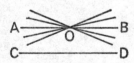

7. In the same figure find the values of ∠*AOC*, ∠*AOD*, ∠*BOE*, and ∠*FOB*.
8. How many degrees are there in: (*a*) ¾ of a rt. ∠; (*b*) ⅔ rt. ∠; (*c*) ½ rt. ∠;
(*d*) ⅓ rt. ∠; (*e*) ¼ rt. ∠?
9. Find the complement of: (*a*) 68°; (*b*) 45°; (*c*) 55°; (*d*) 32°; (*e*) 5°; (*f*) 33° 30'.
10. What is the supplement of: (*a*) 25°; (*b*) 125°; (*c*) 44°; (*d*) 88°; (*e*) 74° 30';
(*f*) 78° 30'?

Parallel Lines

Postulates Concerning Parallels

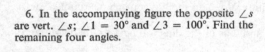

1. Through a given point only one line can be drawn parallel to a given line.

In the diagram the only line that can be drawn ∥ to *CD* through point *O* is *AB*.

2. Two intersecting lines cannot both be parallel to a third straight line.

**3. Two straight lines in the same plane, if produced, either will intersect or else
are parallel.**

Definitions

A **transversal** is a line that intersects two or more other lines.

When a **transversal** cuts two parallel or intersecting lines various angles are
formed. The names and relative positions of these angles are important. The

relationship of angles as shown in the following diagram should be memorized:

∠s 1, 2, 3, 4 are termed **exterior** angles.
∠s 5, 6, 7, 8 are termed **interior** angles.
∠s 1 and 4 ⎱ ⎰ are pairs of **alternate exterior**
∠s 2 and 3 ⎰ ⎱ angles.
∠s 5 and 8 ⎱ ⎰ are pairs of **alternate interior**
∠s 6 and 7 ⎰ ⎱ angles.
∠s 1 and 7 ⎱
∠s 2 and 8 ⎰ are pairs of **corresponding**
∠s 5 and 3 ⎰ angles.
∠s 6 and 4 ⎰

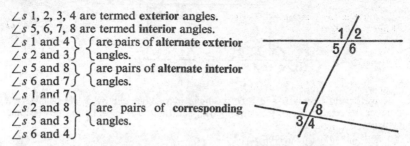

Theorem 1. *If two straight lines are parallel to a third straight line, they are parallel to each other.*

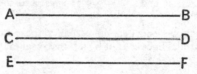

Given: *AB* and *EF* ∥ to *CD*.
To prove: AB ∥ *EF*.

If *AB* is not ∥ to *EF* the two lines would intersect and they would then be two intersecting lines parallel to a third straight line. But this is impossible according to Parallel Postulate 2. Hence *AB* must be parallel to *EF*.

Relationships Formed by Parallels and a Transversal

If two parallel lines are cut by a transversal certain definite relationships will always be found to exist among the angles that are formed by the parallel lines and the transversal.

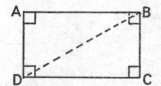
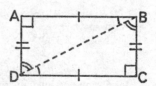

If we take the rectangle *ABCD* we know that the opposite sides are parallel and equal and that all the angles are right angles. If we then draw the diagonal *DB* we have formed two triangles, △*DAB* and △*DCB*.

In △s *DAB* and *DCB* we know *AD* = *CB*, *AB* = *DC*, and ∠*A* = ∠*C*. As will be shown in the section on triangles, when two sides and the included ∠ of one △ are equal to two sides and the included ∠ of another, the two triangles are said to be congruent. This means that all their corresponding sides and angles are equal. (In the diagram the corresponding sides and angles of each triangle are marked with matched check marks.)

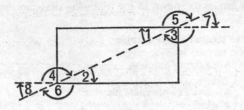

If we extend lines *AB* and *CD*, we have two ∥ lines cut by a transversal. We number the related angles for convenience, and the following relationships become evident.

∠1 = ∠2 (Corr. ∠*s* of cong. △*s*.)

∠1 = ∠7, and ∠2 = ∠8 (Vert. ∠*s* are equal.)

∴ ∠7 = ∠8 = ∠1 = ∠2 (Things = to the same thing are = to each other.)

∠5 is supp. ∠7 (Ext. sides form a st. ∠.)

∴ ∠6 = ∠4, and ∠3 = ∠5 (Vert. ∠*s* are equal.)

∴ ∠3 = ∠6, ∠5 = ∠6, and ∠3 = ∠4 (Things = to the same thing are = to each other (Ax. 1).)

Presenting the above conclusions verbally, the angle relationships that occur when two parallel lines are cut by a transversal may be stated as follows.

1. The alternate interior angles are equal.
    ∠1 = ∠2, and ∠3 = ∠4

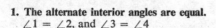

2. The alternate exterior angles are equal.
    ∠5 = ∠6, and ∠7 = ∠8

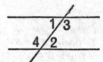

3. The corresponding angles are equal.
    ∠4 = ∠5, ∠3 = ∠6, ∠2 = ∠7, ∠1 = ∠8

4. The two interior angles on the same side of a transversal are supplementary.
    ∠1 supp. ∠4, and ∠3 supp. ∠2

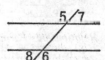
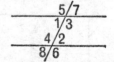

5. The two exterior angles on the same side of a transversal are supplementary.
    ∠5 supp. ∠8, and ∠7 supp. ∠6

These angle relationships may now be employed to prove that certain straight lines are parallel. Such proofs are represented by the *converses* of statements 1–5, in the form of the following theorems:

Theorems on Parallel Lines

Two lines are parallel if:
Theorem 2. *A transversal to the lines makes a pair of alternate interior angles equal.*

Theorem 3. *A transversal to the lines makes a pair of alternate exterior angles equal.*

Theorem 4. *A transversal to the lines makes a pair of corresponding angles equal.*

Theorem 5. *A transversal to the lines makes a pair of interior angles on the same side of the transversal supplementary.*

Theorem 6. *A transversal to the lines makes a pair of exterior angles of the same side of the transversal supplementary.*

A *corollary* that follows from these theorems is the following:

Corollary 1. *If two lines are perpendicular to a third line they are parallel.*
This can be easily proved by showing alt. int. ∠s equal as ∠1 = ∠2, or corr. ∠s equal, as ∠1 = ∠3, etc.

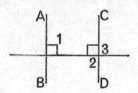

We may summarize the relationships of the angles formed by parallel lines cut by a transversal as follows:

(a) *The four acute angles formed are equal.*
(b) *The four obtuse angles formed are equal.*
(c) *Any one of the acute angles is the supplement of any one of the obtuse angles;*
that is, their sum equals 180°.

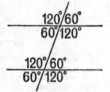

Practice Exercise No. 67

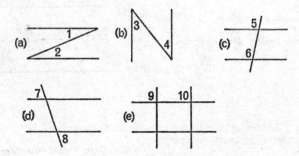

1. In the above diagram identify the kinds of angles indicated.

2. If ∠3 = 50°, what is the value of ∠1, ∠2, and ∠4?

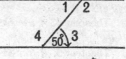

3. If ∠5 = 40°, what is the value of ∠6, ∠7, and ∠8?

4. *AB* is ⊥ to *CD*. Why would any other line that makes a 90° angle with *CD* be ∥ to *AB*?

5. Tell why *AB* ∥ *CD* if given:
   (a) ∠3 = ∠6;   (b) ∠1 = ∠5;   (c) ∠2 = ∠7.

6. If given ∠1 = ∠2, prove that *AB* ∥ *CD*.

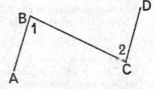

7. Given ∠1 = 65° and ∠4 = 115°, prove that the two horizontal lines are ∥.

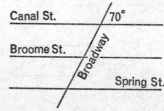

8. If Broadway cuts across Canal Street at an angle of 70°, at what angle does it cut across Broome and Spring Streets, which are ∥ to Canal Street?

9. Given ∠*ABC* = 60°, construct a line ∥ to *BC* using the principle of corresponding angles being equal.

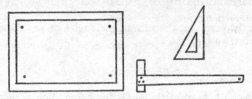

10. Using the drawing-board, T-square, and triangle pictured, how would you construct two angles the sides of which are ‖ to each other?

TRIANGLES

A **triangle** is a plane three-sided figure, the sides of which are straight lines. If you close off any angle a triangle is formed.

Triangles are classified according to their sides as *scalene*, *isosceles*, and *equilateral*.

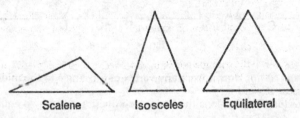

Scalene   Isosceles   Equilateral

A **scalene triangle** is one in which no two sides are equal. An **isosceles triangle** is one in which two sides are equal. An **equilateral triangle** is one with three sides equal.

Triangles may also be classified with respect to their angles as *right*, *acute*, and *obtuse*.

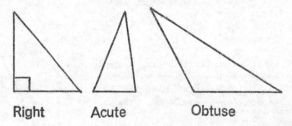

Right   Acute   Obtuse

A **right triangle** (or *right-angled triangle*) contains one right angle (often indicated by placing a small square in the 90° angle).

An **acute triangle** is one in which all angles are less than right angles.

An **obtuse triangle** has one angle greater than a right angle.

Note that a *right* triangle may be *scalene* or *isosceles*; an *acute* triangle may be *scalene*, *isosceles*, or *equilateral*; an *obtuse* triangle may be *scalene* or *isosceles*.

Note also that either the scalene or the isosceles triangle may be right, acute, or obtuse.

It is a basic theorem that the sum of the angles of any triangle is equal to 180°. (See Theorem 14, p. 156.)

Triangular Measurements

The **height** or **altitude** of a triangle is the perpendicular distance from the base to the vertex of the opposite angle. In Fig. 4, *AC* represents height or altitude of the triangles.

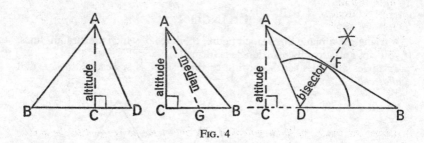

FIG. 4

A **median** is a line drawn from any vertex of a triangle to the middle of the opposite side. *AG* in Fig. 4.

The **bisector** of an angle is the line which divides it into two equal angles. *DF* bisects ∠ *BDA* in Fig. 4.

The **perimeter** of any figure is *the entire distance around the figure*.

Rule: *The area of a triangle equals one-half the product of the base and the height.*

Expressed as a **formula:**

$$A = \tfrac{1}{2}\, bh \text{ or } A = \frac{bh}{2}$$

EXAMPLE 1: Find the area of the triangle shown, with
 $h = 8$ m and $b = 6$ m.

SOLUTION: $A = \dfrac{bh}{2} = \dfrac{6 \times 8}{2} = 24$ m², ANS.

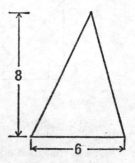

EXAMPLE 2: What is the height of a triangle if its area is 144 m² and its base 16 m?

SOLUTION: $A = \dfrac{bh}{2} \therefore h = \dfrac{2A}{b} = \dfrac{2 \times 144}{16} = 18$ m, ANS.

Facts about Right Triangles

The **hypotenuse** of a right triangle is the side opposite the right angle.

In the figure below it is shown that the square drawn on the hypotenuse of a right triangle is equal in area to the sum of the areas of the squares drawn on the other two sides.

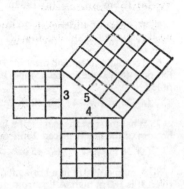

$(3 \text{ m})^2 = 9 \text{ m}^2$
$(4 \text{ m})^2 = 16 \text{ m}^2$
$(5 \text{ m})^2 = 25 \text{ m}^2$
$9 + 16 = 25$
$3^2 + 4^2 = 5^2$

Rule: *The square of the hypotenuse of a right triangle is equal to the sum of the squares of the other two sides.*

From this there arise several self-evident formulae with reference to the right triangle.

Let c = hypotenuse, a = altitude, b = base; then:

Formula 1: $c^2 = a^2 + b^2$

Formula 2: $c = \sqrt{a^2 + b^2}$. (Taking the square root of both sides of the first equation.)

Formula 3: $a^2 = c^2 - b^2$; or, by transposition, $b^2 = c^2 - a^2$.

EXAMPLE 3: Find the hypotenuse of a right triangle whose base is 18 m and altitude 26 m.

$$c = \sqrt{a^2 + b^2} \quad \text{(formula)}$$
$$= \sqrt{(18)^2 + (26)^2} \quad \text{(substituting)}$$
$$= \sqrt{324 + 676} \quad \text{(squaring)}$$
$$= \sqrt{1000} \quad \text{(adding)}$$
$$= 31 \cdot 62, \quad \text{ANS., extracting the square root.}$$

Practice Exercise No. 68

1. A derrick standing perpendicular to the ground is 15 m high, and is tied to a stake in the ground by a cable 17 m long. How far is the foot of the derrick from the stake?

(A) 68 m ......   (B) 6 m ......   (C) 24 m ......   (D) 8 m ......

2. The base of a triangle is 18 m; the altitude is $3\frac{1}{2}$ times the base. What is the area?

(A) 1296 m² ......   (B) 600 m² ......   (C) 567 m² ......   (D) 648 m² ......

3. How much will it cost to fence off an isosceles-shaped lot if one side is 75 m and the base is 50 m? Fencing costs £2 a metre.

(A) £40 ......   (B) £400 ......   (C) £25 ......   (D) £50 ......

4. In a square field it is 90 m between adjacent corners. How far in a straight line is it between opposite corners?

(A) 127 m ......   (B) 180 m ......   (C) 120 m ......   (D) 135 m ......

5. The base of a triangle is 20 m; the altitude is $\frac{1}{2}$ the base. What is the area?

(A) 80 m^2 ......   (B) 100 m^2 ......   (C) 120 m^2 ......   (D) 200 m^2 ......

6. To hold a telephone mast in position a 26-m wire is stretched from the top of the pole to a stake in the ground 10 m from the foot of the pole. How tall is the pole?

(A) 24 m ......   (B) 40 m ......   (C) 12 m ......   (D) 36 m ......

7. What must be the length of a line to reach to the top of a cliff 40 m high, if the bottom of the line is placed 9 m from the cliff?

(A) 36 m ......   (B) 45 m ......   (C) 41 m ......   (D) 54 m ......

8. A mast is 100 m in a horizontal line from a river and its base is 20 m above the river. It is 160 m high. A line from its top to the opposite shore of the river measures 500 m. How wide is the river?

(A) 250·93 m ......   (B) 366·47 m ......   (C) 342·89 m ......   (D) 329·65 m ......

Demonstrating the Congruence of Triangles

In demonstrating some fundamental relationships between lines and angles of triangles, a method of proving triangles to be *congruent* is employed.

Congruent figures are those which can be made to coincide or fit on one another. Thus, if two triangles can be made to coincide in all their parts they are said to be congruent.

The symbol for congruence is $\cong$.

In triangles that are congruent the respective equal angles and equal sides that would coincide if one figure were placed on top of the other are termed **corresponding** angles and *corresponding* sides.

Corresponding parts are also called *homologous* parts. From what has been said it follows that corresponding parts of congruent figures are equal.

In geometry the corresponding or homologous parts of corresponding figures are frequently indicated by using *corresponding check marks* on the respective parts. For example, the corresponding parts in the congruent triangles below are marked with check marks of the same kind.

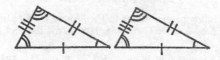

Seven Theorems on Congruence

Theorem 7. *Two triangles are congruent if two sides and the included angle of one are equal respectively to two sides and the included angle of the other.*

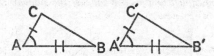

According to this theorem you are given $\angle ABC \cong \triangle A'B'C'$, with $AC = A'C'$, $AB = A'B'$, and $\angle A = \angle A'$.

If you construct the figures with the given equal parts and then place $\triangle ABC$ on $\triangle A'B'C'$ so that the given equal parts correspond, it will be seen that the third line, CB, coincides with $C'B'$, making the triangles congruent at all points. Thus, all the corresponding parts not given may also be assumed to be respectively equal.

For example, construct AC and $A'C'$ to equal 1 cm; $\angle A$ and $\angle A' = 60°$; AB and $A'B' = 2$ cm. Then measure the distances between CB and $C'B'$, and you will find them to be equal. If you measure $\angle s$ C and C' and $\angle s$ B and B' you will find these pairs to be equal as well.

Proving congruence by this theorem is known as the *side included angle side* method. It is abbreviated *s.a.s. = s.a.s.*

By employing a similar approach you can readily verify the following theorems on the correspondence of triangles.*

Theorem 8. *Two triangles are congruent if two angles and the included side of one are equal respectively to two angles and the included side of the other.*

This is known as the *angle side angle* theorem, and is abbreviated *a.s.a = a.s.a.*

Theorem 9. *Two triangles are congruent if the sides of one are respectively equal to the sides of the other.*

 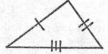

This is known as the *side side side* theorem, and is abbreviated *s.s.s. = s.s.s.*

Theorem 10. *Two triangles are congruent if a side and any two angles of one are equal to the corresponding side and two angles of the other.*

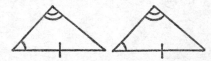

This is known as the *side angle angle* theorem, and is abbreviated *s.a.a. = s.a.a.*

---

* Formal proofs employing geometric axioms, postulates, and theorems to illustrate these cases of congruent triangles are given in regular school textbooks on geometry. The student interested in academic study should refer to such books.

Theorem 11. *Two right triangles are equal if the sides of the right angles are equal respectively.*

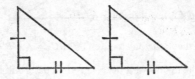

Since the included right angles are equal, this theorem is really a special case of *s.a.s.* = *s.a.s.*

Theorem 12. *Two right triangles are equal if the hypotenuse and an acute angle of one are equal to the hypotenuse and an acute angle of the other.*

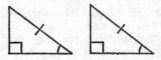

Since the right angles are equal, this theorem is a special case of *s.a.a.* = *s.a.a.*

Theorem 13. *Two right triangles are congruent if a side and an acute angle of one are equal to a side and corresponding acute angle of the other.*

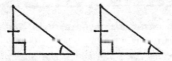

Since the right angles are equal, this is again a special case of *s.a.a.* = *s.a.a.*

Practice Exercise No. 69

Note: Mark corresponding parts with corresponding check marks as previously explained. Use the method of demonstration shown under Theorem 14, following lines of reasoning similar to that used in connexion with Theorem 7.

1. *Given* $AB = AD$
             $\angle 1 = \angle 2$
  Prove $\triangle ABC \cong \triangle ADC$

2. *Given* $BD \perp AC$
             D is the mid-point of AC
  Prove $\triangle ABD \cong \triangle CBD$

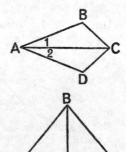

3. *Given* $\angle 3 = \angle 5$ AE is the bisector of BD
    Prove $\triangle ABC \cong \triangle EDC$

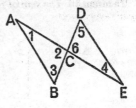

4. *Given* AD and CE bisect each other
    Prove $AE \parallel CD$

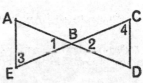

5. *Given* $AD = BC$
           $AC = BD$
    Prove $\triangle BAD \cong \triangle ABC$ and $\angle 1 = \angle 2$

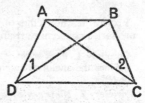

6. *Given* $AB = CB$
           $AD = CD$
    Prove $\angle 1 = \angle 2$
     Hint: Draw BD and then extend it to meet
AC at E.

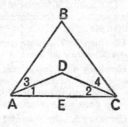

7. *Given* $AB = EF$
           $AB \parallel EF$
           $BC \parallel DE$
    Prove $BC = DE$

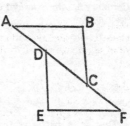

Facts about Triangles in General

The general properties of the triangle not only form the foundation of trigonometry but also find a wide application in the analysis and measurement of straight-sided plane figures of every kind.

One of the most important facts about triangles in general is that, regardless of the shape or size of any triangle, *the sum of the three angles of a triangle is equal to a straight angle, or* 180°. Presented as a theorem, this proposition is easily proved.

Theorem 14. *The sum of the angles of a triangle is equal to a straight angle.*

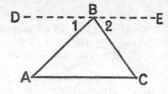

Given: △*ABC*.
To prove: ∠*A* + ∠*B* + ∠*C* = a straight angle.

Steps	Reasons
1. Through *B* draw *DE* ∥ *AC*.	1. Parallel postulate No. 1.
2. ∠1 = ∠*A*.	2. Alt. int. ∠s of ∥ lines are =.
3. ∠2 = ∠*C*.	3. Same reason as 2.
4. ∠1 + ∠*B* + ∠2 = a straight angle.	4. By definition, since the exterior sides lie in a straight line.
5. ∴ ∠*A* + ∠*B* + ∠*C* = a straight angle.	5. Substituting ∠*A* and ∠*C* for ∠1 and ∠2 in step 4 by Axiom 7.

From this knowledge of the sum of the angles of a triangle the following corollaries concerning triangles in general become self-evident.

Corollary 1. *Each angle of an equiangular triangle is 60°.*
Since the angles of an equiangular triangle are equal, each angle = 180° ÷ 3, or 60°.

Corollary 2. *No triangle may have more than one obtuse angle or right angle.*
180° minus 90° or more leaves 90° or less, to be divided between the two remaining angles, and therefore each of the two remaining angles must be acute, i.e. less than 90°.

Corollary 3. *The acute angles of a right triangle are complementary.*
180° minus 90° leaves two angles whose sum equals 90°.

Corollary 4. *If two angles of one triangle are equal respectively to two angles of another the third angles are equal.*
This truth is supported by Ax. 3 (p. 134), namely, that if equals are subtracted from equals the remainders are equal.

Corollary 5. *Any exterior* angle of a triangle is equal to the sum of the two remote interior angles.*

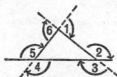

Thus, in △*ABC* if you extend *AC* to *D* and draw *CE* ∥ *AB* you have the two ∥ lines *AB* and *CE* cut by the transversal *AD*. ∴ ∠1 = ∠*B* and ∠2 = ∠*A*, so that ∠1 + ∠2, or ∠*BCD* = ∠*A* + ∠*B*.

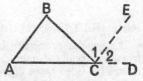

A few characteristic properties of special triangles frequently used are worth noting at this point.

* An exterior angle of a triangle is the angle formed by a side and the extension of its adjacent side. Every triangle has six exterior angles, as shown in the diagram.

Theorem 15. *The base angles of an isosceles triangle are equal.*

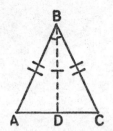

By definition two sides of an isosceles triangle are equal.

∴ if you draw the bisector *BD* of ∠*B* it is readily seen that △*ABD* ≃ △*CBD* by *s.a.s.* = *s.a.s.* Hence ∠*A* = ∠*C*.

This theorem may be stated in another way, namely:

Theorem 16. *If two sides of a triangle are equal, the angles opposite those sides are equal.*

The following corollaries may readily be seen to follow from this theorem.

Corollary 1. *If two sides of a triangle are equal, the angles opposite these sides are equal and the triangle is isosceles.*

Corollary 2. *The bisector of the apex angle of an isosceles triangle is perpendicular to the base, bisects the base, and is the altitude of the triangle.*

Corollary 3. *An equilateral triangle is equiangular.*

Theorem 17. *If one acute angle of a right triangle is double the other, the hypotenuse is double the shorter side.* Or

In a 30°–60° right triangle the hypotenuse equals twice the shorter side.

The following properties of bisectors, altitudes, and medians of triangles are frequently applied in the practical problems of geometric design and construction that arise in shop and office.

Theorem 18. *Every point in the perpendicular bisector of a line is equidistant from the ends of that line.*

If   *CD* is ⊥ bisector of *AB*
Then *DA* = *DB*
       FA = *FB*, etc.

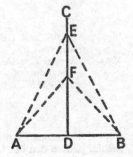

Theorem 19. *Every point in the bisector of an angle is equidistant from the sides of the angle.*

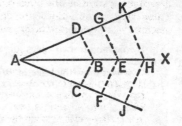

If   *AX* is the bisector of $\angle A$
Then *BC* = *BD*, *EF* = *EG*, *HJ* = *HK*, etc.

Theorem 20. *The perpendicular is the shortest line that can be drawn from a point to a given line.*

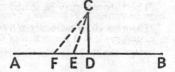

If   *CD* $\perp$ *AB*
Then *CD* < *CE*, *CD* < *CF*, etc.

Theorem 21. *The three bisectors of the sides of a triangle meet in one point which is equidistant from the three vertices of the triangle.*

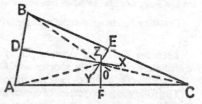

If   *DX*, *EY*, and *FZ* are bisectors of the sides *AB*, *BC*, and *CA*
Then *AO* = *BO* = *CO*, and is equal to the radius of the circle circumscribing $\triangle ABC$.

Note. This fact is often used as a method for finding the centre of a circular object. The procedure consists in inscribing a triangle in the circle and constructing the bisectors of the sides. The point at which they meet is the centre of the circle.

Theorem 22. *The three bisectors of the angles of a triangle meet in one point which is equidistant from the three sides of the triangle.*

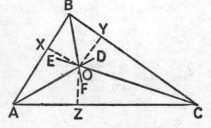

If   *AD*, *BF*, and *CE* are bisectors respectively of $\angle s$ *A*, *B*, and *C*
Then *OX* = *OY* = *OZ*, and is equal to the radius of the circle inscribed in $\triangle ABC$.

Note: This geometric theorem is employed as a method for determining the largest circular pattern that can be cut out of a triangular piece of material.

For practical purposes you should carry out the constructions involved in the theorems of this section. Check the accuracy of your constructions by determining whether the constructed parts fit the hypothesis of the theorem. These very constructions are daily applied in architecture, carpentry, art, machine work, manufacturing, etc.

Practice Exercise No. 70

1. Two angles of a triangle are 62° and 73°. What does the third angle equal?
2. How many degrees are there in the sum of the angles of a quadrilateral?

   Hint: Draw the figure and then construct a diagonal.
3. What is the value of an exterior angle of an equilateral triangle?
4. In a certain right triangle the acute angles are $2x$ and $7x$. What is the size of each angle?
5. An exterior angle at the base of an isosceles triangle equals 116°. What is the value of the vertex angle?
6. In a certain triangle one angle is twice as large as another and three times as large as the third. How many degrees are there in each angle?
7. Draw an equilateral triangle and by it find the ratio between the diameter of the inscribed circle and the radius of the circumscribed circle.

   Hint: Refer to Theorems 21 and 22.

8. Given $\angle 1 = \angle 4$, prove that $\triangle ABC$ is isosceles.

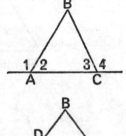

9. Given $BA = BC$ and $DE \parallel BC$, prove that $DE = DA$.

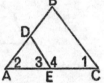

POLYGONS

A **polygon** is a plane geometric figure bounded by three or more sides. Any triangle, for instance, is a polygon.

The **vertices** of a polygon are the angle points where two sides meet.

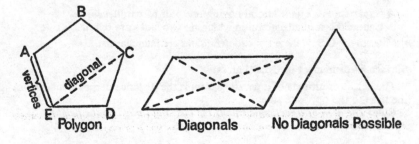

Polygon   Diagonals   No Diagonals Possible

A **diagonal** of a polygon joins two non-consecutive vertices. How many diagonals has a triangle? None. How many diagonals can a four-sided figure have? Two.

Polygons derive their names from the number of and nature of the sides and the types of angles included. A polygon is called regular if all its angles are equal and all its sides are of equal length.

Quadrilaterals are polygons with four sides. There are six types of quadrilaterals: the *rectangle*, the *square* (a special form of rectangle), the *parallelogram*, the *rhombus*, the *trapezoid*, and the *trapezium*.

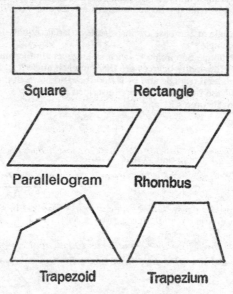

TYPES OF QUADRILATERALS

A **parallelogram** is a quadrilateral in which the opposite sides are parallel.

A **square** is a rectangle having two adjacent sides equal.

A **rectangle** is a parallelogram, one angle of which is a right angle.

A **rhombus** is a parallelogram having two adjacent equal sides but no right angles.

A **trapezium** is a quadrilateral having one pair of parallel sides.

A **trapezoid** is a quadrilateral in which no two sides are parallel.

Note: In the U.S.A. these last two definitions are interchanged.

Surface Measurement of Quadrilaterals

The **height** or **altitude** of a **parallelogram** is the distance perpendicular from the base to the opposite side.

Rule: *The area of a rectangle equals the base multiplied by the height*.

Formula: $A = bh$.

EXAMPLE 1: Find the area of a rectangle that is 3 m high with a 4-m base.

SOLUTION:

$A = bh$, formula.
$A = 4 \times 3 = 12$.
12 m², ANS.

Note: The diagram has been drawn to scale. There are 4 columns and 3 rows. The number of squares by count is seen to be 12. The area is thus 12 square metres.

EXAMPLE 2: Find the height of a rectangle with a 16-m base and an area of 80 m².

SOLUTION:

$$A = bh, \therefore h = \frac{A}{b}$$

$$= \tfrac{80}{16} = 5 \text{ m}, \text{ ANS.}$$

Rule: *The area of a square is equal to the square of one of its sides.*
Formula: $A = S^2$.

EXAMPLE 3: Find the side of a square whose area is 121 m².

SOLUTION:

$A = S^2. \therefore \sqrt{A} - S.$
$= \sqrt{121} = 11 \text{ m}, \text{ ANS.}$

Rule: *The perimeter of a square is equal to four times the square root of the area.*
Formula: $P = 4\sqrt{A}$ or $P = 4S$, where $P = $ perimeter, $A = $ area, and $S = $ side of a square.

EXAMPLE 4: Find the perimeter of a square whose area is 144 m².

SOLUTION:

$P = 4\sqrt{A}, P = 4 \times 12 = 48 \text{ m}, \text{ ANS.}$

Rule: *The diagonal of a square equals the square root of twice the area.*

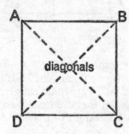

Formula: $D = \sqrt{2A}$

Note: Check back on your right-triangle formula.

EXAMPLE 5: Find the diagonal of a square if the area is 49 cm².

SOLUTION:

$D = \sqrt{2A}, D = \sqrt{98} = 9 \cdot 899, \text{ ANS.}$

SOLUTION by right-triangle formula: $c^2 = a^2 + b^2$, in which c represents the diagonal or hypotenuse, while a and b are the sides. Then

$c^2 = 7^2 + 7^2, c^2 = 98$
$c = \sqrt{98} = 9 \cdot 899 \text{ cm}, \text{ ANS.}$

Any parallelogram can be converted to a rectangle without changing its area. This is shown in the following diagram:

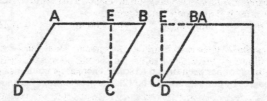

By taking the triangle *EBC* from the figure at the left and changing its position as shown in the figure at the right, we create a rectangle without adding to or deducting from the total area. Hence—

Rule: *The area of a parallelogram is equal to the product of the base times the height.*

EXAMPLE 6: Find the area of a parallelogram whose base is 12 cm and whose height is 8 cm.

SOLUTION: $A = bh. \therefore A = 12 \times 8 = 96$ cm².

Rule: *The area of a trapezium equals half the sum of the parallel sides multiplied by the height.*

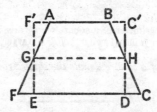

PROOF:
Make a rectangle of the trapezium *ABCF* by drawing a line *GH* ∥ to the two ∥ sides and midway between them. The length of this line is the average of the two ∥ sides *AB* and *FC*. Perpendiculars from the midline *GH* to the larger base *FC* cut off triangles that are exactly equal to the triangles needed above the midline to form a rectangle of the new figure.

Formula: A of trapezium $= \dfrac{B + b}{2} \times h$, in which h is the ⊥ height and B, b are the parallel sides.

EXAMPLE 7: Find the area of a trapezium whose bases are equal to 20 cm and 30 cm and whose height is 15 cm.

SOLUTION: $A = \dfrac{B + b}{2} \times h; A = \dfrac{30 + 20}{2} \times 15$

$= 25 \times 15 = 375$ cm², Ans.

Practice Exercise No. 71

1. An apartment house is rectangular in shape. If its front is 50 m and it goes back 90 m, how far is it all around the house?

 (A) 940 m ......   (B) 280 m ...   (C) 880 m ......   (D) 100 m ......

2. A rectangular hangar is to house an aeroplane. What must its area be if you desire a 6 m allowance on all sides and if the plane is 40 m wide by 30 m long?

 (A) 2184 m² ......   (B) 3064 m² ......   (C) 2964 m² ......   (D) 1998 m² ......

3. How much would it cost to resurface a square plot 75 m long at a cost of 20c a square metre?

 (A) $6000 ......   (B) $1600 ......   (C) $1000 ......   (D) $1125 ......

4. A square field whose area is 1024 m² is to be completely covered by flowerbeds 4 m square. How many beds will be needed to cover the field?

 (A) 32 ......   (B) 56 ......   (C) 64 ......   (D) 84 ......

5. How much barbed wire would be needed to go diagonally across a rectangular piece of land that is 66 m wide by 88 m long?

 (A) 90 m ......   (B) 100 m ......   (C) 110 m ......   (D) 120 m ......

6. Find the length of the diagonal of a square which has an area of 288 m².

 (A) 12 m ......   (B) 17 m ......   (C) 21 m ......   (D) 24 m ......

7. If a moulding cost £0·06 per metre how much would it cost to put a border of moulding round a square room which has an area of 81 m²?

 (A) £2·16 ......   (B) £1 ......   (C) £1·5 ......   (D) £1·8 ......

8. What is the area of the figure below if all the lengths are measured in metres?

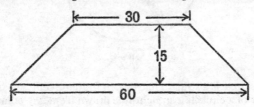

 (A) 450 m² ......   (B) 675 m² ......   (C) 750 m² ......   (D) 2700 m² ......

9. What is the area of the figure below if the lengths are measured in metres?

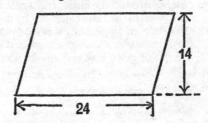

 (A) 168 m² ......   (B) 336 m² ......   (C) 76 m² ......   (D) 206 m² ......

10. What is the area of the figure below if the lengths are measured in metres?

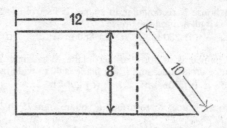

(A) 120 m² ......   (B) 140 m² ......   (C) 160 m² ......   (D) 180 m² ......

CIRCLES

A **circle** is a curved line on which every point is equally distant from a point within called the **centre.**

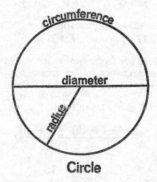

Circle

A **radius** of a circle is a straight line drawn from the centre to the outer edge

The **diameter** of a circle is a straight line drawn from any point on the outer edge through the centre to the outer edge on the opposite side. It is equal to twice any radius.

The **circumference** of a circle is the line representing its outer edge and is equal to the complete distance around the circle. It is analogous to perimeter.

Pi, written π, is the name given to the ratio expressed by dividing the circumference of any circle by its diameter. In quantity it is a constant approximately equal to $3\frac{1}{7}$ or $3 \cdot 1416$. If you measure the distance around any circle, and its diameter, and then divide the distance by the diameter you will always get a result of approximately $3\frac{1}{7}$.

Formula: $\pi = \dfrac{C}{d}$, where C = circumference and d = diameter; or $\pi = \dfrac{C}{2r}$, where r = radius.

Rule: *To find the circumference of a circle multiply the diameter by π.*

Formula: $C = \pi d$; or $C = 2\pi r$.

EXAMPLE 1: The spoke of a wheel is 21 cm. Find its circumference.

SOLUTION: $C = 2\pi r$

$$= 2 \times \frac{22}{7} \times \overset{3}{\underset{1}{21}} = 132 \text{ cm, Ans.}$$

EXAMPLE 2: The circumference of a pulley is 33 cm. What is its diameter?

SOLUTION: $C = \pi d, \therefore d = \dfrac{C}{\pi}$.

$$d = \frac{33}{22} = \overset{3}{\cancel{33}} \times \frac{7}{\underset{2}{22}} = \frac{21}{2} = 10 \cdot 5 \text{ cm, Ans.}$$

Area of a Circle

Rule: *The area of a circle equals one-half the product of the circumference and the radius.*

This can be reasoned informally as follows. Any circle can be cut to form many narrow triangles, as shown in Fig. 5. The altitude of each triangle

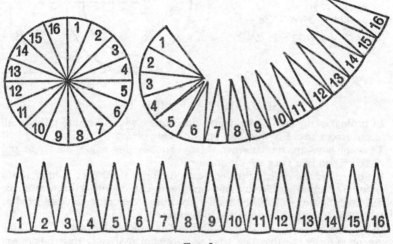

FIG. 5

would be equal to a radius r. The base would be a part of the circumference C. We know the area of each triangle to be equal to $\frac{1}{2}$ the base times the altitude. Since r is the altitude, and the sum of the bases equals the circumference, the area $= \frac{1}{2}r \times C$. Since $C = 2\pi r$, $A = \frac{1}{2}r \times 2\pi r$. $\therefore A = r \times \pi r = \pi r^2$.

Rule: *The area of a circle in terms of the radius is π times the radius squared.*

Formula: $A = \pi r^2$.

EXAMPLE 3: Find the area of a circle that has a 6-metre radius.

SOLUTION: $$A = \pi r^2 = 3 \cdot 1416 \times (6)^2$$
$$= 113 \cdot 10 \text{ m}^2, \text{ Ans.}$$

EXAMPLE 4: The area of a circle is 396 m². Find its radius.

SOLUTION: $A = \pi r^2, \quad \dfrac{A}{\pi} = r^2, \quad \sqrt{\dfrac{A}{\pi}} = r,$

$$r = \sqrt{\frac{396}{\frac{22}{7}}} = \sqrt{396 \times \tfrac{7}{22}} = \sqrt{126} = 11 \cdot 22 \text{ m}$$

Rule: *The area of a circular ring equals the area of the outside circle minus the area of the inside circle.*

Formula: $A = \pi R^2 - \pi r^2$, where $R =$ radius of larger circle and $r =$ radius of smaller circle.

EXAMPLE: In a circular ring the outside diameter is 8 cm and the inside diameter is 6 cm. What is the area of a cross-section of the ring?

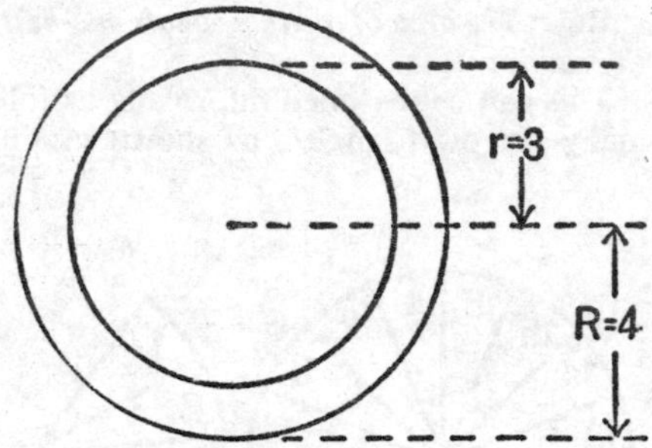

SOLUTION:

$A = \pi R^2 - \pi r^2.$
$D = 8, \therefore R = 4.$
$d = 6, \therefore r = 3.$
$\therefore A = \pi(4^2 - 3^2).$
$A = \tfrac{22}{7}(4^2 - 3^2) = \tfrac{22}{7}(16 - 9)$
$\quad = \dfrac{22}{7} \times 7 = 22 \text{ cm}^2, \text{ Ans.}$

r=3

R=4

SIMILAR PLANE FIGURES

In ordinary language plane figures are similar when they are alike in all respects except size. For instance, all circles are obviously similar.

Two polygons are similar when their corresponding angles are equal and their corresponding sides are proportional.

If the *consecutive* order of the angles is the same it makes no difference if they follow each other clockwise in one figure and counter-clockwise in the other. Such figures will still be similar, because either may be considered as having been reversed like an image in a mirror.

In the case of triangles it is impossible *not* to arrange the angles in the same consecutive order, so that two triangles are similar if only their angles are equal.

In the following diagram all three triangles are similar because they all have the same angles.

The two rhombuses are similar, though the direction of the lines in one reverses that in the other.

All *regular* polygons with a given number of sides are similar.

Rule 1: *If two figures are similar the ratio of any line in one to the corresponding line in the other applies to all the lines that correspond in the two figures.*

Rule 2: *If two figures are similar the ratio of their areas is that of the squares of corresponding lines.*

These rules apply not only to simple geometric figures, but also to drawings, photographs, engravings, blueprints, etc., presenting the greatest complexity

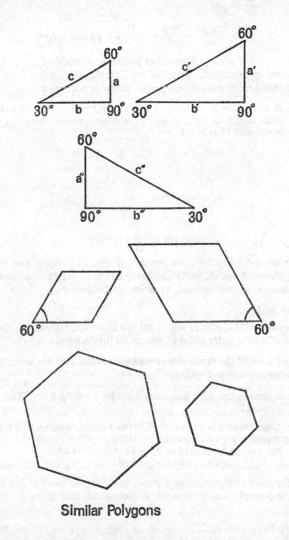

Similar Polygons

of lines. Because of the broad applicability of the rules governing similar polygons, we have generalized the whole subject.

To find the length of any line in a plane figure that is similar to another plane figure, *apply the ratio that exists between any other two corresponding lines.*

EXAMPLE: In a rhombus measuring 4 cm on a side the longer diagonal is 5·5 cm. How long would this diagonal be in a similar rhombus measuring 7 cm on a side?

SOLUTION: $D : d :: S : s.$

$$\frac{D}{5\frac{1}{2}} = \frac{7}{4},$$

$$D = \frac{7 \times 5\frac{1}{2}}{4} = \frac{38\frac{1}{2}}{4} = \frac{77}{2 \times 4} = 9\frac{5}{8} \text{ cm}, \quad \text{ANS.}$$

To find the area of a plane figure that is similar to another plane figure having a known area, *determine the ratio of any two corresponding lines in the two figures and make the required area proportional to the squares of these lines.*

EXAMPLE: A trapezium in which one of the sides measures 6 m has an area of 54 m². What would be the area of a similar trapezium in which a corresponding side measured 15 m?

SOLUTION: $A' : A :: S^2 : s^2,$

$$\frac{A'}{54} = \frac{15^2}{6^2},$$

$$A' = \frac{225 \times 54}{36} = \frac{225 \times 3}{2} = 337 \cdot 5 \text{ m}^2, \quad \text{ANS.}$$

SOLID GEOMETRY

Plane geometry treats of surfaces or of figures having *two* dimensions, namely *length* and *breadth*. **Solid geometry** treats of **solids** or of **bodies** having *three* dimensions, namely *length*, *breadth*, and *thickness*.

Rectangular Solids

A **rectangular solid** is one in which all the faces are rectangles. The **cube** is a special type of rectangular solid in which all the faces are equal.

To find the area of the faces of a rectangular solid, *add the areas of the three different forms of face and multiply by 2.*

EXAMPLE: A rectangular solid measures 6 cm by 4 cm by 3 cm. What is the total area of its faces?

SOLUTION: It has two faces measuring 6 cm by 4 cm, two measuring 6 cm by 3 cm, and two measuring 4 cm by 3 cm.
$$(6 \times 4) + (6 \times 3) + (4 \times 3) = 54 \text{ cm}^2$$
$$2 \times 54 = 108 \text{ cm}^2, \quad \text{ANS.}$$

To find the area of the faces of a cube, *multiply the area of one face by six.*
To find the cubical contents of a rectangular solid, *multiply together the three dimensions.*

EXAMPLE: What are the cubical contents of a box measuring 110 mm by 60 mm by 45 mm?

SOLUTION: $110 \times 60 \times 45 = 297\,000 \text{ mm}^3, \quad \text{ANS.}$

With solids other than rectangular ones we consider the surfaces and areas of the sides as distinct from those of the bottom and top (if any). We call the area of the sides the **lateral area** and speak of the top as well as the bottom as **bases.**

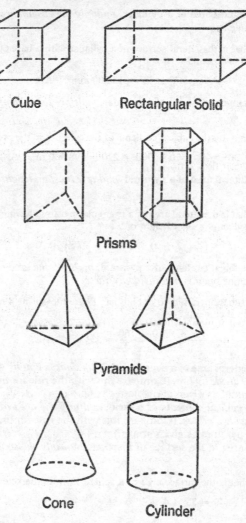

Cube   Rectangular Solid

Prisms

Pyramids

Cone   Cylinder

To find the lateral area of a prism, *multiply the perimeter of one of the bases by the height.*

EXAMPLE: A prism 6 cm high has as its base an equilateral triangle measuring $1\frac{1}{2}$ cm on a side. What is its lateral area?

SOLUTION: $(1\frac{1}{2} + 1\frac{1}{2} + 1\frac{1}{2}) \times 6 = 27 \text{ cm}^2$, ANS.

To find the cubical contents of a prism, *multiply the area of one of the bases by the height.*

EXAMPLE: What are the cubical contents of a prism 8 cm high if the area of one of the bases is $3 \cdot 75 \text{ cm}^2$?

SOLUTION: $3 \cdot 75 \times 8 = 30 \text{ cm}^3$, ANS.

To find the lateral area of a cylinder, *multiply the circumference of one of the bases by the height.*

EXAMPLE: What is the lateral surface of a cylinder with a base 60 mm in diameter if its height is 70 mm?

SOLUTION: $60 \times \frac{22}{7} \times 70 = 13\,200$ mm², ANS.

To find the cubical contents of a cylinder, *multiply the area of one of the bases by the height.*

EXAMPLE: What are the cubical contents of the cylinder in the preceding example?

SOLUTION $30^2 \times \frac{22}{7} \times 70 = 900 \times 220 = 198\,000$ m³, ANS.

To find the lateral area of a pyramid, *multiply its slant height by the perimeter and divide by two.*

EXAMPLE: What is the lateral area of a triangular pyramid having a base measuring 2 m on a side and a slant height of 9 m?

SOLUTION: $(2 + 2 + 2) \times 9 \div 2 = 27$ m², ANS.

To find the cubical contents of a pyramid, *multiply the area of the base by the* altitude (not slant height) *and divide by three.*

EXAMPLE: A square pyramid 10 m high has a base measuring 4 m on a side. What are its cubical contents?

SOLUTION: $\dfrac{4 \times 4 \times 10}{3} = \dfrac{160}{3} = 53\frac{1}{3}$ m³, ANS.

To find the lateral area of a cone, *multiply its slant height by the circumference of the base and divide by two.* (Compare this with the rule for finding the lateral area of a pyramid as given above.)

To find the cubical contents of a cone, *multiply the area of the base by the altitude and divide by three.* (Compare this with the rule for finding the cubical contents of a pyramid as given above.)

To find the area of the surface of a sphere, *multiply the square of the radius by* 4π.

EXAMPLE: What is the surface area of a sphere 12 m in diameter?

SOLUTION: $6^2 \times 4\pi = 36 \times 4 \times \frac{22}{7} = \frac{3168}{7} = 452\frac{4}{7}$ m², ANS.

To find the cubical contents of a sphere, *multiply the cube of the radius by* $\dfrac{4\pi}{3}$.

EXAMPLE: What are the cubical contents of a sphere 12 m in diameter?

SOLUTION: $6^3 \times \dfrac{4\pi}{3} = \dfrac{216 \times 4 \times 22}{3 \times 7} = \dfrac{6336}{7} = 905\frac{1}{7}$ m³, ANS.

Practice Exercise No. 72

1. The diameter of a wheel is 28 cm. What is its circumference?

(A) 66 cm ......   (B) 77 cm ......   (C) 88 cm ......   (D) 99 cm ......

2. The circumference of a wheel is 110 mm. How long is one of its spokes?

(A) 35 mm ......   (B) 17½ mm ......   (c) 15 mm ......   (D) 12½ mm ......

3. To make a circular coil for a magnet you need 49 turns of wire. How much wire will you need if the diameter of the coil is 4 cm?

(A) 12⁴⁄₇ cm ......   (B) 84⁴⁄₇ cm ......   (c) 324 cm ......   (D) 616 cm ......

4. How many square metres of tin are needed for the top of a tank that is 1·4 m in diameter?

(A) 6·16 m² ......   (B) 3·08 m² ......   (c) 4·62 m² ......   (D) 1·54 m² ......

5. You have a circular grazing field 96 m in diameter, which is roped around. Concentric with that you have a circular trotting track 128 m in diameter. How much will it cost to regravel the trotting track at a price of £0·10 per square metre?

(A) £426 ......   (B) £563·2 ......   (c) £826·4 ......   (D) £968·2 ......

6. The area of canvas needed to just cover the muzzle of a cannon is 50¼ cm². What is the diameter of the muzzle?

(A) 12 cm ......   (B) 8 cm ......   (c) 6 cm ......   (D) 5 cm ......

7. If the radius of a circle is twice as great as the radius of a smaller circle, how many times as large will the area of the greater circle be than the area of the smaller circle?

(A) 2 ......   (B) 4 ......   (c) 6 ......   (D) 8 ......

8. You have 4 circular garden plots, each having a 1·4 m radius. What must the radius be of one large circular plot that will have as much area as the four combined?

(A) 2·8 m ......   (B) 2·1 m ......   (c) 2·0 m ......   (D) 6·4 m ......

9. How much will it cost to resurface a circular swimming tank that has a diameter of 5·6 m if surfacing costs £25 a square metre?

(A) £154 ......   (B) £308 ......   (c) £462 ......   (D) £616 ......

10. If you wish to convert a circular field that has a diameter of 56 m to a square field with the same area, how long will a side of the square be?

(A) 28 m ......   (B) 36·8 m ......   (c) 49·6 m ......   (D) 46 m ......

CHAPTER FIFTEEN

TRIGONOMETRY

Trigonometry is the branch of mathematics that deals with the measurement of triangles. (The word *trigonometry* comes from the Greek and means *to measure a triangle*.) Trigonometry enables us to find the unknown parts of triangles by arithmetical processes. For this reason it is constantly used in surveying, mechanics, navigation, engineering, physics and astronomy.

From geometry you learned that there are many shapes of triangles. For our purpose we can start with the simple case of a right triangle. Starting from this, you will eventually be able to work with all types of triangles, because any triangle can be broken down into two right triangles.

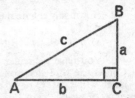

In the right triangle *BAC* you know from geometry that

(a) $\angle A + \angle B = 90$
(b) $c^2 = a^2 + b^2$

From equation (*a*) you can find one of the acute angles if the other is given, and from equation (*b*) you can determine the length of any side if the other two are given. But as yet you do not have a method for finding angle *A* if given the two sides *a* and *b*, even though by geometry you could construct the triangle with this information. And this is where trigonometry makes its contribution. It gives you a method for calculating the angles if you know the sides or for calculating the sides if you know the angles.

Trigonometric Functions of an Angle

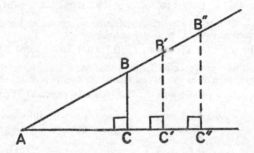

If we take the triangle in the previous figure and extend lines *AB* and *AC*, and then drop perpendiculars from points *B'* and *B''* to *AC*, we form three similar triangles:

$$\triangle CAB, \quad \triangle C'AB', \quad \text{and} \quad \triangle C''AB''$$

When two triangles are similar the ratio of any two sides of one triangle equals the ratio of corresponding sides of the second triangle. Thus, in the three triangles of the figure

$$\frac{BC}{AC} = \frac{B'C'}{AC'} = \frac{B''C''}{AC''}$$

or

$$\frac{BC}{AB} = \frac{B'C'}{AB'} = \frac{B''C''}{AB''}$$

Similar equalities hold for the ratios between the other sides of the triangles.

These equalities between the ratios of the corresponding sides of similar triangles illustrate the fact that *no matter how the size of a right triangle may vary, the values of the ratios of the sides remain the same so long as the acute angles are unchanged.* In other words, each of the above ratios is a **function** of angle *A*.

From algebra and geometry we learn that a variable quantity which depends upon another quantity for its value is called a **function** of the latter value.

Therefore in the above figure the value of the ratio $\dfrac{BC}{AC}$ is a function of the magnitude of angle *A*; and as long as the magnitude of angle *A* remains the same, the value of the ratio $\dfrac{BC}{AC}$ will be the same.

Description of the Tangent Function

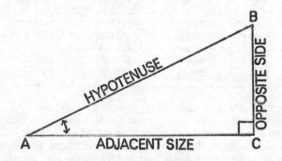

The constant ratio or function, $\dfrac{BC}{AC}$, is termed the **tangent** of angle *A*. It will be noted that this function represents the ratio of the side *opposite* angle *A* divided by the side next to angle *A*, called the *adjacent* side—that is, the side next to it other than the hypotenuse. Accordingly,

$$\text{tangent } \angle A = \frac{\textbf{opposite side}}{\textbf{adjacent side}}$$

or

$$\tan A = \frac{\textbf{opp.}}{\textbf{adj.}}$$

Making a Table of Trigonometric Functions

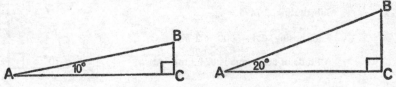

If you construct $\angle A$ equal to 10° and measure BC and AC and then compute the value of $\dfrac{BC}{AC}$, you will find it to be 0·176. Then if you construct $\angle A$ to equal 20° you will find $\dfrac{BC}{AC}$ equal to 0·364. For $\angle A$ at 30° you will find $\dfrac{BC}{AC}$ equal to 0·577. This means that thereafter you will know that the *tangent* of any angle of 10° in a right triangle is equal to 0·176, and the tangent of any angle of 20° is equal to 0·364. Thus, by computing the values of the ratios of $\dfrac{BC}{AC}$ for all angles from 1° to 90° you would obtain a complete table of tangent values. A sample of such a table is shown below.

Sample Table of Trigonometric Functions

Angle	Sine	Cosine	Tangent
68	0·9272	0·3746	2·4751
69	0·9336	0·3584	2·6051
70	0·9397	0·3420	2·7475
71	0·9455	0·3256	2·9042
72	0·9511	0·3090	3·0777

This sample table and the more complete table on p. 179 give the tangents of angles to four decimal places. For instance, in the table above, to find the value of the tangent of an angle of 69° you first look in the column head *Angle* and find 69°. Then on the same horizontal line in the column headed *Tangent* you find the value 2·6051. This means that tan 69° = 2·6051.

The following example will show how you can solve problems in trigonometry by the use of the table of tangents.

EXAMPLE: An aeroplane is sighted by two observers. One observer at A indicates it to be directly overhead. The other observer at B, 300 m due west of A, measures its angle of elevation (*see below*) at 70°. What is the altitude of the aeroplane?

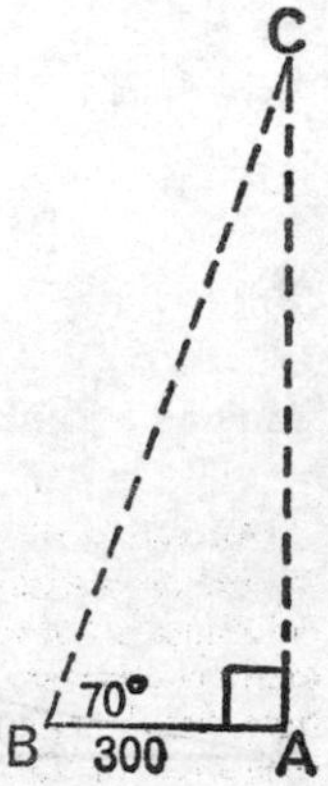

SOLUTION:

$$\tan \angle B = \frac{\text{(opp. side)}}{\text{(adj. side)}} = \frac{CA}{BA}$$

Since $\quad \angle B = 70°$
$\tan \angle B = 2·7475$
(*see table above*)

Substituting, $2·7475 = \dfrac{CA}{300}$.

Transposing, $CA = 300 \times 2·7475$
$\qquad\qquad = 824·25$ m
Altitude of aeroplane is 824·3 m, ANS.

Practical Observation of Angles

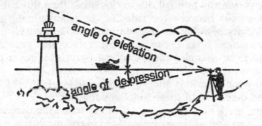

The *angle of elevation or depression* of an object is the angle made between a line from the eye to the object and a horizontal line in the same vertical plane. If the object is above the horizontal line it makes an *angle of elevation*; if below the horizontal line it makes an *angle of depression*.

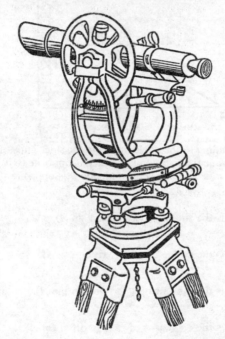

Courtesy of Keuffel & Esser Co., New York

For measuring both vertical and horizontal angles out of doors an engineer's *transit* or *theodolite* is used. As may be seen from the illustration, the instrument combines a telescope with a horizontal and a vertical plate, each of which is graduated by degrees, minutes, and seconds. By moving the telescope to right or left, horizontal angles can be measured on the horizontal plate. Vertical angles are measured on the vertical disc by moving the telescope up and down.

The Six Trigonometric Functions

As has been previously pointed out, ratios other than those involved in the *tangent function* exist between the sides of the triangle, and have, like the tangent, an equality of value for a given magnitude of angle, irrespective of the size of the triangle. It is to be expected, therefore, that problems involving the solution of right triangles can be solved by other known trigonometric ratios or functions of the self-same angle. As a matter of fact, there exist six important ratios or functions for any acute angle of a right triangle. The description and definition of these functions follows.

The sides and angles of triangle *CAB* in the following diagram have been marked in the manner traditionally employed in trigonometry. It is the custom to have the angles represented by capital letters and the sides indicated by the

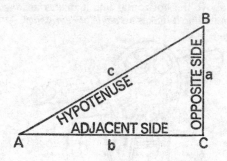

small letter corresponding to the angle opposite the side. Thus, the right angle is designated by *C*, while the hypotenuse, which is opposite to it, is designated by *c*. Similarly, side *a* is opposite $\angle A$, and side *b* is opposite $\angle B$. Thus, we have these six ratios:*

$\dfrac{a}{c}$ is the **sine** of $\angle A$ (written **sin** *A*).

$\dfrac{b}{c}$ is the **cosine** of $\angle A$ (written **cos** *A*).

$\dfrac{a}{b}$ is the **tangent** of $\angle A$ (written **tan** *A*).

$\dfrac{b}{a}$ is the **cotangent** of $\angle A$ (written **cot** *A*).

$\dfrac{c}{b}$ is the **secant** of $\angle A$ (written **sec** *A*).

$\dfrac{c}{a}$ is the **cosecant** of $\angle A$ (written **cosec** *A*).

---

* Two additional functions which are little used are: **versed sine** $\angle A = 1 - \cos A$ (written **vers** *A*), and **coversed sine** $\angle A = 1 - \sin A$ (written **covers** *A*).

Using self-explanatory abbreviations, we thus have by definition:

$$\sin A = \frac{\text{opp.}}{\text{hyp.}} = \frac{a}{c}, \qquad \cos A = \frac{\text{adj.}}{\text{hyp.}} = \frac{b}{c},$$

$$\tan A = \frac{\text{opp.}}{\text{adj.}} = \frac{a}{b}, \qquad \cot A = \frac{\text{adj.}}{\text{opp.}} = \frac{b}{a},$$

$$\sec A = \frac{\text{hyp.}}{\text{adj.}} = \frac{c}{b}, \qquad \text{cosec } A = \frac{\text{hyp.}}{\text{opp.}} = \frac{c}{a},$$

This table of definitions of the trigonometric functions should be committed to memory.

Practice Exercise No. 73

1. In the preceding figure, $\tan B = \frac{b}{a}$. Write the other five functions of $\angle B$.
2. Which is greater, $\sin A$ or $\tan A$?
3. Which is greater, $\cos A$ or $\cot A$?
4. Which is greater, $\sec A$ or $\tan A$?
5. Which is greater, $\text{cosec } A$ or $\cot A$?
6. $\sin A = \frac{3}{5}$. What is the value of $\cos A$? *Hint:* Use rt. $\triangle$ formula $c^2 = a^2 + b^2$ to find side b.
7. $\tan A = \frac{3}{4}$. What is the value of $\sin A$?
8. $\sin A = \frac{8}{17}$. Find $\cos A$.
9. $\cot A = \frac{15}{8}$. Find $\sec A$.
10. Find the value of the other five functions of A if $\sin A = \frac{5}{13}$.

Relations between Functions of Complementary Angles

If you observe the relations between the functions of the two acute angles of the same right triangle you will note that every function of each of the two acute angles is equal to a different function of the other acute angle. These correspondences of value are demonstrated in the following.

$$\sin A = \frac{a}{c} \quad \text{and} \quad \cos B = \frac{a}{c},$$

$$\cos A = \frac{b}{c} \quad \text{and} \quad \sin B = \frac{b}{c},$$

$$\tan A = \frac{a}{b} \quad \text{and} \quad \cot B = \frac{a}{b}, \text{ etc.}$$

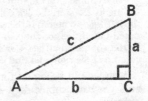

Thus we have:

$\sin A = \cos B$	$\cot A = \tan B$
$\cos A = \sin B$	$\sec A = \text{cosec } B$
$\tan A = \cot B$	$\text{cosec } A = \sec B$

From these equalities it will be evident that any function of an acute angle of a right triangle equals the co-function of the complement of that angle.*

For example, $\tan 40° = \cot 50°$; $\sin 70° = \cos 20°$; $\text{cosec } 41° 20' = \sec 48° 40'$.

---

* The name *cosine* means *complement's sine*. It is a contraction from the Latin *complementi sinus*. The words *cotangent* and *cosecant* were derived in the same manner.

Since angles A and B are complementary, another way of writing these equations is as follows:

$$\sin (90° - A) = \cos A \qquad \cot (90° - A) = \tan A$$
$$\cos (90° - A) = \sin A \qquad \sec (90° - A) = \csc A$$
$$\tan (90° - A) = \cot A \qquad \csc (90° - A) = \sec A$$

Practice Exercise No. 74

Fill in the blanks in examples 1–6 with the equivalent co-functions

1. $\sin 26° =$
2. $\tan 43° =$
3. $\cos 24° \, 28' =$
4. $\cot 88° \, 50' =$
5. $\sec 6° \, 10' =$
6. $\csc 77\frac{1}{2}° =$
7. How many degrees must $\angle A$ be if $90° - A = 5A$?
8. What is the value of $\angle A$ if $\tan A = \cot A$?
9. Find A if $90° - A = A$.
10. Find A if $\cos A = \sin 2A$.

How to Use a Table of Trigonometric Functions

From the foregoing it becomes apparent that you can easily compute the functions of any angle greater than 45° if you know the functions of all angles between 0° and 45°. Therefore in a table of trigonometric functions, such as appears on the next page, it is only necessary to have a direct table of functions for angles from 0° to 45°, since the function of any angle above 45° is equal to the co-function of its complement.

To find the functions of angles from 0° to 45° read the table from the top down, using the values of angles at the left and the headings at the top of the table. To find the functions of angles from 45° to 90° read from the bottom up, using the values of angles at the right and the function designations at the bottom of the table.

If you know the value of the function of an angle and wish to find the angle, look in the body of the table in the proper column and then read the magnitude of the angle in the corresponding row of one or the other of the angle columns.

For example, you are told that the sine of a certain angle is ·5000 and wish to find the angle. Look in the *Sin* column, locate ·5000 and read the angle value (30°) from the left *Angle* column. If this value had been given to you as a cosine you would have noted that it does not appear in the column headed *Cos* at the top, but does appear in the column that has *Cos* at the bottom. Hence you would then use the *Angle* column at the right and find ·5000 to be the cosine of 60°.

You should become thoroughly familiar with the use of the table. To this end you can supplement the following exercise by making up your own examples.

Further familiarity with the table of functions will indicate the following about variations of the trigonometric functions.

As an angle increases from 0° to 90°, its:

sine	*increases* from	0 to 1,
cosine	*decreases* from	1 to 0,
tangent	*increases* from	0 to ∞,
cotangent	*decreases* from	∞ to 0,
secant	*increases* from	1 to ∞,
cosecant	*decreases* from	∞ to 1.

Table VII. Table of Natural Trigonometric Functions
For explanation of the use of this table see preceding page.

Angle	Sin	Cos	Tan	Cot	Sec	Cosec	
0°	·0000	1·0000	·0000	∞	1·0000	∞	90°
1	·0175	·9998	·0175	57·2900	1·0002	57·2987	89
2	·0349	·9994	·0349	28·6363	1·0006	28·6537	88
3	·0523	·9986	·0524	19·0811	1·0014	19·1073	87
4	·0698	·9976	·0699	14.3007	1·0024	14·3356	86
5°	·0872	·9962	·0875	11·4301	1·0038	11·4737	85°
6	·1045	·9945	·1051	9·5144	1·0055	9·5668	84
7	·1219	·9925	·1228	8·1443	1·0075	8·2055	83
8	·1392	·9903	·1405	7·1154	1·0098	7·1853	82
9	·1564	·9877	·1584	6·3138	1·0125	6·3925	81
10°	·1736	·9848	·1763	5·6713	1·0154	5·7588	80°
11	·1908	·9816	·1944	5·1446	1·0187	5·2408	79
12	·2079	·9781	·2126	4·7046	1·0223	4·8097	78
13	·2250	·9744	·2309	4·3315	1·0263	4·4454	77
14	·2419	·9703	·2493	4·0108	1·0306	4·1336	76
15°	·2588	·9659	·2679	3·7321	1·0353	3·8637	75°
16	·2756	·9613	·2867	3·4874	1·0403	3·6280	74
17	·2924	·9563	·3057	3·2709	1·0457	3·4203	73
18	·3090	·9511	·3249	3·0777	1·0515	3·2361	72
19	·3256	·9455	·3443	2·9042	1·0576	3·0716	71
20°	·3420	·9397	·3640	2·7475	1·0642	2·9238	70°
21	·3584	·9336	·3839	2·6051	1·0711	2·7904	69
22	·3746	·9272	·4040	2·4751	1·0785	2·6695	68
23	·3907	·9205	·4245	2·3559	1·0864	2·5593	67
24	·4067	·9135	·4452	2·2460	1·0946	2·4586	66
25°	·4226	·9063	·4663	2·1445	1·1034	2·3662	65°
26	·4384	·8988	·4877	2·0503	1·1126	2·2812	64
27	·4540	·8910	·5095	1·9626	1·1223	2·2027	63
28	·4695	·8829	·5317	1·8807	1·1326	2·1301	62
29	·4848	·8746	·5543	1·8040	1·1434	2·0627	61
30°	·5000	·8660	·5774	1·7321	1·1547	2·0000	60°
31	·5150	·8572	·6009	1·6643	1·1666	1·9416	59
32	·5299	·8480	·6249	1·6003	1·1792	1·8871	58
33	·5446	·8387	·6494	1·5399	1·1924	1·8361	57
34	·5592	·8290	·6745	1·4826	1·2062	1·7883	56
35°	·5736	·8192	·7002	1·4281	1·2208	1·7434	55°
36	·5878	·8090	·7265	1·3764	1·2361	1·7013	54
37	·6018	·7986	·7536	1·3270	1·2521	1·6616	53
38	·6157	·7880	·7813	1·2799	1·2690	1·6243	52
39	·6293	·7771	·8098	1·2349	1·2868	1·5890	51
40°	·6428	·7660	·8391	1·1918	1·3054	1·5557	50°
41	·6561	·7547	·8693	1·1504	1·3250	1·5243	49
42	·6691	·7431	·9004	1·1106	1·3456	1·4945	48
43	·6820	·7314	·9325	1·0724	1·3673	1·4663	47
44	·6947	·7193	·9657	1·0355	1·3902	1·4396	46
45°	·7071	·7071	1·0000	1·0000	1·4142	1·4142	45°
	Cos	Sin	Cot	Tan	Cosec	Sec	Angle

Also note that:

> sines and cosines are never > 1,
> secants and cosecants are never < 1,
> tangents and cotangents may have any value from 0 to ∞.*

Practice Exercise No. 75

From the table of trigonometric functions find the values required in examples 1–15:

1. sin 8°	2. sin 42°	3. tan 40°	4. cot 63°	5. sec 22°
6. cos 25°	7. cosec 14°	8. sin 78°	9. cot 69°	10. sec 81°
11. cos 62°	12. tan 56°	13. sin 58°	14. cos 45°	15. sin 30°

16. Find the angle whose sine is 0·2588.
17. Find the angle whose tangent is 0·7002.
18. Find the angle whose cosine is 0·5000.
19. Find the angle whose secant is 2·9238.
20. Find the angle whose cotangent is 5·6713.

Functions of 45°, 30°, and 60° Angles

For some rather common angles the exact values of their functions can be easily found by the application of elementary principles of geometry.

Functions of a 45° Angle

In the isosceles right triangle ACB, if $\angle A = 45°$, then $\angle B = 45°$, and therefore side a = side b. Now if we let side a equal 1 or unity, then from the right triangle formula of

$$a^2 + b^2 = c^2$$

we get

$$c = \sqrt{1 + 1} \text{ or } \sqrt{2}$$

(taking the square root of both sides of the equation). Now since any trigonometric function of an acute angle is equal to the corresponding co-function of its complement, therefore

$$\sin 45° = \frac{1}{\sqrt{2}} \text{ or } \tfrac{1}{2}\sqrt{2} = \cos 45°$$

$$\tan 45° = \frac{1}{1} \text{ or } 1 = \cot 45°$$

$$\sec 45° = \frac{\sqrt{2}}{1} \text{ or } \sqrt{2} = \csc 45°$$

* The symbol ∞ denotes 'infinity' and is used in mathematics to represent a number that is indefinitely large, or larger than any preassignable quantity.

Functions of 30° *and* 60° *Angles*

In the equilateral triangle *ABD* the three sides are equal and the three angles each equal 60°. If we drop a perpendicular from *B* to *AD* it bisects ∠*B* and the base *AD* at *C*.

If we let the length of each of the sides equal 2 units, then *AC* = *CD* = 1; and in the right triangle *ACB*

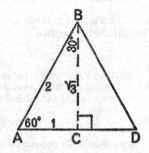

∠*B* = 30°, ∠*C* = 90°, ∠*A* = 60°
AC = 1, *AD* = 2

Then, since $(AB)^2 = (AC)^2 + (BC)^2$, it follows that $(BC)^2 = 3$ and $BC = \sqrt{3}$.
Thus in the right triangle *ACB*

$$\sin 30° = \tfrac{1}{2} \qquad\qquad = \cos 60°$$

$$\tan 30° = \frac{1}{\sqrt{3}} \quad \text{or} \quad \tfrac{1}{3}\sqrt{3} = \cot 60°$$

$$\sec 30° = \frac{2}{\sqrt{3}} \quad \text{or} \quad \tfrac{2}{3}\sqrt{3} = \operatorname{cosec} 60°$$

$$\cos 30° = \frac{\sqrt{3}}{2} \qquad\qquad = \sin 60°$$

$$\cot 30° = \frac{\sqrt{3}}{1} \quad \text{or} \quad \sqrt{3} = \tan 60°$$

$$\operatorname{cosec} 30° = \frac{2}{1} \quad \text{or} \quad 2 \quad = \sec 60°$$

It is an advantage to know the values of the 30°, 45°, and 60° angles by heart. To help yourself memorize them, fill in the outline of the table below with the proper values of the functions.

Function	30°	60°	45°
Sine			
Cosine			
Tangent			
Cotangent			
Secant			
Cosecant			

Clearly this table is inadequate when values such as sin 42° 33′, cot 26° 14′, etc., are required. We shall therefore use tables for angles which included minutes of arc.

Since cos A = sin (90–A), a sine table will do for cosines.
Since cot A = tan (90–A), a tangent table will do for cotangents.
Since cosec A = sec (90–A), a secant table will do for cosecants.

Consequently, three sets of tables will give the values of all the trigonometric ratios we need.

EXAMPLE 1: To find sin 42° 33′.

SOLUTION: Locate 42° in the degree column, move across the table into the column headed 30′. We have so far found sin 42° 30′ = 0·6756. To obtain the final result move across the table into the column headed 3′ and *add* the number found, i.e. 6.

$$\therefore \text{ sin } 42° 33′ = 0·6762, \quad \text{Ans.}$$

EXAMPLE 2: To find cot 26° 14′.

SOLUTION: We are using tangent tables and therefore the result cot 26° 14′ = tan 63° 46′. To find tan 63° 46′ locate 63 in the degree column and move across the table into the column headed 42′. We have so far found tan 63° 42′ = 2·0233 (notice that the darker type warns us that the whole number has increased by 1). To obtain the final result move into the column headed 4′ and *add* the number found, i.e. 58.

$$\therefore \text{ tan } 63° 46′ = 2·0291$$
$$\text{Hence cot } 26° 14′ = 2·0291, \quad \text{Ans.}$$

EXAMPLE 3: To find cosec 14° 23′.

SOLUTION: We are using secant tables, and therefore the result sec 75° 37′ = cosec 14° 23′. To find sec 75° 37′ locate 75 in the degree column, move across the table to the column headed 36′. We have so far found

$$\text{sec. } 75° 36′ = 4·0211$$

Moving into the column headed 1′, we obtain

$$\text{sec. } 75° 37′ = 4·0256$$
$$\text{Hence cosec } 14° 23′ = 4·0256, \quad \text{Ans.}$$

EXAMPLE 4: Find the value of A when cos A = 0·9419.

SOLUTION: We are using sine tables, therefore we need the result cos A = sin (90–A) = 0·9419. Locate within the body of the table the number nearest to 0·9419 either equal to or less than 0·9419.

This gives an angle of 70° 18′ so far, because 0·9415 is the number located. We need a 4 which occurs in the 4′ column.

$$\therefore \text{ sin } 70° 22′ = 0·9419$$
$$\text{Hence cos } 19° 38′ = 0·9419$$
$$\text{and } \angle A = 19° 38′, \quad \text{Ans.}$$

Practice Exercise No. 76

Find the values of the functions in examples 1–5:

1. sin 15° 30′
2. cos 25° 40′
3. tan 47° 10′
4. cot 52° 30′
5. sec 40° 30′

Find the value of $\angle A$ to the nearest minute in examples 6–10:

6. $\sin A = 0.0901$ 7. $\tan A = 0.3411$ 8. $\cos A = 0.4173$
9. $\cot A = 0.8491$ 10. $\operatorname{cosec} A = 2.3662$

Reciprocals among the Functions

If you inspect the ratios of the six functions of $\angle A$ you will readily note that they are not independent of each other. In fact, if you line them up as follows:

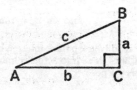

$$\sin A = \frac{a}{c} \qquad \operatorname{cosec} A = \frac{c}{a}$$

$$\cos A = \frac{b}{c} \qquad \sec A = \frac{c}{b}$$

$$\tan A = \frac{a}{b} \qquad \cot A = \frac{b}{a}$$

it becomes obvious that *the sine is the reciprocal of the cosecant, the cosine is the reciprocal of the secant, and the tangent is the reciprocal of the cotangent.* Accordingly,

$$\sin A = \frac{1}{\operatorname{cosec} A} \qquad \cos A = \frac{1}{\sec A}$$

$$\tan A = \frac{1}{\cot A} \qquad \operatorname{cosec} A = \frac{1}{\sin A}$$

$$\sec A = \frac{1}{\cos A} \qquad \cot A = \frac{1}{\tan A}$$

Therefore:
$$\sin A \times \operatorname{cosec} A = 1 \qquad \cos A \times \sec A = 1$$

$$\tan A \times \cot A = 1$$

In accordance with the usual algebraic method of notation (by which ab is equivalent to $a \times b$) these relationships are usually written:

$$\sin A \operatorname{cosec} A = 1 \qquad \cos A \sec A = 1$$

$$\tan A \cot A = 1$$

To illustrate such a relation, find, for example, in the table of functions the tangent and the cotangent of $30°$.

$$\tan 30° = 0{\cdot}5774, \qquad \cot 30° = 1{\cdot}7321$$

$$\tan 30° \cot 30° = 0{\cdot}5774 \times 1{\cdot}7321$$
$$= 1{\cdot}00011454$$
$$= 1{\cdot}0001 \text{ 4 dec. pls.}$$

Interrelations among the Functions

Since $\tan A = \dfrac{a}{b}$, $\sin A = \dfrac{a}{c}$, and $\cos A = \dfrac{b}{c}$, it follows that

$$\tan A = \frac{\sin A}{\cos A}, \quad \text{and } \sin A = \tan A \cos A.$$

The student will the more readily grasp these interrelations if instead of considering only abstract values he translates these into actual numbers. The 3–4–5 right triangle in the diagram will serve this purpose.

From the interrelations of sine, cosine, and tangent it follows that if we know two of these values we can always find the third.

From the Pythagorean theorem of the right triangle we know that $a^2 + b^2 = c^2$. If we divide both sides of this equation by c^2 we get

$$\frac{a^2}{c^2} + \frac{b^2}{c^2} = 1$$

Since $\dfrac{a}{c} = \sin A$ and $\dfrac{b}{c} = \cos A$, it follows that

(1) $\sin^2 A + \cos^2 A = 1$*

Therefore

(2) $\sin A = \sqrt{1 - \cos^2 A}$ and

(3) $\cos A = \sqrt{1 - \sin^2 A}$

Making Practical Use of the Functions

With the information on trigonometry outlined in the previous pages you will be able to solve many triangles if you know three parts, one of which is a side. And in the case of the right triangle, since the right angle is a part of it, you need only to know two other parts, one of which must be a side.

As will be brought out in the practice exercises that follow, these trigonometric methods of solving triangles are used daily in handling problems that arise in military operations, engineering, navigation, shop-work, physics, surveying, etc.

* $(\sin A)^2$ is customarily written as $\sin^2 A$, and likewise for the other functions.

You should adopt a planned method of procedure in solving problems. One such method is as follows:

1. After reading the problem, draw a figure to a convenient scale, and in it show those lines and angles which are given and those which are to be found.

2. Write down all the formulae that apply to the particular problem.

3. Substitute the given data in the proper formulae, and solve for the unknowns.

4. Check your results.

Incidentally we would suggest that you work with a hard lead pencil or a fine-pointed pen. Nothing is of greater help to accuracy in mathematics than neatness of work, and neatness is next to impossible if you use writing instruments that make thick lines and sprawly figures.

Applying the Sine Function,

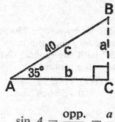

$$\sin A = \frac{\text{opp.}}{\text{hyp.}} = \frac{a}{c}$$

EXAMPLE 1: In the accompanying figure $c = 40$ and $\angle A = 35°$. Find a.

SOLUTION:
$$\frac{a}{c} = \sin A, \quad a = c \sin A,$$

$$\sin 35° = 0\cdot 5736, \quad c = 40,$$
$$c \sin A = 40 \times 0\cdot 5736 = 22\cdot 944,$$
$$a = 22\cdot 944, \quad \text{ANS.}$$

CHECK:
$$\frac{a}{c} = \sin A.$$

$$\frac{22\cdot 944}{40} = 0\cdot 5736, \text{ which is } \sin 35°.$$

EXAMPLE 2: Given $c = 48$ and $\angle B = 22°$, find a by means of the sine formula.

SOLUTION:
$$\frac{a}{c} = \sin A, \quad a = c \sin A,$$

$$\angle A = 90° - \angle B, \quad \angle A = 90° - 22° = 68°,$$
$$\sin 68° = 0\cdot 9272, \quad c = 48,$$
$$c \sin A = 48 \times 0\cdot 9272 = 44\cdot 5056,$$
$$a = 44\cdot 50 +, \quad \text{ANS.}$$

CHECK:
$$\frac{a}{c} = \sin A.$$

$$\frac{44\cdot 5056}{48} = 0\cdot 9272, \text{ which is } \sin 68°.$$

Practice Exercise No. 77

The problems in this exercise should be solved by using the sine function. Answers need be accurate only to the first decimal place.

1. Given $c = 100$, $\angle A = 33°$, find a.
2. Given $c = 10$, $\angle A = 20°$, find a.
3. Given $a = 71$, $c = 78$, find $\angle A$.
4. Given $a = 14$, $\angle A = 28°$, find c.
5. Given $c = 50$, $a = 36$, find $\angle A$.

6. An aeroplane is 405 m above a landing field when the pilot cuts out his motor. He glides to a landing at an angle of 13° with the field. How far will he glide in reaching the field?

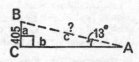

(A) 300 m ......   (B) 1248 m ......   (C) 1800 m ......   (D) 1641 m ......

7. An ascension balloon is moored by a rope 150 m long. A wind blowing in an easterly direction keeps the rope taut and causes it to make an angle of 50° with the ground. What is the vertical height of the balloon from the ground?

(A) 180 m ......   (B) 114·9 m ......
(C) 177·5 m ......   (D) 189·4 m ......

8. A carpenter has to build a ramp to be used as a loading platform for a carrier aeroplane. The height of the loading door is 1·2 m, and the required slope or gradient of the ramp is to be 18°. How long must the ramp be?

(A) 2·4 m   (B) 3·883 m ......   (C) 4·842 m ......   (D) 10·14 m ......

9. The fire department has a new 20 m ladder. The greatest angle at which it can be placed against a building with safety is at 71° with the ground. What is the maximum vertical height that the ladder can reach?

(A) 18·91 m ......   (B) 20·94 m ......
(C) 30 m ......   (D) 16·23 m ......

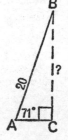

10. A road running from the bottom of a hill to the top is 625 m long. If the hill is 54½ m high, what is the angle of elevation of the road?

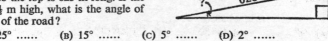

(A) 25° ......   (B) 15° ......   (C) 5° ......   (D) 2° ......

Applying the Cosine Function,

$$\cos A = \frac{\text{adj.}}{\text{hyp.}} = \frac{b}{c}$$

EXAMPLE 1: In the accompanying figure $c = 36$ and $\angle A = 40°$. Find b.

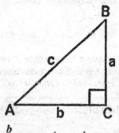

SOLUTION:
$$\frac{b}{c} = \cos A, \qquad b = c \cos A.$$

$$\cos 40° = 0.7660$$
$$c = 36$$
$$c \cos A = 36 \times 0.7660 = 27.576$$
$$b = 27.58, \quad \text{Ans.}$$

CHECK:
$$\frac{b}{c} = \cos A, \qquad \frac{27.576}{36} = 0.7660 \text{ or } \cos 40°.$$

EXAMPLE 2: Given $b = 26$ and $\angle A = 22°$; find c.

SOLUTION:
$$\frac{b}{c} = \cos A, \quad c = \frac{b}{\cos A},$$

$$b = 26$$
$$\cos 22° = 0.9272$$

$$\frac{b}{\cos A} = 26 \div 0.9272 = 28.04$$

$$c = 28.04, \quad \text{Ans.}$$

CHECK:
$$\frac{b}{c} = \cos A, \qquad \frac{26}{28.04} = 0.9272 \text{ which is } \cos 22°.$$

Practice Exercise No. 78

Use the cosine function in solving the problems in this exercise.

1. Given $c = 400$, $b = 240$; find $\angle A$.
2. Given $c = 41$, $\angle A = 39°$; find b.
3. Given $c = 67.7$, $\angle A = 23° 30'$; find b.
4. Given $c = 187$, $b = 93\frac{1}{2}$; find $\angle A$.
5. Given $b = 40$, $\angle A = 18°$; find c.
6. A roof truss is to be 30 m wide. If the rafters are 17 m long, at what angle will the rafters be laid at the eaves?

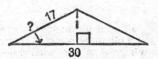

(A) 34° ......   (B) 19° 30' ......   (C) 28° 05' ......   (D) 42° 10' ......

7. Desiring to measure distance across a pond, a surveyor standing at point A sighted on a point B across the pond. From A he ran a line AC, making an angle of 27° with AB. From B he ran a line perpendicular to AC. He measured the line AC to be 681 m. What is the distance across the pond from A to B?

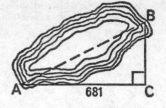

(A) 100 m ......   (B) 764 m ......   (C) 681 m ......   (D) 862·8 m ......

8. A scout on a hill 125 m above a lake sights a boat on the water at an angle of depression of 10° as shown. What is the exact distance from the scout to the boat?

(A) 240·5 m ......   (B) 720 m ......   (C) 468·4 m ......   (D) 1020 m ......

9. A mountain climber stretches a cord from the rocky ledge of a sheer cliff to a point on a horizontal plane, making an angle of 50° with the ledge. The cord is 84 m long. What is the vertical height of the rocky ledge from its base?

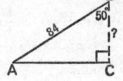

(A) 45 m ......   (B) 82 m ......
(c) 76·8 m ......   (D) 54 m ......

10. In $\triangle ABC$, angle $ACB = 90°$, $AB = 100$ mm, $AC = 16·5$ mm. Calculate angle BAC.

(A) 65° ......   (B) 25° 40′ ......
(c) 80° 30′ ......   (D) 72° 20′ ......

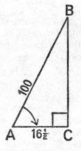

Applying the Tangent Function,

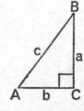

$$\tan A = \frac{\text{opp.}}{\text{adj.}} = \frac{a}{b}$$

EXAMPLE 1: In the accompanying figure $a = 40$ and $b = 27$. Find $\angle A$.

SOLUTION:
$$\frac{a}{b} = \tan A, \quad a = 40, \quad b = 27,$$

$$\frac{40}{27} = 1{\cdot}4815,$$

$$\tan A = 1{\cdot}4815. \quad \angle A = 55° \ 59', \quad \text{Ans.}$$

CHECK: $a = b \tan A$; $27 \times 1{\cdot}4815 = 40$ which is a.

EXAMPLE 2: Given angle $A = 28°$ and $a = 29$. Find b.

SOLUTION:
$$\frac{a}{b} = \tan A, \quad b = \frac{a}{\tan A},$$

$$a = 29, \quad \tan 28° = 0{\cdot}5317, \quad \frac{29}{0{\cdot}5317} = 54{\cdot}54,$$

$$b = 54{\cdot}54, \quad \text{Ans.}$$

CHECK: $\dfrac{a}{b} = \tan A$, $\quad \dfrac{29}{54{\cdot}54} = 0{\cdot}5317$ which is $\tan A$.

Practice Exercise No. 79

Use the tangent function in solving the problems in this exercise.

1. Given $a = 18$, $b = 24$; find $\angle A$.
2. Given $b = 64$, $\angle A = 45°$; find a.
3. Given $b = 62$, $\angle A = 36°$; find a.
4. Given $\angle A = 70°$, $a = 50$; find b.
5. Given $\angle A = 19° \ 36'$, $b = 42$; find a.

6. An engineer desires to learn the height of a cone-shaped hill. He measures its diameter to be 280 m. From a point on the circumference of the base he determines that the angle of elevation is 43°. What is the altitude?

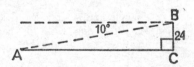

    (A) 130·55 m ......   (B) 260 m ......
    (C) 125·45 m ......   (D) 560 m ......

7. From a look-out tower 24 m high an enemy tank division is sighted at an angle of depression which is measured to be 10°. How far is the enemy away from the look-out tower if they are both on the same level?

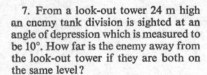

    (A) 136·1 m ......   (B) 64·23 m ......   (C) 86·6 m ......   (D) 243·4 m ......

8. The upper deck of a ship stands 30 m above the level of its dock. A runway to the deck is to be built having an angle of inclination of 20°. How far from the boat should it start?

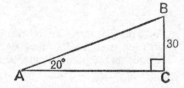

    (A) 60 m ......   (B) 76·25 m ......
    (C) 82·42 m ......   (D) 42·30 m ......

9. From a boat-house 10 m above the level of a lake two rowing crews were sighted racing in the direction of the boat-house. The boats were directly in a line with each other. The leading boat was sighted at an angle of depression equal to 15°, and the other at 14°. How far apart were the boats?

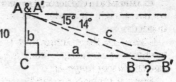

(A) 37·32 m ......   (B) 2·787 m ......   (C) 6·414 m ......   (D) 40·11 m ......

10. A clock on the tower of a building is observed from two points which are on the same level and in the same straight line with the foot of the tower. At the nearer point the angle of elevation to the clock is 60°, and at the farther point it is 30°. If the two points are 30 m apart, what is the height of the clock?

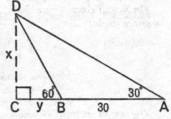

(A) 13·08 m ......   (B) 40 m ......
(C) 25·98 m ......   (D) 36·04 m ......

The Oblique Triangle

As previously stated, you can use right-triangle methods to solve most oblique triangles by introducing perpendiculars and resolving the oblique triangle into two right triangles.

For example:

1. Triangle *ABC* can be resolved into right triangles *ADC* and *BDC* by introducing the perpendicular *CD*.

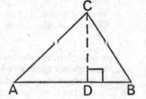

2. Triangle *DEF* can be resolved into right triangles *DGF* and *EGF* by extending *DE* and dropping the perpendicular *FG*.

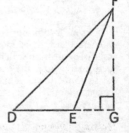

3. Triangle *HJK* can be resolved into right triangles *HLJ* and *KLJ* by introducing the perpendicular *JL*.

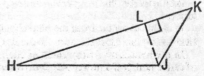

In practical problems, however, it is often impossible or too cumbersome to use a right triangle, and in such cases formulae for oblique angles are needed.

There are three important formulae that may be used in the solution of triangles of any shape. They are known as the *law of sines*, the *law of cosines*, and the *law of tangents*.

For our purposes it will be sufficient to state the law, give the corresponding formulae and show the application of the law to the solution of problems involving oblique triangles.

The law of sines: *The sides of a triangle are proportional to the sines of their opposite angles:*

$$\frac{a}{\sin A} = \frac{b}{\sin B} = \frac{c}{\sin C}$$

or $\frac{a}{b} = \frac{\sin A}{\sin B}, \frac{b}{c} = \frac{\sin B}{\sin C}, \frac{a}{c} = \frac{\sin A}{\sin C}$

The law of cosines: *The square of any side of a triangle is equal to the sum of the squares of the other two sides minus twice their product times the cosine of the included angle.*

$$a^2 = b^2 + c^2 - 2bc \cos A$$
$$b^2 = a^2 + c^2 - 2ac \cos B$$
$$c^2 = a^2 + b^2 - 2ab \cos C$$

or $a = \sqrt{b^2 + c^2 - 2bc \cos A}$
$b = \sqrt{a^2 + c^2 - 2ac \cos B}$
$c = \sqrt{a^2 + b^2 - 2ab \cos C}$

The law of tangents: *The difference between any two sides of a triangle is to their sum as the tangent of half the difference between their opposite angles is to tangent of half their sum.*

$$\frac{a - b}{a + b} = \frac{\tan \frac{1}{2}(A - B)}{\tan \frac{1}{2}(A + B)}$$

$$\frac{a - c}{a + c} = \frac{\tan \frac{1}{2}(A - C)}{\tan \frac{1}{2}(A + C)}$$

$$\frac{b - c}{b + c} = \frac{\tan \frac{1}{2}(B - C)}{\tan \frac{1}{2}(B + C)}$$

or if $b > a$, then

$$\frac{b - a}{b + a} = \frac{\tan \frac{1}{2}(B - A)}{\tan \frac{1}{2}(B + A)}$$

Solving Oblique Triangles

Any triangle has six parts, namely, three angles and the sides opposite the angles.

In order to solve a triangle three independent parts must be known in addition to the fact that the sum of the angles of any triangle equals 180°.

In problems involving triangles there occur the following four combinations of parts which if known will determine the size and form of the triangle.

I. *One side and two angles are known*
II. *Two sides and the included angle are known*
III. *Three sides are known*
IV. *Two sides and the angle opposite one of them is known.**

Applying the Laws of Sine, Tangent, and Cosine to Oblique Triangles

Case I: One side and two angles are known

EXAMPLE: Given $\angle A = 56°$, $\angle B = 69°$, and $a = 467$; find b and c.

SOLUTION: We use the law of sines.

Formulae needed:

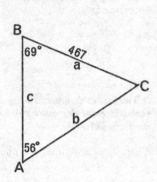

1. $C = 180° - (\angle A + \angle B)$,

2. $\dfrac{b}{a} = \dfrac{\sin B}{\sin A}$, $\therefore b = \dfrac{a \sin B}{\sin A}$.

3. $\dfrac{c}{a} = \dfrac{\sin C}{\sin A}$, $\therefore c = \dfrac{a \sin C}{\sin A}$.

Substituting:

1. $\angle C = 180° - (56° + 69°) = 55°$.

2. $b = \dfrac{467 \times 0.9336}{0.8290} = 525.9$

3. $c = \dfrac{467 \times 0.8192}{0.8290} = 461.5$ $\Bigg\}$ ANS.

Case II: Two sides and the included angle are known

EXAMPLE: Given $a = 17$, $b = 12$, and $\angle C = 58°$; find $\angle A$, $\angle B$, and c.

SOLUTION: We use the law of tangents to obtain $\angle A$ and $\angle B$ and the law of sines to obtain c.

Formulae needed:

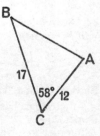

1. $A + B = 180° - C$ and $\frac{1}{2}(A + B) = \frac{1}{2}(180° - C)$.

When $\frac{1}{2}(A + B)$ has been determined $\frac{1}{2}(A - B)$ is found by the following:

2. $\dfrac{a - b}{a + b} = \dfrac{\tan \frac{1}{2}(A - B)}{\tan \frac{1}{2}(A + B)}$,

$\therefore \tan \frac{1}{2}(A - B) = \dfrac{a - b}{a + b} \times \tan \frac{1}{2}(A + B)$.

3. $\angle A = \frac{1}{2}(A + B) + \frac{1}{2}(A - B)$,
in which the Bs cancel out.

* This combination is considered an ambiguous case because it is often possible to form more than one triangle to satisfy the given conditions.

4. $\angle B = \frac{1}{2}(A + B) - \frac{1}{2}(A - B)$,
   in which the *A*s cancel out.

5. $\dfrac{c}{a} = \dfrac{\sin C}{\sin A}$, $\therefore c = \dfrac{a \sin C}{\sin A}$.

Substituting:

1. $\frac{1}{2}(A + B) = \frac{1}{2}(180° - 58°) = 61°$.

2. $\tan \frac{1}{2}(A - B) = \dfrac{17 - 12}{17 + 12} \times \tan 61° = 0.3110$,
   which is the tan of 17° 16′ and equal to $\frac{1}{2}(A - B)$.

$$
\left.
\begin{array}{l}
\text{3. } \angle A = 61° + 17° 16′ = 78° 16′ \\
\text{4. } \angle B = 61° - 17° 16′ = 43° 44′ \\
\text{5. } c = \dfrac{17 \times \sin 58°}{\sin 78° 16′} = 14.7
\end{array}
\right\} \text{ Ans.}
$$

This example could also be solved by the use of the law of cosines by first finding *c* ($c = \sqrt{a^2 + b^2 - 2ab \cos C}$). When the three sides and $\angle C$ are known the law of sines can be employed to find $\angle A$ and $\angle B$. For purposes of a check, do this example by the second method.

Case III: Three sides are known

EXAMPLE: Given $a = 5$, $b = 6$, and $c = 7$; find $\angle A$, $\angle B$, and $\angle C$.

SOLUTION: We use the law of cosines and the law of sines.

Formulae needed:

1. $a^2 = b^2 + c^2 - 2bc \cos A$,

$$\therefore \cos A = \frac{b^2 + c^2 - a^2}{2bc}.$$

2. $\dfrac{a}{b} = \dfrac{\sin A}{\sin B}$, $\therefore \sin B = \dfrac{b \sin A}{a}$.

3. $\angle C = 180° - (A + B)$.

Substituting:

$$\cos A = \frac{36 + 49 - 25}{2(6 \times 7)} = 0.7143,$$

which is the cos of 44° 25′.

$$\sin B = \frac{6 \times 0.6999}{5} = 0.8399,$$

which is the sin of 57° 8′

$$
\left.
\begin{array}{l}
\angle C = 180° - (44° 25′ + 57° 8′) = 78° 27′ \\
\angle A = 44° 25′ \\
\angle B = 57° 8′ \\
\angle C = 78° 27′
\end{array}
\right\} \text{Ans.}
$$

Case IV (*the ambiguous case*): **Two sides and the angle opposite one of them are known**

When given two sides of a triangle and the angle opposite one of them, there is often a possibility of two solutions unless one of the solutions is excluded by the statement of the problem.

This fact may be clarified by the next figure. It will be seen in the triangle *ABC* that if $\angle A$ and sides *a* and *b* are given, either of the triangles *ABC* or *AB'C* meet the given conditions.

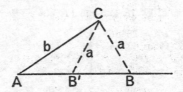

By varying the relative lengths of *a* and *b* and the magnitude of $\angle A$, the following possibilities can be recognized.

If $a > b$, $\angle A > \angle B$, which makes $\angle B$ less than 90°, and allows for only one solution.

If $a = b$, $\angle A = \angle B$; both angles are less than 90° and only an isosceles triangle can be formed.

If $a < b$ and $\angle A$ is acute two triangles are possible.

If $a = b \sin A$ the figure is a right triangle and only one solution is possible.

If $a < b \sin A$ no triangle is possible.

Before doing a problem of this type you can generally determine the number of possible solutions by making an approximate small-scale drawing of the given parts.

In the cases where there are two possible solutions and the unknown parts are $\angle B$, $\angle C$, and side *c*, the second set of unknown parts should be designated as $\angle B'$, $\angle C'$, and side *c'*. They will then be found as follows:

$$B' = 180° - B$$

because when an angle is determined by its sine, it has two possible values that are supplementary to each other.

$$C' = 180° - (A + B')$$

$$c' = \frac{a \sin C'}{\sin A}$$

EXAMPLE: Given $a = 5$, $b = 8$, and $\angle A = 30°$; find $\angle B$, $\angle C$, and side *c*.

Here $a < b$ and $\angle A$ is acute. ∴ two triangles are possible.

Formulae needed for $\triangle ABC$:

1. $\dfrac{b}{a} = \dfrac{\sin B}{\sin A}$,　　　　　　∴ $\sin B = \dfrac{b \sin A}{a}$.

2. $\angle C = 180° - (A + B)$.

3. $\dfrac{c}{a} = \dfrac{\sin C}{\sin A}$,　　　　　　∴ $c = \dfrac{a \sin C}{\sin A}$.

Substituting:

1. $\sin B = \dfrac{8 \times 0.5000}{5} = 0.8000,$

which is the sin of 53° 8′.

2. $\angle C = 180° - (30° + 53° 8′) = 96° 52′.$

3. $c = \dfrac{5 \times 0.9929}{0.5000} = 9.929.$

$$\left. \begin{array}{l} \angle B = 53° 8′ \\ \angle C = 96° 52′ \\ c = 9.929 \end{array} \right\} \text{ Ans.}$$

To find $\angle B′$, $\angle C′$, and $c′$:

$\angle B′ = 180° - B = 126° 52′,$ Ans.

$\angle C′ = 180° - (A + B′) = 23° 8′,$ Ans.

$$c′ = \frac{a \sin C′}{\sin A} = \frac{5 \times 3.928}{5} = 3.928, \text{ Ans.}$$

Practice Exercise No. 80

In working out the problems in this exercise apply the principles for solving oblique triangles.

1. Given $\angle A = 45°$, $\angle B = 60°$, and $c = 9.562$; find a and b.
2. Given $a = 43$, $\angle A = 43°$, and $\angle B = 68°$; find $\angle C$, b, and c.
3. Given $a = 22$, $b = 13$, and $\angle C = 68°$; find $\angle A$, $\angle B$, and c.
4. Given $a = 27$, $b = 26$, $c = 34$; find $\angle A$, $\angle B$, and $\angle C$.
5. Given $a = 8$, $b = 5$, and $\angle A = 21°$; find c, $\angle B$, and $\angle C$.

6. Two aeroplane spotters, A and B, are 1·83 km apart on the same level of ground. B is due north of A. At the same instant they both spot an aeroplane to the north, which makes an angle of elevation of 67° 31′ at A and 82° 16′ at B. What is the altitude of the aeroplane from the ground?

(A) 2·5 km ......   (B) 6·6 km ......   (C) 4 km ......   (D) 3·2 km ......

7. An observer on a boat anchored off shore sights on two points, A and B, on the shore. He determines the distance from himself to point A to be 98·5 m, and the distance between A and B as 146 m. The angle to the observer subtended by the points on shore is 64° 20′. How far is it from the observer to point B?

(A) 158·6 m ......   (B) 124·3 m ......   (C) 176 m ......   (D) 92·77 m ......

8. An observer at a fire tower spots a fire in a forest area extending across a stretch of land from point A to point B. The distance from the tower to A is 5 km, and to B, 5·5 km. The angle subtended by the stretch of land to the tower is 50°. What is the distance across which the fire extends? (*Note:* For practice purposes solve by the tangent law.)

(A) 6 km ......   (B) 4·46 km ......   (C) 3·42 km ......   (D) 8·5 km ......

9. Two scouts start from a point C at the same time and branch out at an angle of 33° to each other. If one scout travels at the rate of 1 km per hour while the other travels at the rate of 3 km per hour, how far apart will they be at the end of 2 hours? (*Note:* Solve by cosine law.)

(A) 3 km ......   (B) 5·42 km ......   (C) 4·46 km ......   (D) 8·56 km ......

Table VIII. Natural Sines

	0'	6'	12'	18'	24'	30'	36'	42'	48'	54'	1'	2'	3'	4'	5'
0°	·0000	0017	0035	0052	0070	0087	0105	0122	0140	0157	3	6	9	12	15
1	·0175	0192	0209	0227	0244	0262	0279	0297	0314	0332	3	6	9	12	15
2	·0349	0366	0384	0401	0419	0436	0454	0471	0488	0506	3	6	9	12	15
3	·0523	0541	0558	0576	0593	0610	0628	0645	0663	0680	3	6	9	12	15
4	·0698	0715	0732	0750	0767	0785	0802	0819	0837	0854	3	6	9	12	14
5	·0872	0889	0906	0924	0941	0958	0976	0993	1011	1028	3	6	9	12	14
6	·1045	1063	1080	1097	1115	1132	1149	1167	1184	1201	3	6	9	12	14
7	·1219	1236	1253	1271	1288	1305	1323	1340	1357	1374	3	6	9	12	14
8	·1392	1409	1426	1444	1461	1478	1495	1513	1530	1547	3	6	9	12	14
9	·1564	1582	1599	1616	1633	1650	1668	1685	1702	1719	3	6	9	11	14
10	·1736	1754	1771	1788	1805	1822	1840	1857	1874	1891	3	6	9	11	14
11	·1908	1925	1942	1959	1977	1994	2011	2028	2045	2062	3	6	9	11	14
12	·2079	2096	2113	2130	2147	2164	2181	2198	2215	2233	3	6	9	11	14
13	·2250	2267	2284	2300	2317	2334	2351	2368	2385	2402	3	6	8	11	14
14	·2419	2436	2453	2470	2487	2504	2521	2538	2554	2571	3	6	8	11	14
15	·2588	2605	2622	2639	2656	2672	2689	2706	2723	2740	3	6	8	11	14
16	·2756	2773	2790	2807	2823	2840	2857	2874	2890	2907	3	6	8	11	14
17	·2924	2940	2957	2974	2990	3007	3024	3040	3057	3074	3	6	8	11	14
18	·3090	3107	3123	3140	3156	3173	3190	3206	3223	3239	3	6	8	11	14
19	·3256	3272	3289	3305	3322	3338	3355	3371	3387	3404	3	5	8	11	14
20	·3420	3437	3453	3469	3486	3502	3518	3535	3551	3567	3	5	8	11	14
21	·3584	3600	3616	3633	3649	3665	3681	3697	3714	3730	3	5	8	11	14
22	·3746	3762	3778	3795	3811	3827	3843	3859	3875	3891	3	5	8	11	13
23	·3907	3923	3939	3955	3971	3987	4003	4019	4035	4051	3	5	8	11	13
24	·4067	4083	4099	4115	4131	4147	4163	4179	4195	4210	3	5	8	11	13
25	·4226	4242	4258	4274	4289	4305	4321	4337	4352	4368	3	5	8	11	13
26	·4384	4399	4415	4431	4446	4462	4478	4493	4509	4524	3	5	8	10	13
27	·4540	4555	4571	4586	4602	4617	4633	4648	4664	4679	3	5	8	10	13
28	·4695	4710	4726	4741	4756	4772	4787	4802	4818	4833	3	5	8	10	13
29	·4848	4863	4879	4894	4909	4924	4939	4955	4970	4985	3	5	8	10	13
30	·5000	5015	5030	5045	5060	5075	5090	5105	5120	5135	3	5	8	10	13
31	·5150	5165	5180	5195	5210	5225	5240	5255	5270	5284	2	5	7	10	12
32	·5299	5314	5329	5344	5358	5373	5388	5402	5417	5432	2	5	7	10	12
33	·5446	5461	5476	5490	5505	5519	5534	5548	5563	5577	2	5	7	10	12
34	·5592	5606	5621	5635	5650	5664	5678	5693	5707	5721	2	5	7	10	12
35	·5736	5750	5764	5779	5793	5807	5821	5835	5850	5864	2	5	7	9	12
36	·5878	5892	5906	5920	5934	5948	5962	5976	5990	6004	2	5	7	9	12
37	·6018	6032	6046	6060	6074	6088	6101	6115	6129	6143	2	5	7	9	12
38	·6157	6170	6184	6198	6211	6225	6239	6252	6266	6280	2	5	7	9	11
39	·6293	6307	6320	6334	6347	6361	6374	6388	6401	6414	2	4	7	9	11
40	·6428	6441	6455	6468	6481	6494	6508	6521	6534	6547	2	4	7	9	11
41	·6561	6574	6587	6600	6613	6626	6639	6652	6665	6678	2	4	7	9	11
42	·6691	6704	6717	6730	6743	6756	6769	6782	6794	6807	2	4	6	9	11
43	·6820	6833	6845	6858	6871	6884	6896	6909	6921	6934	2	4	6	8	11
44	·6947	6959	6972	6984	6997	7009	7022	7034	7046	7059	2	4	6	8	10
	0'	6'	12'	18'	24'	30'	36'	42'	48'	54'	1'	2'	3'	4'	5'

Table VIII. Natural Sines—contd.

	0′	6′	12′	18′	24′	30′	36′	42′	48′	54′	1′	2′	3′	4′	5′
45°	·7071	7083	7096	7108	7120	7133	7145	7157	7169	7181	2	4	6	8	10
46	·7193	7206	7218	7230	7242	7254	7266	7278	7290	7302	2	4	6	8	10
47	·7314	7325	7337	7349	7361	7373	7385	7396	7408	7420	2	4	6	8	10
48	·7431	7443	7455	7466	7478	7490	7501	7513	7524	7536	2	4	6	8	10
49	·7547	7559	7570	7581	7593	7604	7615	7627	7638	7649	2	4	6	8	9
50	·7660	7672	7683	7694	7705	7716	7727	7738	7749	7760	2	4	6	7	9
51	·7771	7782	7793	7804	7815	7826	7837	7848	7859	7869	2	4	5	7	9
52	·7880	7891	7902	7912	7923	7934	7944	7955	7965	7976	2	4	5	7	9
53	·7986	7997	8007	8018	8028	8039	8049	8059	8070	8080	2	3	5	7	9
54	·8090	8100	8111	8121	8131	8141	8151	8161	8171	8181	2	3	5	7	8
55	·8192	8202	8211	8221	8231	8241	8251	8261	8271	8281	2	3	5	7	8
56	·8290	8300	8310	8320	8329	8339	8348	8358	8368	8377	2	3	5	6	8
57	·8387	8396	8406	8415	8425	8434	8443	8453	8462	8471	2	3	5	6	8
58	·8480	8490	8499	8508	8517	8526	8536	8545	8554	8563	2	3	5	6	8
59	·8572	8581	8590	8599	8607	8616	8625	8634	8643	8652	1	3	4	6	7
60	·8660	8669	8678	8686	8695	8704	8712	8721	8729	8738	1	3	4	6	7
61	·8746	8755	8763	8771	8780	8788	8796	8805	8813	8821	1	3	4	6	7
62	·8829	8838	8846	8854	8862	8870	8878	8886	8894	8902	1	3	4	5	7
63	·8910	8918	8926	8934	8942	8949	8957	8965	8973	8980	1	3	4	5	6
64	·8988	8996	9003	9011	9018	9026	9033	9041	9048	9056	1	3	4	5	6
65	·9063	9070	9078	9085	9092	9100	9107	9114	9121	9128	1	2	4	5	6
66	·9135	9143	9150	9157	9164	9171	9178	9184	9191	9198	1	2	3	5	6
67	·9205	9212	9219	9225	9232	9239	9245	9252	9259	9265	1	2	3	4	6
68	·9272	9278	9285	9291	9298	9304	9311	9317	9323	9330	1	2	3	4	5
69	·9336	9342	9348	9354	9361	9367	9373	9379	9385	9391	1	2	3	4	5
70	·9397	9403	9409	9415	9421	9426	9432	9438	9444	9449	1	2	3	4	5
71	·9455	9461	9466	9472	9478	9483	9489	9494	9500	9505	1	2	3	4	5
72	·9511	9516	9521	9527	9532	9537	9542	9548	9553	9558	1	2	3	4	4
73	·9563	9568	9573	9578	9583	9588	9593	9598	9603	9608	1	2	2	3	4
74	·9613	9617	9622	9627	9632	9636	9641	9646	9650	9655	1	2	2	3	4
75	·9659	9664	9668	9673	9677	9681	9686	9690	9694	9699	1	1	2	3	4
76	·9703	9707	9711	9715	9720	9724	9728	9732	9736	9740	1	1	2	3	3
77	·9744	9748	9751	9755	9759	9763	9767	9770	9774	9778	1	1	2	3	3
78	·9781	9785	9789	9792	9796	9799	9803	9806	9810	9813	1	1	2	2	3
79	·9816	9820	9823	9826	9829	9833	9836	9839	9842	9845	1	1	2	2	3
80	·9848	9851	9854	9857	9860	9863	9866	9869	9871	9874	0	1	1	2	2
81	·9877	9880	9882	9885	9888	9890	9893	9895	9898	9900	0	1	1	2	2
82	·9903	9905	9907	9910	9912	9914	9917	9919	9921	9923	0	1	1	2	2
83	·9925	9928	9930	9932	9934	9936	9938	9940	9942	9943	0	1	1	1	2
84	·9945	9947	9949	9951	9952	9954	9956	9957	9959	9960	0	1	1	1	1
85	·9962	9963	9965	9966	9968	9969	9971	9972	9973	9974	0	0	1	1	1
86	·9976	9977	9978	9979	9980	9981	9982	9983	9984	9985	0	0	1	1	1
87	·9986	9987	9988	9989	9990	9990	9991	9992	9993	9993	0	0	0	1	1
88	·9994	9995	9995	9996	9996	9997	9997	9997	9998	9998					
89	·9998	9999	9999	9999	9999	1·000	1·000	1·000	1·000	1·000					
	0′	6′	12′	18′	24′	30′	36′	42′	48′	54′	1′	2′	3′	4′	5′

Table VIIIa. Natural Tangents

	0'	6'	12'	18'	24'	30'	36'	42'	48'	54'	1'	2'	3'	4'	5'
0	0·0000	0017	0035	0052	0070	0087	0105	0122	0140	0157	3	6	9	12	1
1	0·0175	0192	0209	0227	0244	0262	0279	0297	0314	0332	3	6	9	12	1
2	0·0349	0367	0384	0402	0419	0437	0454	0472	0489	0507	3	6	9	12	1
3	0·0524	0542	0559	0577	0594	0612	0629	0647	0664	0682	3	6	9	12	1
4	0·0699	0717	0734	0752	0769	0787	0805	0822	0840	0857	3	6	9	12	1
5	0·0875	0892	0910	0928	0945	0963	0981	0998	1016	1033	3	6	9	12	1
6	0·1051	1069	1086	1104	1122	1139	1157	1175	1192	1210	3	6	9	12	1
7	0·1228	1246	1263	1281	1299	1317	1334	1352	1370	1388	3	6	9	12	1
8	0·1405	1423	1441	1459	1477	1495	1512	1530	1548	1566	3	6	9	12	1
9	0·1584	1602	1620	1638	1655	1673	1691	1709	1727	1745	3	6	9	12	1
10	0·1763	1781	1799	1817	1835	1853	1871	1890	1908	1926	3	6	9	12	1
11	0·1944	1962	1980	1998	2016	2035	2053	2071	2089	2107	3	6	9	12	1
12	0·2126	2144	2162	2180	2199	2217	2235	2254	2272	2290	3	6	9	12	1
13	0·2309	2327	2345	2364	2382	2401	2419	2438	2456	2475	3	6	9	12	1
14	0·2493	2512	2530	2549	2568	2586	2605	2623	2642	2661	3	6	9	12	1
15	0·2679	2698	2717	2736	2754	2773	2792	2811	2830	2849	3	6	9	13	1
16	0·2867	2886	2905	2924	2943	2962	2981	3000	3019	3038	3	6	9	13	1
17	0·3057	3076	3096	3115	3134	3153	3172	3191	3211	3230	3	6	10	13	1
18	0·3249	3269	3288	3307	3327	3346	3365	3385	3404	3424	3	6	10	13	1
19	0·3443	3463	3482	3502	3522	3541	3561	3581	3600	3620	3	7	10	13	1
20	0·3640	3659	3679	3699	3719	3739	3759	3779	3799	3819	3	7	10	13	1
21	0·3839	3859	3879	3899	3919	3939	3959	3979	4000	4020	3	7	10	13	1
22	0·4040	4061	4081	4101	4122	4142	4163	4183	4204	4224	3	7	10	14	1
23	0·4245	4265	4286	4307	4327	4348	4369	4390	4411	4431	3	7	10	14	1
24	0·4452	4473	4494	4515	4536	4557	4578	4599	4621	4642	4	7	11	14	1
25	0·4663	4684	4706	4727	4748	4770	4791	4813	4834	4856	4	7	11	14	1
26	0·4877	4899	4921	4942	4964	4986	5008	5029	5051	5073	4	7	11	15	1
27	0·5095	5117	5139	5161	5184	5206	5228	5250	5272	5295	4	7	11	15	1
28	0·5317	5340	5362	5384	5407	5430	5452	5475	5498	5520	4	8	11	15	1
29	0·5543	5566	5589	5612	5635	5658	5681	5704	5727	5750	4	8	12	15	1
30	0·5774	5797	5820	5844	5867	5890	5914	5938	5961	5985	4	8	12	16	20
31	0·6009	6032	6056	6080	6104	6128	6152	6176	6200	6224	4	8	12	16	20
32	0·6249	6273	6297	6322	6346	6371	6395	6420	6445	6469	4	8	12	16	20
33	0·6494	6519	6544	6569	6594	6619	6644	6669	6694	6720	4	8	13	17	21
34	0·6745	6771	6796	6822	6847	6873	6899	6924	6950	6976	4	9	13	17	21
35	0·7002	7028	7054	7080	7107	7133	7159	7186	7212	7239	4	9	13	18	22
36	0·7265	7292	7319	7346	7373	7400	7427	7454	7481	7508	5	9	14	18	22
37	0·7536	7563	7590	7618	7646	7673	7701	7729	7757	7785	5	9	14	18	22
38	0·7813	7841	7869	7898	7926	7954	7983	8012	8040	8069	5	9	14	19	2
39	0·8098	8127	8156	8185	8214	8243	8273	8302	8332	8361	5	10	15	20	2
40	0·8391	8421	8451	8481	8511	8541	8571	8601	8632	8662	5	10	15	20	2
41	0·8693	8724	8754	8785	8816	8847	8878	8910	8941	8972	5	10	16	21	2
42	0·9004	9036	9067	9099	9131	9163	9195	9228	9260	9293	5	11	16	21	27
43	0·9325	9358	9391	9424	9457	9490	9523	9556	9590	9623	6	11	17	22	28
44	0·9657	9691	9725	9759	9793	9827	9861	9896	9930	9965	6	11	17	23	29
	0'	6'	12'	18'	24'	30'	36'	42'	48'	54'	1'	2'	3'	4'	5'

Table VIIIa. Natural Tangents—contd.

°	0'	6'	12'	18'	24'	30'	36'	42'	48'	54'	1'	2'	3'	4'	5'
45	1·0000	0035	0070	0105	0141	0176	0212	0247	0283	0319	6	12	18	24	30
46	1·0355	0392	0428	0464	0501	0538	0575	0612	0649	0686	6	12	18	25	31
47	1·0724	0761	0799	0837	0875	0913	0951	0990	1028	1067	6	13	19	25	32
48	1·1106	1145	1184	1224	1263	1303	1343	1383	1423	1463	7	13	20	26	33
49	1·1504	1544	1585	1626	1667	1708	1750	1792	1833	1875	7	14	21	28	34
50	1·1918	1960	2002	2045	2088	2131	2174	2218	2261	2305	7	14	22	29	36
51	1·2349	2393	2437	2482	2527	2572	2617	2662	2708	2753	8	15	23	30	38
52	1·2799	2846	2892	2938	2985	3032	3079	3127	3175	3222	8	16	24	31	39
53	1·3270	3319	3367	3416	3465	3514	3564	3613	3663	3713	8	16	25	33	41
54	1·3764	3814	3865	3916	3968	4019	4071	4124	4176	4229	9	17	26	34	43
55	1·4281	4335	4388	4442	4496	4550	4605	4659	4715	4770	9	18	27	36	45
56	1·4826	4882	4938	4994	5051	5108	5166	5224	5282	5340	10	19	29	38	48
57	1·5399	5458	5517	5577	5637	5697	5757	5818	5880	5941	10	20	30	40	50
58	1·6003	6066	6128	6191	6255	6319	6383	6447	6512	6577	11	21	32	43	53
59	1·6643	6709	6775	6842	6909	6977	7045	7113	7182	7251	11	23	34	45	56
60	1·7321	7391	7461	7532	7603	7675	7747	7820	7893	7966	12	24	36	48	60
61	1·8040	8115	8190	8265	8341	8418	8495	8572	8650	8728	13	26	38	51	64
62	1·8807	8887	8967	9047	9128	9210	9292	9375	9458	9542	14	27	41	55	68
63	1·9626	9711	9797	9883	9970	**0057**	**0145**	**0233**	**0323**	**0413**	15	29	44	58	73
64	2·0503	0594	0686	0778	0872	0965	1060	1155	1251	1348	16	31	47	63	78
65	2·1445	1543	1642	1742	1842	1943	2045	2148	2251	2355	17	34	51	68	85
66	2·2460	2566	2673	2781	2889	2998	3109	3220	3332	3445	18	37	55	73	91
67	2·3559	3673	3789	3906	4023	4142	4262	4383	4504	4627	20	40	60	79	99
68	2·4751	4876	5002	5129	5257	5386	5517	5649	5782	5916	22	43	65	87	108
69	2·6051	6187	6325	6464	6605	6746	6889	7034	7179	7326	24	47	71	95	119
70	2·7475	7625	7776	7929	8083	8239	8397	8556	8716	8878	26	52	78	104	130
71	2·9042	9208	9375	9544	9714	9887	**0061**	**0237**	**0415**	**0595**	29	58	87	116	144
72	3·0777	0961	1146	1334	1524	1716	1910	2106	2305	2506	32	64	97	129	161
73	3·2709	2914	3122	3332	3544	3759	3977	4197	4420	4646	36	72	108	144	180
74	3·4874	5105	5339	5576	5816	6059	6305	6554	6806	7062	41	81	122	163	203
75	3·7321	7583	7848	8118	8391	8667	8947	9232	9520	9812	46	93	139	186	232
76	4·0108	0408	0713	1022	1335	1653	1976	2303	2635	2972	53	107	160	214	267
77	4·3315	3662	4015	4373	4737	5107	5483	5864	6252	6646	62	124	186	248	310
78	4·7046	7453	7867	8288	8716	9152	9594	**0045**	**0504**	**0970**	73	146	220	293	366
79	5·1446	1929	2422	2924	3435	3955	4486	5026	5578	6140	87	175	263	350	438
80	5·671	5·730	5·789	5·850	5·912	5·976	6·041	6·107	6·174	6·243					
81	6·314	6·386	6·460	6·535	6·612	6·691	6·772	6·855	6·940	7·026					
82	7·115	7·207	7·300	7·396	7·495	7·596	7·700	7·806	7·916	8·028					
83	8·144	8·264	8·386	8·513	8·643	8·777	8·915	9·058	9·205	9·357					
84	9·51	9·68	9·84	10·02	10·20	10·39	10·58	10·78	10·99	11·20		Differences			
85	11·43	11·66	11·91	12·16	12·43	12·71	13·00	13·30	13·62	13·95		untrustworthy			
86	14·30	14·67	15·06	15·46	15·89	16·35	16·83	17·34	17·89	18·46		here			
87	19·08	19·74	20·45	21·20	22·02	22·90	23·86	24·90	26·03	27·27					
88	28·64	30·14	31·82	33·69	35·80	38·19	40·92	44·07	47·74	52·08					
89	57·29	63·66	71·62	81·85	95·49	114·6	143·2	191·0	286·5	573·0					
	0'	6'	12'	18'	24'	30'	36'	42'	48'	54'	1'	2'	3'	4'	5'

The black type indicates that the integer changes.

Table VIIIb. Natural Secants

°	0'	6'	12'	18'	24'	30'	36'	42'	48'	54'	1'	2'	3'	4'	5'
0	1·0000	0000	0000	0000	0000	0000	0001	0001	0001	0001					
1	1·0002	0002	0002	0003	0003	0003	0004	0004	0005	0006					
2	1·0006	0007	0007	0008	0009	0010	0010	0011	0012	0013					
3	1·0014	0015	0016	0017	0018	0019	0020	0021	0022	0023					
4	1·0024	0026	0027	0028	0030	0031	0032	0034	0035	0037	0	0	0	0	1
5	1·0038	0040	0041	0043	0045	0046	0048	0050	0051	0053	0	1	1	1	1
6	1·0055	0057	0059	0061	0063	0065	0067	0069	0071	0073	0	1	1	1	2
7	1·0075	0077	0079	0082	0084	0086	0089	0091	0093	0096	0	1	1	2	2
8	1·0098	0101	0103	0106	0108	0111	0114	0116	0119	0122	0	1	1	2	2
9	1·0125	0127	0130	0133	0136	0139	0142	0145	0148	0151	0	1	1	2	2
10	1·0154	0157	0161	0164	0167	0170	0174	0177	0180	0184	1	1	2	2	3
11	1·0187	0191	0194	0198	0201	0205	0209	0212	0216	0220	1	1	2	3	3
12	1·0223	0227	0231	0235	0239	0243	0247	0251	0255	0259	1	1	2	3	3
13	1·0263	0267	0271	0276	0280	0284	0288	0293	0297	0302	1	1	2	3	3
14	1·0306	0311	0315	0320	0324	0329	0334	0338	0343	0348	1	2	2	3	4
15	1·0353	0358	0363	0367	0372	0377	0382	0388	0393	0398	1	2	3	3	4
16	1·0403	0408	0413	0419	0424	0429	0435	0440	0446	0451	1	2	3	4	4
17	1·0457	0463	0468	0474	0480	0485	0491	0497	0503	0509	1	2	3	4	5
18	1·0515	0521	0527	0533	0539	0545	0551	0557	0564	0570	1	2	3	4	5
19	1·0576	0583	0589	0595	0602	0608	0615	0622	0628	0635	1	2	3	4	5
20	1·0642	0649	0655	0662	0669	0676	0683	0690	0697	0704	1	2	3	5	6
21	1·0711	0719	0726	0733	0740	0748	0755	0763	0770	0778	1	2	4	5	6
22	1·0785	0793	0801	0808	0816	0824	0832	0840	0848	0856	1	3	4	5	7
23	1·0864	0872	0880	0888	0896	0904	0913	0921	0929	0938	1	3	4	6	7
24	1·0946	0955	0963	0972	0981	0989	0998	1007	1016	1025	1	3	4	6	7
25	1·1034	1043	1052	1061	1070	1079	1089	1098	1107	1117	2	3	5	6	8
26	1·1126	1136	1145	1155	1164	1174	1184	1194	1203	1213	2	3	5	6	8
27	1·1223	1233	1243	1253	1264	1274	1284	1294	1305	1315	2	3	5	7	8
28	1·1326	1336	1347	1357	1368	1379	1390	1401	1412	1423	2	4	5	7	9
29	1·1434	1445	1456	1467	1478	1490	1501	1512	1524	1535	2	4	6	8	9
30	1·1547	1559	1570	1582	1594	1606	1618	1630	1642	1654	2	4	6	8	10
31	1·1666	1679	1691	1703	1716	1728	1741	1753	1766	1779	2	4	6	8	10
32	1·1792	1805	1818	1831	1844	1857	1870	1883	1897	1910	2	4	7	9	11
33	1·1924	1937	1951	1964	1978	1992	2006	2020	2034	2048	2	5	7	9	12
34	1·2062	2076	2091	2105	2120	2134	2149	2163	2178	2193	2	5	7	10	12
35	1·2208	2223	2238	2253	2268	2283	2299	2314	2329	2345	3	5	8	10	13
36	1·2361	2376	2392	2408	2424	2440	2456	2472	2489	2505	3	5	8	11	13
37	1·2521	2538	2554	2571	2588	2605	2622	2639	2656	2673	3	6	8	11	14
38	1·2690	2708	2725	2742	2760	2778	2796	2813	2831	2849	3	6	9	12	15
39	1·2868	2886	2904	2923	2941	2960	2978	2997	3016	3035	3	6	9	12	16
40	1·3054	3073	3093	3112	3131	3151	3171	3190	3210	3230	3	7	10	13	16
41	1·3250	3270	3291	3311	3331	3352	3373	3393	3414	3435	3	7	10	14	17
42	1·3456	3478	3499	3520	3542	3563	3585	3607	3629	3651	4	7	11	14	18
43	1·3673	3696	3718	3741	3763	3786	3809	3832	3855	3878	4	8	11	15	19
44	1·3902	3925	3949	3972	3996	4020	4044	4069	4093	4118	4	8	12	16	20
	0'	6'	12'	18'	24'	30'	36'	42'	48'	54'	1'	2'	3'	4'	5'

Table VIIIb. Natural Secants—contd.

	0'	6'	12'	18'	24'	30'	36'	42'	48'	54'	1'	2'	3'	4'	5'
45	1·4142	4167	4192	4217	4242	4267	4293	4318	4344	4370	4	8	13	17	21
46	1·4396	4422	4448	4474	4501	4527	4554	4581	4608	4635	4	9	13	18	22
47	1·4663	4690	4718	4746	4774	4802	4830	4859	4887	4916	5	9	14	19	23
48	1·4945	4974	5003	5032	5062	5092	5121	5151	5182	5212	5	10	15	20	25
49	1·5243	5273	5304	5335	5366	5398	5429	5461	5493	5525	5	10	16	21	26
50	1·5557	5590	5622	5655	5688	5721	5755	5788	5822	5856	6	11	17	22	28
51	1·5890	5925	5959	5994	6029	6064	6099	6135	6171	6207	6	12	18	23	29
52	1·6243	6279	6316	6353	6390	6427	6464	6502	6540	6578	6	12	19	25	31
53	1·6616	6655	6694	6733	6772	6812	6852	6892	6932	6972	7	13	20	26	33
54	1·7013	7054	7095	7137	7179	7221	7263	7305	7348	7391	7	14	21	28	35
55	1·7434	7478	7522	7566	7610	7655	7700	7745	7791	7837	7	15	22	30	37
56	1·7883	7929	7976	8023	8070	8118	8166	8214	8263	8312	8	16	24	32	40
57	1·8361	8410	8460	8510	8561	8612	8663	8714	8766	8818	8	17	25	34	42
58	1·8871	8924	8977	9031	9084	9139	9194	9249	9304	9360	9	18	27	36	45
59	1·9416	9473	9530	9587	9645	9703	9762	9821	9880	9940	10	19	29	39	49
60	2·0000	0061	0122	0183	0245	0308	0371	0434	0498	0562	10	21	31	42	52
61	2·0627	0692	0757	0824	0890	0957	1025	1093	1162	1231	11	22	34	45	56
62	2·1301	1371	1441	1513	1584	1657	1730	1803	1877	1952	12	24	36	48	61
63	2·2027	2103	2179	2256	2333	2412	2490	2570	2650	2730	13	26	39	52	65
64	2·2812	2894	2976	3060	3144	3228	3314	3400	3486	3574	14	28	43	57	71
65	2·3662	3751	3841	3931	4022	4114	4207	4300	4395	4490	15	31	46	62	77
66	2·4586	4683	4780	4879	4978	5078	5180	5282	5384	5488	17	34	50	67	84
67	2·5593	5699	5805	5913	6022	6131	6242	6354	6466	6580	18	37	55	73	92
68	2·6695	6811	6927	7046	7165	7285	7407	7529	7653	7778	20	40	60	81	101
69	2·7904	8032	8161	8291	8422	8555	8688	8824	8960	9099	22	44	67	89	111
70	2·9238	9379	9521	9665	9811	9957	**0106**	**0256**	**0407**	**0561**	25	49	74	98	123
71	3·0716	0872	1030	1190	1352	1515	1681	1848	2017	2188	27	55	82	110	137
72	3·2361	2535	2712	2891	3072	3255	3440	3628	3817	4009	31	61	92	123	153
73	3·4203	4399	4598	4799	5003	5209	5418	5629	5843	6060	35	69	104	138	173
74	3·6280	6502	6727	6955	7186	7420	7657	7897	8140	8387	39	79	118	157	196
75	3·8637	8890	9147	9408	9672	9939	**0211**	**0486**	**0765**	**1048**	45	90	135	180	225
76	4·1336	1627	1923	2223	2527	2837	3150	3469	3792	4121	52	104	156	207	260
77	4·4454	4793	5137	5486	5841	6202	6569	6942	7321	7706	61	121	182	242	303
78	4·8097	8496	8901	9313	9732	**0159**	**0593**	**1034**	**1484**	**1942**	72	143	215	287	359
79	5·2408	2883	3367	3860	4362	4874	5396	5928	6470	7023	86	172	258	344	431
80	5·759	5·816	5·875	5·935	5·996	6·059	6·123	6·188	6·255	6·323					
81	6·392	6·464	6·537	6·611	6·687	6·765	6·845	6·927	7·011	7·097					
82	7·185	7·276	7·368	7·463	7·561	7·661	7·764	7·870	7·979	8·091					
83	8·206	8·324	8·446	8·571	8·700	8·834	8·971	9·113	9·259	9·411		Differences untrustworthy here			
84	9·57	9·73	9·90	10·07	10·25	10·43	10·63	10·83	11·03	11·25					
85	11·47	11·71	11·95	12·20	12·47	12·75	13·03	13·34	13·65	13·99					
86	14·34	14·70	15·09	15·50	15·93	16·38	16·86	17·37	17·91	18·49					
87	19·11	19·77	20·47	21·23	22·04	22·93	23·88	24·92	26·05	27·29					
88	28·65	30·16	31·84	33·71	35·81	38·20	40·93	44·08	47·75	52·09					
89	57·30	63·66	71·62	81·85	95·49	114·6	143·2	191·0	286·5	573·0					
	0'	6'	12'	18'	24'	30'	36'	42'	48'	54'	1'	2'	3'	4'	5'

The black type indicates that the integer changes.

10. A cannon is placed in position at point A to fire upon an enemy fort located on a mountain. The airline distance from the gun to the fort has been determined as 5 km. The distance on a horizontal plane from the gun to a point C at the base of the mountain is 3½ km. From this point at the base to the fort itself the distance is 1·8 km. (*a*) At what angle of elevation will the cannon have to be set in order to score a direct hit upon the fort? (*b*) What is the angle of depression from the fort to the cannon?

(*a*) (A) 27° 21′ ......   (B) 38° 59′ ......   (C) 13° 40′ ......   (D) 16° 8′ ......

(*b*) (A) 38° 59′ ......   (B) 27° 21′ ......   (C) 22° 16′ ......   (D) 13° 40′ ......

CHAPTER SIXTEEN

SCALES AND GRAPHS

SCALES

A **scale** is a convenient representation of one quantity or magnitude in terms of another. A scale thus expresses a ratio.

For instance, what covers km in reality, occupies only mm on a map. Hence, a map has a scale based on the ratio between distances *mapped* and distances *on* the map. This scale is usually shown on the map by a diagram, as here illustrated.

KILOMETRES

1   0   1   2   3   4
Map Scale

The *speedometer* in a car has a scale based on the ratio between circular displacements on its dial and the speed at which the car is travelling.

EXAMPLE: In making a speedometer, if a displacement equivalent to a 10° angle represented a speed of 5 km/h, what angle would be needed to represent a speed of 30 km/h?

SOLUTION: If 10° = 5 km/h, then 2° = 1 km/h, and 30 km/h = 30 × 2 or 60°, ANS.

A *thermometer* has a scale based on the ratio between the height of a column of mercury and the temperature of air or of some other medium or material, as measured in Centigrade or Fahrenheit degrees.

A *blueprint* is 'drawn to scale'. This means that the drawing on the blueprint represents a scale ratio. For instance, in the blue print of the hull of a ship, if the scale is 2 mm to 1 m, then every line 1 cm in length on the blueprint represents 5 m of the hull.

EXAMPLE: If a beam of the hull is to be 100 m, how long a line would be needed on the blueprint just described in order to represent this beam?

SOLUTION: If 2 mm = 1 m, then 1 cm = 5 m. 100 ÷ 5 = 20 cm, ANS.

GRAPHS

A *graph* is a diagram showing relationships between two or more factors. Most graphs have two scales, as the following population graph indicates:

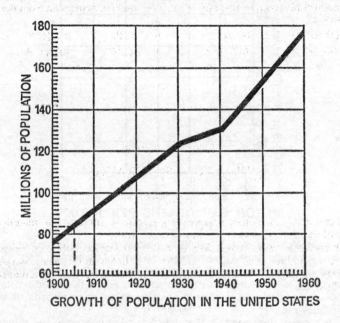

GROWTH OF POPULATION IN THE UNITED STATES

Parts of a Graph

In the preceding example the graph has a vertical scale, in which a unit of space on the paper represents 20 million population, and a horizontal scale, in which the same unit represents a period of 10 years. The **vertical scale** is sometimes called the **vertical** axis or **ordinate**, while the **horizontal scale** is called the **horizontal axis** or **abscissa**.

Reading the Graph

The graph just presented shows the population of the United States in millions from 1900 to 1960. If you could assume that the population growth was fairly uniform for any ten-year period, then you could estimate from the graph the population for the years 1905, 1906, 1907, etc. For example, 1905 is midway between 1900 and 1910. At this point move up vertically until you reach the graph line. From this point follow the broken line horizontally. It intersects the population scale about one-fifth of the way between 80 and 100. The population estimate would therefore be about 84 million in 1905.

Types of Graphs

The **broken-line graph** next shown illustrates the variations in deaths due to automobile accidents between the years 1925 and 1955. It is called a line graph, because a line is used to represent the factors in the two scales.

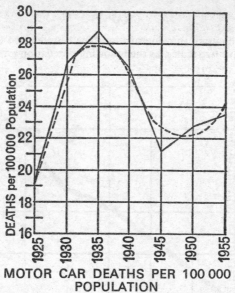

MOTOR CAR DEATHS PER 100 000
POPULATION

The unbroken line in this graph is plotted from the actual number of accidents. The light dotted line averages the path of the broken line.

Broken-line graphs are especially suitable for recording so-called *historical* data involving a factor subject to constant and severe fluctuation. This type of graph, accordingly, is preferred for presenting price records of stocks, commodities, etc.

The **bar graph** illustrated in the right-hand column gives the number of persons examined in a certain city during the years 1935–60. The height of the

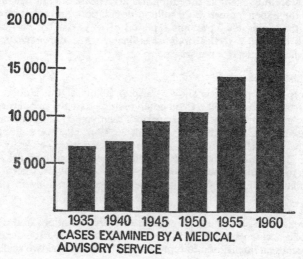

CASES EXAMINED BY A MEDICAL
ADVISORY SERVICE

black bar, indicating the number of cases, represents the vertical scale, while its base position, indicating the specific year, represents the horizontal scale.

The **circle graph** is usually on a percentage basis: the whole circle represents 100%, while fractional percentages are indicated by proportionate segments of the circle. The following figure illustrates such a graph:

HOW A MANUFACTURER'S EXPENSES
ARE APPORTIONED BEFORE TAXES

Mathematical Graphs

The graphs presented up to this point were merely graphic representations of statistical facts about items that are related but not according to any well-

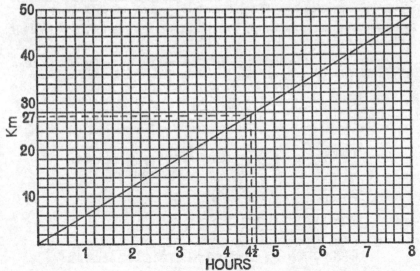

GRAPH SHOWING RELATION BETWEEN WALKER'S TIME AND DISTANCE

defined rules or formulae. The items were not causally related, that is, a change in one did not always produce a change in the other. However, the accompanying graph is a line graph representing quantities that change together. Such graphs may be made from formulae or equations, and vice versa. These may be called mathematical graphs.

EXAMPLE: A man walks at a rate of 6 km/h. Show graphically the relation between the distance he walks and the number of hours he walks.

	Hours	Distance
SOLUTION: Formula $D = 6T$.	1	6
	2	12
Make a table of hours and distances.	3	18
	4	24
Plot a graph by placing distances	5	30
on one scale and hours on another.	6	36
(See diagram on p. 205.)	7	42
	8	48

The range of numbers is decided according to the purpose of the graph. For *hours* the range selected is 8; for *distance* the range is 48. Next choose your unit intervals for each scale. For the scale of hours, since the range is 8, you can choose one large or unit box on the graph paper as equal to one hour. For the scale of distance this would not be practical, since you would need too much space. Therefore you choose a convenient interval. In this case let one small unit or square equal a distance of 2 km. Then proceed to plot each point according to the pairs of items in the table. Thus, the first point on the graph represents 1 hour and a distance of 6 km, the second point represents 2 hours and a distance of 12 km, etc. These points are called co-ordinates. When all the points are plotted draw your line, which in this case proves to be a straight line, showing that there must be a direct relationship between the distance covered and the time spent in walking.

Note: In plotting a graph which represents a formula or direct relation it is not necessary to plot all the points. In a straight-line graph three or four points are sufficient, two to determine the line, and one or two more as a check for accuracy.

Using or Reading the Graph. From such a graph you may now read off directly the distance walked in, say, $4\frac{1}{2}$ hours.

METHOD: $4\frac{1}{2}$ would be midway between four and five on the hours scale. Draw a line from this point to where it meets the line of the graph. From the point of the intersection draw a line out to the distance scale. This line is seen to cut the distance scale at 27, which is the answer. Check this by arithmetic in your formula. Substituting in $D = 6T$, $D = 6 \times \frac{9}{2} = \frac{54}{2} = 27$.

Picture graphs are used to illustrate statistical information, but the unit is

| PUBLIC UTILITIES | | Each symbol represents 200 000 men |

THE MAN-POWER SITUATION IN "COUNTRY X"

not a distance measure on a page as in the line and bar graphs, but rather a symbol of the fact being illustrated. The next figure illustrates such a graph in which several symbols are used.

Practice Exercise No. 81

1. Construct a broken-line graph from the data given in the bar graph on p. 204.
2. From the data given in the broken-line graph construct a bar graph giving the number of deaths due to motor-car accidents during the years 1925, 1930, 1935, 1940, 1945, 1950, and 1955.
3. If a certain type of timber cost 8 cents a metre, make a table and graph showing the relationship between the cost and the number of metres of timber bought. The range in metres should be from 15 to 55 m. Use intervals of 10 m and 80 cents. By arithmetic compute the price of $36\frac{1}{2}$ m of timber. Now find the price of $36\frac{1}{2}$ m of timber from the graph by the method indicated for graphic reading.
4. From the graph on page 205 find out how long it would take to walk 39 km and check by arithmetic.
5. From your graph in Problem 3 above, find out how many metres of timber you can buy for $256. Check by arithmetic.
6. A young man's day is divided as follows: work, 7 hours; eating and dressing, $3\frac{1}{2}$ h; study, 3 h; travel, $1\frac{1}{2}$ h; recreation and miscellaneous, 1 h; sleep, 8 h. Figure out the percentages for these divisions and draw a circular graph to illustrate.

CHAPTER SEVENTEEN

COMBINATIONS AND PERMUTATIONS

PRELIMINARY EXPLANATIONS

When a cricket captain picks *11* men out of a *squad* of *20* to form a team, mathematically he may be said to be making a *combination* of *20* 'things' taken *11 at a time*. When the captain assigns the members of such a team the *sequence* of a particular batting *order*, mathematically he may be said to be making a *permutation* of the *same* 20 'things' 11 at a time.

Combinations, we thus see, are *groupings* of things *without* regard to their order. *Permutations* are *arrangements* of the same in the *sequences* of *particular orders*. And the numerical aspects of both topics are important, since our workaday world is obviously one of much grouping and arranging.

Common symbols for the *total number* of *all possible combinations* of *n things taken t at a time* are

$$C(n, t), \quad C_t^n, \quad {}_nC_t, \quad \text{or} \quad \binom{n}{t}$$

Of these we shall here use only the first two, as when we designate all the possible combinations of 20 things taken 11 at a time, by the symbols $C(20, 11)$ or C_{11}^{20}.

The corresponding *symbols* for the *total number* of *all possible permutations* of *n* things taken *t* at a time are

$$P(n, t), \quad P_t^n, \quad \text{or} \quad {}_nP_t$$

Of these we shall here use only the first, as when we designate *all* the possible permutations of 20 things, taken 11 at a time, by the symbol P(20, 11).

For any given pair of numerical values of *n* and *t* in these formulae there are in general many *more* possible *permutations* than there are possible *combinations*.

As a simple illustration, suppose that we have any 3 objects—such as a set of three paintings designated *A, B, C*. The only possible *grouping* of *all 3* of these which we can select at a time *without* regard to sequence is the *combination*,

$$C(3, 3) = 1: \quad \overparen{A \text{ and } B \text{ and } C.}$$

But we can *arrange* this one combination—by hanging the paintings on adjacent wall panels, for instance—in any of the *sequences* of the possible *permutations* of 3 things taken all 3 at a time,

$$P(3, 3) = 6: \quad \begin{matrix} ABC, & BAC, & CAB, \\ ACB, & BCA, & CBA. \end{matrix}$$

Moreover, we can select *groupings* from these same 3 objects, *2* at a time without regard to sequence, in any of the *combinations*,

$$C(3, 2) = 3: \quad \begin{matrix} \overparen{A \text{ and } B,} \\ \overparen{B \text{ and } C,} \\ \overparen{C \text{ and } A.} \end{matrix}$$

But we can *arrange* these *combinations* in the *sequences* of the possible *permutations* of 3 things taken 2 at a time,

$$P(3, 2) = 6: \quad \begin{matrix} AB, & BA, & CA, \\ AC, & BC, & CB. \end{matrix}$$

However, in the *special case* that we select *only 1* of these 3 objects at a time, we *cannot* permute any single grouping from its original order of *1 thing standing by itself*. In other words, the numbers of possible combinations and permutations are *the same*,

$$C(3, 1) = P(3, 1) = 3: \quad \begin{matrix} A \text{ (alone)}, \\ B \text{ (alone)}, \\ C \text{ (alone)}. \end{matrix}$$

This illustrates the *general fact* that, for *any number* of objects, *n*,

$$P(n, 1) = C(n, 1) = n$$

EXAMPLE 1: By forming arrays of all the possibilities, as above, find C(5, 2) and P(5, 2).

SOLUTION: Letting *A, B, C, D, E* represent the *5 things*, we can systematically *combine* them 2 at a time in any of the ways,

$$\overbrace{A \& B,}$$

C(5, 2): $\overbrace{A \& C,}$ $\overbrace{B \& C,}$

$\overbrace{A \& D,}$ $\overbrace{B \& D,}$ $\overbrace{C \& D,}$

$\overbrace{A \& E,}$ $\overbrace{B \& E,}$ $\overbrace{C \& E,}$ $\overbrace{D \& E.}$

And we can systematically *permute* these combinations in the ways,

AB,	BA,	CA,	DA,	EA,
AC,	BC,	CB,	DB,	EB,
AD,	BD,	CD,	DC,	EC,
AE,	BE,	CE,	DE,	ED.

P(5, 2):

By actual count of these arrays, as in the preceding illustrations, we now find that

$$C(5, 2) - 10, \quad P(5, 2) - 20, \quad \text{Ans.}$$

In other words: there are 10 possible *combinations* of 5 things taken 2 at a time; and there are 20 possible *permutations* of 5 things taken 2 at a time.

Unless *n* and *t* have very small values, it is tedious to find $C(n, t)$ and $P(n, t)$ by the method of the above illustrations, in which all the separate possibilities are written out in an *array* and then counted. Hence, in actual practice, *short-method formulae* are used.

One might, perhaps, expect the derivation of these formulae to begin with the simpler concept of a combination, which does not involve the idea of arrangement in different sequences. Nevertheless, in most instances it is mathematically simpler to compute $P(n, t)$ than to compute $C(n, t)$; and for this reason we begin with permutations formulae.

PERMUTATIONS FORMULAE

Returning to the above illustration of P(3, 3) = 6, the 6 possible permutations of 3 things taken 3 at a time—*ABC, ACB*, etc.—note now that each arrangement in the array begins with one of the *3 possibilities, A, B, C*. But once this first permutational position has been filled, only the 2 remaining possibilities can be assigned to the second permutational position (*B* or *C* after *A*, and *A* or *C* after *B*, etc.). And once both of the first 2 permutational positions have been filled, only the *one* remaining possibility can be assigned to the third permutational position (*C* after *A* and *B*, etc.).

Thus, we see that P(3, 3) must be the *product* of 3 possible entries for the *first* permutational position, times $3 - 1 = 2$ possible entries for the *second* permutational position, times $3 - 2 = 1$ possible entry for the *third* permutational position; or $3(2)(1) = 6$, which was the actual count.

The product in this form is called **factorial three**. The symbol for it is 3!, meaning **3 times 2 times 1,** which is a special case of

$$n! = n(n - 1)(n - 2) \ . \ . \ . \ (2)(1)$$

Return next to the illustration in Example 1 of $P(5, 2) = 20$, the 20 possible permutations of 5 things taken 2 at a time—*AB, AC, AD*, etc. We can similarly see that $P(5, 2)$ must be the product of 5 possible entries for the *first* permutational position, times $5 - 1 = 4$ possible entries for the *second* permutational position; or $5(4) = 20$, which again was the actual count.

Note in this last case, however, that $5(4) = 5!/3! = 5!/(5 - 2)!$, since $3!$ $= (5 - 2)!$ in the denominator of this fraction cancels out the $(3)(2)(1)$ of $5!$ in its numerator. Therefore, *generalizing* the reasoning of the two above examples to the case of *all the possible permutations* of any *n* things taken *t* at a time, we arrive at the *formula*,

$$P(n, t) = \frac{n!}{(n - t)!}$$

In the **special case** where $t = n$, this becomes

$$P(n, n) = n!$$

since the denominator of the $P(n, t)$ formula is then *eliminated*. (NOTE especially here that when $t = n$, then $(n - n)!$ or $(0)!$ in the denominator is, by definition, *no number at all*. Therefore, it is NOT equal to *zero*, which would make the first permutation formula *meaningless!*)

Permutations formulae are applicable to the solution of many different types of problems involving arrangements in different sequences.

EXAMPLE 2: In how many different orders may 5 of 7 different wires from one piece of electrical equipment be attached to 5 different posts on another?

SOLUTION: Since the question of *sequence* (order) is involved in this problem, it is one of *permutations* with $n = 7$ and $t = 5$. Hence

$$P(7, 5) = \frac{7!}{(7 - 5)!} = \frac{7!}{2!}$$

$$= \frac{7 \cdot 6 \cdot 5 \cdot 4 \cdot 3 \cdot \overset{1}{\cancel{2}} \cdot 1}{\underset{1}{\cancel{2}} \cdot 1} = 2520, \quad \text{ANS.}$$

Since it is obvious from the arithmetic of this solution that the *1* (one) factor of *n!* never affects the result, we shall hereafter leave it implied (not written) in numerical computations.

EXAMPLE 3: In how many different arrangements may 5 persons be seated along the head of a banquet table?

SOLUTION: Here $n = t = 5$, and
$$P(5, 5) = 5! = 5 . 4 . 3 . 2 = 120, \quad \text{ANS.}$$

When *special conditions* are attached to permutation problems, the *same formulae* may often still be used, but they must be *adapted* to the problems' different requirements.

EXAMPLE 4: How many different arrangements are there possible in the preceding example if one particular pair of persons must always be seated next to each other?

SOLUTION: Temporarily considering the special pair as *a unit*, we may think of the first part of our problem as one of permuting *this unit* with the other three persons. *Then n = t = 4*, and

$$P(4, 4) = 4! = 4 \cdot 3 \cdot 2 = 24, \quad \text{PARTIAL ANS.}$$

But in *each* such permutation, the 2 members of the special pair may sit next to each other in *either* of the 2 possible orders, *AB* or *BA*. Hence, the total number of possible permutations is really *twice P(4,4)*, or

$$2P(4, 4) = 2(24) = 48, \quad \text{ANS.}$$

EXAMPLE 5: Answer the same question if the special pair must always be seated *separated from each other*.

SOLUTION: The entire group can be seated with no special conditions in $P(5, 5)$ different ways (*Example 3*), and with the special pair together in $2P(4, 4)$ different ways (*Example 4*). Hence, the number of ways in which they can be seated with the special pair separated must be the *difference* between these two figures, or

$$P(5, 5) - 2P(4, 4) = 120 - 48 = 72, \quad \text{ANS.}$$

EXAMPLE 6: Recompute the answer to Example 2 if the 5 persons are to be seated, with no others, about a *round* table.

SOLUTION: Now no one person is actually 'first', 'last', or 'centred', in a linear ('lined up') arrangement. However, we may arbitrarily think of any *one* person's position as 'fixed' for purposes of reference and then permute the positions of the other *4* with respect to each other, clockwise or counter-clockwise from this point of reference. Thus $n = 5 - 1 = 4$, and as further above,

$$P(4, 4) = 4! = 24, \quad \text{ANS.}$$

Arrangements of things in a *closed chain* or *ring*, as in the last example, are called *circular permutations*. Generalizing the method of reasoning just illustrated, therefore, we derive for P_c, *the total number of circular permutations possible for any n things taken n at a time*, the *formula*

$$P_c(n, n) = (n - 1)!$$

Thus the preceding solution could have been arrived at directly, by substitution in this formula, as

$$P_c(5, 5) = (5 - 1)! = 4!, \text{ etc.}$$

EXAMPLE 7: In Example 3 suppose 3 of the 5 persons to be men (represented by M's) and 2 to be women (represented by W's). In how many different arrangements can they be seated *according to gender* rather than as individuals?

SOLUTION: Let our answer be P_a, the total number of permutations possible for 3 (indistinguishable) M's and 2 (indistinguishable) W's, taken all 5 at a time as in the typical case of $MMMWW$. Then in order to express $P(5, 5)$ in terms of P_a, we would have to multiply the latter by $P(3, 3) = 3!$ to obtain the number of permutations in which the 3 M's are distinguishable as M_1, M_2, M_3, and by $P(2, 2) = 2!$ to obtain the number of cases in which the 2 W's are also distinguishable as W_1 and W_2. But we already know that $P(5, 5) = 5!$ Hence, by substitution

$$3!2!P_a = P(5, 5) = 5!$$

And therefore

$$P_a = \frac{5!}{3!2!} = \frac{\overset{2.1.1}{5.4.3.2}}{\underset{1.1.1}{3.2.2}} = 10, \quad \text{Ans.}$$

These 10 possible ways, 6 beginning with a man and 4 with a woman, are illustrated as follows:

$$6 \begin{cases} MMMWW, \\ MMWMW, \\ MMWWM, \\ MWMMW, \\ MWMWM, \\ MWWMM, \end{cases} \left.\begin{matrix} WMMMW, \\ WMMWM, \\ WMWMM, \\ WWMMM. \end{matrix}\right\} 4$$

Note that the preceding problem requires us to treat its 'men' as M's 'all alike' and its 'women' as W's 'all alike'. This is doubtless sociologically superstitious and matrimonially foolish. But the accompanying mathematical procedure is nevertheless sound. Generalizing the solution's method of reasoning, we derive for P_a, *the total number of* ***indistinguishable permutations*** *possible for* ***n*** *things taken* ***n*** *at a time when* ***n_1*** *are all alike, and* ***n_2*** *are all alike,* etc., the *formula*,

$$P_a = \frac{n!}{n_1!n_2!\,\ldots}$$

EXAMPLE 8: As for purposes of cryptography, how many indistinguishable 9-letter words can be formed by permutations of the letters in the code word, 'TENNESSEE'?

SOLUTION: Here $n = 9$, $n_1 = 4$ for the E letters, and $n_2 = n_3 = 2$ for the N and S letters. Hence,

$$P_a = \frac{9!}{4!2!2!} = \frac{\overset{2}{9.8.7.6.5.}\overset{1.1.1}{4.3.2}}{\underset{1.1.1.1.1}{4.3.2.2.2}} = 3780, \quad \text{Ans.}$$

EXAMPLE 9: In how many different ways can a captain assign speakers to the first and second positions of a 2-man debating team from 12 candidates?

SOLUTION: Here $N = 12$, $t = 2$, and

$$P(12, 2) = \frac{12!}{(12 - 2)!} = \frac{12!}{10!}$$

$$= \frac{\overset{1.1.1.1.1.1.1.1.1}{12.11.10.9.8.7.6.5.4.3.2}}{\underset{1.1.1.1.1.1.1.1.1}{10.9.8.7.6.5.4.3.2}}$$

$$= 132, \quad \text{Ans.}$$

Or, with the second line written more briefly,

$$= 12 \cdot 11 \cdot \frac{10!}{10!} = 132, \quad \text{THE SAME ANS.}$$

since $10!/10! = 1$.

From this example we see that *it is always correct, and often arithmetically more convenient, to treat*

$$\frac{n!}{(n-t)!} = n(n-1)(n-2) \ldots (n-t+1)$$

in the *basic formula* for $P(n, t)$. From now on, therefore, we shall write $12!/(12-2)!$ or $12!/10!$ directly as $12 \cdot 11$, etc., whenever we do not wish to use the other factors of the numerator for cancellations with other denominator factors in the same term.

COMBINATIONS FORMULAE

From the very beginning of this chapter we have seen that *permutations are simply different arrangements of the things grouped in corresponding combinations.* More particularly, each *set* of *permutations* of any given t things in $P(n, t)$ consists simply of *different arrangements* of the *same* t things in a *corresponding combination* of $C(n, t)$.

But now we know from our basic permutations formula that *each combination* of t things in $C(n, t)$ can have $P(t, t) = t!$ possible *permutations*. Hence,

$$t! C(n, t) = P(n, t)$$

Dividing by $t! = t!$, we then obtain the second important *basic relationship*,

$$C(n, t) = \frac{P(n, t)}{t!}$$

and by substitution for $P(n, t)$ from our basic permutations formula, this gives us for the **number of all possible combinations of n *things taken t at a time*,** the *basic formula*,

$$C(n, t) = \frac{n!}{t!(n-t)!}$$

In the *special case* where $t = n$, the denominator factor, $(t - t)!$ or $(0)!$ is again *eliminated* (see *NOTE* on p. 210), and the above formula takes the anticipated form,

$$C(n, n) = n!/n! = 1$$

Combinations formulae can be applied to the solution of many different types of problems in which selections are to be made without regard to the possible permutations of arrangement.

EXAMPLE 10: (*a*) In how many different ways may a coach assign players, from a squad of 12, to the 9 *positions* on a baseball line-up? (*b*) In how many different ways can he select a team of 9 from the same squad *without regard to the positions they are to play*?

SOLUTION: Here $n = 12$ and $t = 9$. Since question (a) involves positions, we must again use the previous derived *permutations formula*,

$$P(12, 9) = \frac{12!}{(12 - 9)!} = \frac{12!}{3!}$$

$$= 12.11.10.9.8.7.6.5.4$$
$$= 79\ 833\ 600, \quad \text{ANS. } (a)$$

But since question (b) concerns *selection without regard to position played*, we must now use our new *combinations* formula,

$$C(12, 9) = \frac{12!}{9!(12 - 9)!} = \frac{12!}{9!3!}$$

Then, dividing numerator and denominator by $9!$ rather than by $3!$, this gives us most simply,

$$\frac{\overset{4}{12}.11.\overset{5}{10}}{\underset{1}{3}.\underset{1}{2}} = 220, \quad \text{ANS. } (b)$$

Note in this numerical instance how very large a number of possible permutations may be as compared to the corresponding number of possible combinations. The latter may also be surprisingly large, however; and then it is important to shorten the computation by the arithmetic device of the above solution.

EXAMPLE 11: In how many different ways can one hand of 5 cards be dealt from a 52-card pack.

SOLUTION: Since the final grouping of the cards, rather than the order in which they are dealt, is important here, the problem is again one of combinations rather than of permutations. Hence, $n = 52, t = 5$, and

$$C(52, 5) = \frac{52!}{5!(52 - 5)!} = \frac{52!}{5!47!}$$

$$= \frac{52.51.\overset{10}{50}.49.\overset{2}{48}}{\underset{1}{5}.\underset{1}{4}.\underset{1}{3}.\underset{1}{2}} = 2\ 598\ 960, \quad \text{ANS.}$$

INTER-GROUP COMBINATIONS

Thus far we have meant by 'combinations' only '*sub-group* combinations'— those of t things taken from the *same* group of n things. But many related problems concern *inter-group* combinations. These may be defined as selections of t_1 things at a time from a *first* group of n_1 things, *along with* t_2 things at a time from a *second* group of n_2 things, *along with* t_3 things at a time from a *third* group of n_3 things, *etc.*

Suppose, for instance, that we wish to combine 1 of 2 photographs, R and S, with 2 of the 3 paintings, A, B, and C, mentioned at the beginning of this chapter (p. 208). Here $n_1 = 2$ (photographs), $t_1 = 1$; $n_2 = 3$ (paintings), $t_2 = 2$; and *the total number of possible inter-group combinations, $N = 6$, is* illustrated by the array:

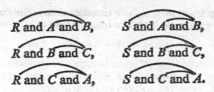

R and A and B, S and A and B,

R and B and C, S and B and C,

R and C and A, S and C and A.

Evidently, here, we may make our *first* choice (of 1 of the 2 photographs, R and S) in either of the 2 ways,

$$C(2, 1) = \frac{2!}{1!(2 - 1)!} = \frac{2!}{1!1!} = \frac{2}{1.1} = 2$$

Also, we may make our *second* choice (of 2 of the 3 paintings, A, B, and C) in any of the 3 ways,

$$C(3, 2) = \frac{3!}{2!(3 - 2)!} = \frac{3!}{2!1!} = \frac{3.2}{2} = 3$$

Hence, the total number, N, of possible ways in which we may combine *both* of these choices (to make up a selection of 1 photograph *and* 2 paintings) is

$$N = C(2, 1)C(3, 2) = 2(3) = 6$$

Applying the same method of reasoning to the general case defined above, therefore, we derive the basic *inter-group combinations* formula,

$$N = C(n_1, t_1) \, C(n_2, t_2) \, . \, . \, . \, C(n_k, t_k)$$

for *the total number of joint selections possible from k different sets of $n_1, n_2 \ldots n_k$, things taken $t_1, t_2 \ldots, t_k$ at a time, respectively. In the special case where $t_1 = t_2 = \ldots t_k = 1$*, this becomes

$$N = n_1 n_2 \, . \, . \, . \, n_k$$

since $C(n, 1) = n$ for *all* possible values of n.

The k different *kinds of things* from which choices are to be *combined* in applications of these formulae may be *objects*, *events*, *placements*, or whatever. They may also occur or be selected in the actual time sequence, $1, 2, 3, \ldots k$, or simultaneously. In the latter case, however, it is convenient to assign some *one arbitrarily selected order* to them. This is to avoid the sort of duplicated count we would get if we were (mistakenly) to regard the combination of photograph R with paintings A and B, in the above instance, as different from the combination of paintings A and B with photograph R (the latter being a different *permutation*, but *not* a different *combination*).

EXAMPLE 12: In planning a plane trip from Boston to San Diego with stops at Chicago and St. Louis, a travel agent finds he can book a seat for his customer on 5 different flights to Chicago, on 4 different flights from there to St. Louis, and on 7 different flights from there on. How many different complete itineraries can he book?

SOLUTION: Here $k = 3$ (different stages of the trip), and it is *common-sense routine* to let $n_1 = 5$ (possible choices for the first stage), $n_2 = 4$ (possible choices for the second stage), and $n_k = n_3 = 7$ (possible choices for the third stage), all in their *actual time sequence*. Then the total number of possible itineraries is

$$N = n_1 n_2 n_3 = 5(4)(7) = 140, \quad \text{Ans.}$$

EXAMPLE 13: An international scientific conference is attended by 5 representatives from North America, 4 from South America, 7 from Europe, 6 from Asia, and 3 from Africa. How many different committees can be chosen from this assemblage with three representatives from each area?

SOLUTION: Here there is no natural time sequence of choice, but we may *arbitrarily* let $n_1 = 5$ (possibilities of choice from North America), $n_2 = 4$ (from South America), etc. Then $t_1 = t_2 = \ldots t_5 = 3$, and

$$N = C(5, 3)C(4, 3)C(7, 3)C(6, 3)C(3, 3)$$

$$= \frac{\overset{2}{5}.4.3}{\underset{1}{3}.2} \cdot \frac{\overset{1.1}{4}.3.2}{\underset{1}{3}.2} \cdot \frac{\overset{1}{7}.6.5}{\underset{1}{3}.2} \cdot \frac{\overset{1}{6}.5.4}{\underset{1}{3}.2} \cdot \frac{\overset{1.1}{3}.2}{\underset{1}{3}.2}$$

$$= (10)(4)(35)(20)(1) = 28\,000, \quad \text{Ans.}$$

EXAMPLE 14: Designate the 'head' and 'tail' faces of 3 coins in any arbitrary order as H_1 and T_1, H_2 and T_2, H_3 and T_3, respectively. If we count $H_1 T_2 T_3$ as *different* from $T_1 H_2 T_3$ or $T_1 T_2 H_3$, etc., in how many different patterns, of their head or tail faces up, can the 3 coins land when tossed?

SOLUTION: Here the 3 coins and their corresponding faces may be identical. However, when we consider them in the proposed *arbitrary* order, $n_1 = 2$ (possibilities of H_1 or T_1 for the *first* coin), $n_2 = 2$ (possibilities of H_2 or T_2 for the *second* coin), etc.; then $k = 3$, and the total number of different possible ways in which the 3 coins can land is

$$N = n_1 n_2 n_3 = 2(2)(2) = 2^3 = 8, \quad \text{Ans.}$$

EXAMPLE 15: Each die of a pair has its faces numbered from 1 to 6. In how many different patterns of their numbered faces up—'different' being defined as in the preceding problem—can the pair come to rest when rolled?

SOLUTION: By the method of the preceding solution, $k = 2$, and $n_1 = n_2 = 6$. Hence,

$$N = n_1 n_2 = 6^2 = 36, \quad \text{Ans.}$$

From the last two examples we see that, in the further special case where $n_1 = n_2 = \ldots n_k = n$, the preceding formula becomes:

$$N = n^k$$

In Example 14, for instance, $k = 3$, $n = 2$, and

$$N = n^k = 2^3 = 8, \text{ as before}$$

In Example 15, also, $k = 2$, $n = 6$, and

$$N = n^k = 6^2 = 36, \text{ as before}$$

But a more general result now follows. . . .

In a fairly common type of combinations problem we are required to find M, the *sum of all the possible combinations* of k things taken *1, 2, 3, . . . and any other number up to, and including, k at a time.* This can always be done 'the long way' by substitution in the defining formula,

$$M = C_1^k + C_2^k + \ldots + C_k^k.$$

Note, however, that we can think of the k things in this formula as the same k as in the preceding formula for $N = n^k$. Also, we can think of $n = 2$ in the latter as the *two possibilities* of each of the same k things being *either* (a) *included* in possible combinations, *or* (b) *excluded* from such combinations. Then

$$N = M + 1, \quad or \quad M = N - 1$$

with the *1* in these equations being the number of the *one* possibility in which *all k* things are *excluded* from any possible combination. Hence, by substitution, we obtain the formula

$$M = C_1^k + C_2^k + \ldots + C_k^k = 2^k - 1$$

(*Note:* Most textbooks derive this formula by a longer method which applies a special case of quite a different principle called *the binomial theorem.* But this method is recommended as shorter, more direct, and less dependent upon other principles or the special device of ingenious substitution.)

EXAMPLE 16: In how many different combinations can we replace 1 or more electronic tubes in a set of 8?

SOLUTION: Since these replacements may be made for any 1, *or* 2, *or* 3, *or* more tubes *up to and including* all 8, our *M formula* applies with $n = 8$. Hence,

$$M = 2^8 - 1 = 256 - 1 = 255, \quad \text{ANS.}$$

Obviously, the *minus one* in this computation *adjusts* the total for the *rejected* possibility that *no* tube be replaced.

As when we apply permutations formulae (p. 209) we may sometimes have to *adapt* the above *combination formulae* to the *special conditions* of particular problems.

Care must be taken, however, not to *insinuate duplications of arrangement in this procedure,* for then we shall obtain *numerical results* which are *too large by some factor of a corresponding permutations formula.* The very real danger of this becomes immediately evident when we recognize that the reasoning by which we originally arrived at our basic permutations formula (p. 210) was itself but an application of our basic inter-group combinations formula (p. 215) to the systematic filling of successive positions in a permutational arrangement!

EXAMPLE 17: Of all the possible hands of cards computed in the solution of Example 11, how many contain '4 of a kind'—defined as 4 cards of 1 rank (all *sevens*, or all *jacks*, etc.)?

SOLUTION: Since there are only 4 cards in each of the 13 ranks (*ace, king, queen, jack, 10, 9, 8, 7, 6, 5, 4, 3, 2*) of a bridge pack, *4* cards of the *same rank* can be dealt from such a pack in only

$$13C_4^4 = 13 \cdot 1 = 13 \text{ different ways}$$

But each such combination can be further combined with any one of the remaining $12C_1^4 = 12 \cdot 4 = 48$ cards of 12 *other* ranks in the pack. Hence, the number of 5-card hands with '4 of a kind' is

$$13C_4^4 12C_1^4 = 13 \cdot 12 \cdot 4 = 624, \quad \text{Ans.}$$

EXAMPLE 18: In how many different ways can a hand be dealt containing only 'one pair'—defined as 2 cards of *1* rank and 3 other cards of ranks *different* from each other and from that of the pair?

SOLUTION: Obviously a pair can be formed in *each* rank in C_2^4 ways, and hence in *all* 13 ranks of the pack in

$$13C_2^4 = 13 \cdot \frac{4 \cdot 3}{2} = 78 \text{ ways}$$

There is a *danger* we may now make the *mistake* of reasoning (erroneously) that each such combination can be further combined with 1 card from each of 3 remaining ranks in

$$12C_1^4 11C_1^4 10C_1^4 = 12 \cdot 4 \cdot 11 \cdot 4 \cdot 10 \cdot 4$$
$$= 84\,480 \text{ different ways}$$

Multiplied by *78*, this would give us a product of *6 589 440 ways*, which is *impossible*, since it is more than twice as great as the total of *all* possible hands, including those with no pairs or with other sub-combinations than pairs (see Example 11)!

The error of the preceding reasoning is that it *permutes*, rather than *combines*, selections from three of the remaining suits. Hence, the product at which it arrives is *too great* by a factor of $3! = 3 \cdot 2 = 6$.

To correct the error, we could divide the obtained product by a compensating factor of $3! = 6$. Or, *to avoid the error altogether*, we can reason that 3 ranks different from the pair's rank can be combined in

$$C_3^{12} = \frac{12 \cdot 11 \cdot 10}{3 \cdot 2} = 220 \text{ ways}$$

Then from each of the latter we can make an inter-group combination of one of $n = 4$ cards from each of $k = 3$ ranks in

$$n^k = 4^3 = 64 \text{ ways}$$

Hence a 5-card hand with only '1 pair' can be dealt in

$$13C_2^4 \, 64C_3^{12} = 78 \cdot 64 \cdot 220$$
$$= 1\,098\,240 \text{ ways}, \quad \text{Ans.}$$

Further illustrations of how to apply permutations and combinations formulae are given in the following chapter on *Probability Theory*.

Practice Exercise No. 82

1. Illustrate in array form, and then count, $C(4, 3)$ and $P(4, 3)$.
2. Compute the same by formula.
3. How many different selections can we make of 5 flags from among 12 different flags?
4. How many different signals can we form by hoisting in a line to a ship's masthead: (*a*) 5 different flags? (*b*) 5 of 12 different flags?
5. (*a*) Derive the formula,

$$P(n, t) = C(n, t)P(t, t)$$

(*b*) By substitution in this formula, check your answers to questions 3 and 4 above.

6. In how many different ways can 8 people be seated at a round table if one particular couple is never to be separated?

7. How many 5-letter words can there be formed, like 'crate', in which no letter is repeated, the third and fifth are from the 5 vowels, *a*, *e*, *i*, *o*, *u*, and the other 3 are from the other 21 letters of the alphabet?

8. In how many different indistinguishable sequences can 3 reds, 6 blues, and 4 greens be arranged?

9. (*a*) How many 2-piece suits can a man select from a wardrobe of 4 jackets and 6 pairs of trousers?

(*b*) How many 3-piece suits if the same wardrobe also includes 3 waistcoats?

10. If 5 coins are tossed, in how many different patterns can they land heads or tails?

11. Answer the corresponding question if 4 dice are rolled.

12. A manufacturer offers cars, with or without automatic transmission, in 6 different body styles, 4 different interior finishes, and 9 different colours. How many models would a dealer have to stock in order to display all possible combinations?

13. The Braille system of writing for the blind is based on raised dots in 6 possible positions. How many different characters are there possible in this system?

14. How many different signals can we form by hoisting, in a line to the masthead of a ship, any 5 of 5 different *kinds* of flags?

15. How many different selections can there be made of 1 or more books from a shelf of 16?

16. In how many different ways can each of the following card hands be dealt:

(*a*) a 'full house'—defined as a pair from 1 rank and 3 of a kind from another?
(*b*) any 'flush'—defined as 5 cards from one suit?
(*c*) any 'straight'—defined as 5 cards of consecutive ranks with ace either high or low?
(*d*) a 'straight flush'?
(*e*) a 'flush' other than a '*straight* flush'?
(*f*) a 'straight' other than a 'straight *flush*'?

Note: Authors such as Hoyle in *The Official Rules of Card Games* usually mean by 'straights' and 'flushes' only those which are not 'straight flushes'. The latter they give a separate, higher listing—above 'full houses' and 'fours of a kind'—because of their relative rarity.

SUMMARY

Combinations are *groupings* of things *without* regard to their sequence. *Permutations* are *arrangements* of the same in *different sequences*.

In theory, at least, we can always find how many combinations and permutations are possible under any stated set of conditions by systematically constructing *arrays* of the separate possibilities and then *counting* their numbers.

In most instances, however, it is more practicable to apply *formulae* such as those recapitulated here for convenient reference:

Formula	*Application*
$P(n, t) = \dfrac{n!}{(n - t)!}$	The number *P* of all possible *permutations* of any *n* things taken *t* at a time.

Formula	Application
$P(n, n) = n!$	The same in the special case when $t = n$.
$P_c(n, n) = (n - 1)!$	The number P_c of all possible *circular permutations* of n things arranged in a *circle* or *closed chain*.
$P_a = \dfrac{n!}{n_1! n_2! \ldots}$	The number P_a of all possible *indistinguishable permutations* of n things when n_1 are all alike, n_2 are all alike, etc.
$C(n, t) = \dfrac{n!}{t!(n - t)!}$	The number of all possible *combinations* of n things taken t at a time.
$C(n, n) = 1$	The same for the special case of $t = n$.
$C(n, t) = \dfrac{P(n, t)}{t!}$ $P(n, t) = C(n, t)P(t, t)$	Basic relationships between P and C.
$N = C(n_1, t_1)C(n_2, t_2)$ $\ldots C(n_k, t_k)$	The number of all possible *inter-group combinations* of n_1 things taken t_1 at a time, etc., up to n_k things taken t_k at a time.
$N = n_1 n_2 \ldots n_k$	The same for the special case of $t_1 = t_2 = \ldots = t_k = 1$.
$N = n^k$	The same for the further special case of $n_1 = n_2 = \ldots n_k = n$.
$M = C_1^k + C_2^k + \ldots C_k^k$ $= 2^k - 1$	The number of all possible combinations of k things taken *1*, or *2*, or $\ldots k$ at a time.

The principal danger to avoid in applying these formulae is that of (erroneously) insinuating *permutation* factors into calculations of numbers of *combinations*.

CHAPTER EIGHTEEN

THEORY OF PROBABILITY

PRELIMINARY EXPLANATIONS

In *everyday* language we call something 'probable' if we believe it *likely to happen*, 'improbable' if we believe it *unlikely to happen*, or 'certain' if we believe it *sure to happen*. In **the mathematical theory of probability** we try to *define* such concepts *more precisely* so as to assign them *measures*, or *numerical indices*, which can be computed, and therefore compared, arithmetically.

As may be expected from its title, the theory deals with the likelihood of things which happen *by chance* or are selected *at random*. By definition, these are *events* which are not influenced to happen one way rather than another by *known* causes. Simple instances are: the landing of a tossed coin heads or tails, or the 'blindfolded' selection of 'any card' from a shuffled pack.

At first approach this sort of subject matter may seem to upset all our previous ideas of what mathematics is about. Previously in this subject we have not made such statements, for instance, as 'the base angles of an isosceles triangle are *probably* equal', or 'the *odds* are *in favour of* the sine of 30° being 0·500'.

When we understand the nature of mathematics, however, we realize that the more precise statements usually made about such matters are just short forms of '*If . . . , then . . .*' hypothetical generalizations. What we really mean, when we take the trouble to repeat it all over again each time, is that '*if* a triangle is isosceles, *then* its base angles are equal', or '*if* an angle is 30°, *then* its sine is 0·500'. These *hypothetical generalizations* follow from the *theoretical assumptions* upon which they are based, *even though* there may be *no* actual physical structure in the entire material universe which is *exactly* isosceles or which has an angle of *precisely* 30°.

The corresponding *theoretical assumption* by which we can put even *un*-certainties on *an exact mathematical* basis is that of something happening in a *definite number* of *equally likely ways*. Instances are: the assumption of a 'well-balanced' coin being *equally likely* to land with *either* heads *or* tails face up, or the assumption of an 'unloaded' die being *equally likely* to stop rolling with *any one* of its 6 faces up.

BASIC DEFINITIONS

If *on any given trial*, we say, an *event E* can *happen* in *h ways*, and *fail to happen* in *f ways*, out of $w = h + f$ *equally likely ways*, then the *mathematical measure* of *the probability of E happening* is, by definition,

$$p = h/w, \text{ or } h \text{ chances out of } w$$

and the *mathematical measure* of *the probability of E failing to happen* (or of the *im*probability of *E* happening) is, also by definition,

$$q = f/w, \text{ or } f \text{ chances out of } w$$

In other words, *p* and *q* are simply *numerical indices* assigned to the *likelihood* and *unlikelihood*, respectively, that *E* will happen *under the assumed theoretical conditions*.

From these definitions it follows at once that, in the *special case* when $h = 0$, then $h/w = 0/w = 0$, and

$$p = 0, \text{ the } measure \text{ of } impossibility$$

Also, in the *special case* when $h = w$, then $h/w = w/w = 1$, and

$$p = 1, \text{ the } measure \text{ of } certainty$$

For instance, let our *trial* be the tossing of a 'well-balanced coin'—by assumption, a coin which can land heads *or* tails with *equal* likelihood but *cannot* stand on edge. And let three *events* be defined as follows:

> *E*—the coin lands heads,
> E_1—the coin stands on edge,
> E_2—the coin lands heads or tails.

Then for event *E*, that the coins land heads,

$$h = 1, \quad f = 1, \quad w = h + f = 1 + 1 = 2$$

and we obtain the probability measures,

$$p = h/w = 1/2 \text{ for } E \text{ } happening$$
$$q = f/w = 1/2 \text{ for } E \text{ } failing \text{ to happen}$$

These *express numerically* the fact that the coin is *equally likely* to land heads *or* not to land heads.

Likewise for event E_1, that the coin stands on edge,

$$h_1 = 0, \quad f_1 = 2$$
$$w_1 = h_1 + f_1 = 0 + 2 = 2$$

and we obtain the *probability measures*,

$$p_1 = h_1/w_1 = 0/2 = 0 \text{ for } E_1 \text{ } happening$$
$$q_1 = f_1/w_1 = 2/2 = 1 \text{ for } E_1 \text{ } failing$$

These *express numerically* the fact that the coin *cannot* stand on edge.

Finally, for event E_2,

$$h_2 = 2, \quad f_2 = 0$$
$$w_2 = h_2 + f_2 = 2 + 0 = 2$$

and we obtain the *probability measures*,

$$p_2 = h_2/w_2 = 2/2 = 1 \text{ for } E_2 \text{ } happening$$
$$q_2 = f_2/w_2 = 0/2 = 0 \text{ for } E_2 \text{ } failing$$

These *express numerically* the fact that the coin *must* certainly land *either* heads *or* tails (not heads) *under the given assumptions*.

From the above definitions it is also clear that *the probability measures, p and q, must always have values from 0 to 1 inclusive*:

$$0 \leqq p \leqq 1, \text{ and } 0 \leqq q \leqq 1$$

This is only reasonable, since we should not expect a measure of probability to be less than that of impossibility (*0*) or greater than that of certainty (*1*).

Finally, since $w = h + f$ by definition, we see that

$$\frac{h}{w} + \frac{f}{w} = \frac{h+f}{w} = \frac{w}{w} = 1$$

Hence, by substitution of $p = h/w$ etc.,

$$p + q = 1, \quad p = 1 - q, \quad q = 1 - p$$

Any one of these last three relationships may serve as a partial check upon the accuracy of work in calculating p and q. (The 'check' works unless pairs of errors have been made exactly compensating for each other.) The last two relationships may also serve for finding either p or q once the other has been computed.

In the preceding illustration of event E, for instance,

$$p + q = \tfrac{1}{2} + \tfrac{1}{2} = 1. \textit{ Partial check}$$

Also, if we had not separately computed q, we could have found it by substitution, as

$$q = 1 - p = 1 - \tfrac{1}{2} = \tfrac{1}{2}, \text{ as before}$$

Values of the probability fractions, p and q, may also be expressed in *decimal* or in percentage form. In the latter case they are called *percentage chances*. From the above illustration again, for instance,

$$p = \tfrac{1}{2} = 0{\cdot}5, \text{ or a } 50\% \textit{ chance of E happening}$$
$$q = \tfrac{1}{2} = 0{\cdot}5, \text{ or a } 50\% \textit{ chance of E failing}$$

In this special case the 'chances' may be said to be '50–50'; or, if betting is involved, it can be said to be based on 'even money' terms.

A different, but equivalent, way of expressing the ideas of probability and improbability is in terms of 'odds'. By definition, *the ratio*,

$$h : f, \text{ or } h \text{ to } f,$$

states the *odds for E* happening; and the *inverse ratio*

$$f : h, \text{ or } f \text{ to } h,$$

states the *odds against E* happening. In the case of the preceding illustration, for instance, the odds are:

$$h : f, \text{ or } 1 \text{ to } 1 \text{ for E happening}$$
$$\text{and} \quad f : h, \text{ or } 1 \text{ to } 1 \text{ against E happening}$$

In this special case, where both ratios are the same, we say 'the odds are even'.

When the *odds for E happening* are *greater than 1:1* they may also be called the odds *in favour* of E.

If the statement of a probability problem is such that you can determine h, f, and w by actual count you can always solve the problem directly by the method of the above illustrations.

The most important **applications of probability** theory which you are likely to encounter later are in *economics, genetics, thermodynamics, nuclear physics, information theory,* and other highly technical sciences which require special backgrounds for an adequate understanding of their problems. However, most essential principles of the probability theory involved in such applications can be more simply illustrated in elementary experiments like the tossing of coins, or in common games of chance. Hence, we shall here make greatest use of examples from the latter sources.

EXAMPLE 1: If 2 well-balanced coins are tossed, what are the probabilities of, and the odds on, the following events:

E_1—both land heads,
E_2—1 lands heads and 1 tails,
E_3—neither lands heads,
E —at least 1 lands heads?

SOLUTION: Designate the respective heads and tails sides of the 2 coins in some arbitrary order as H_1, T_1, and H_2, T_2 (see the preceding chapter, p. 212). Then the assumption that the coins are well balanced means that they are equally likely to land with any 1 of the 4 possible inter-group combinations of their faces up,

$$H_1H_2, \quad H_1T_2, \quad T_1H_2, \quad T_1T_2$$

For event E_1, then, $w_1 = 4$, $h_1 = 1$ (the count of the H_1H_2 possibility), and *the probability of E_1 happening* is

$$p_1 = h_1/w_1 = 1/4, \text{ or 1 chance out of 4}$$

which may also be expressed as

$$p_1 = 0{\cdot}25, \text{ or } a\ 25\%\ chance$$

Alternatively, since $f_1 = w_1 - h_1 = 4 - 1 = 3$, *the odds on E_1 happening* are

$$h_1 : f_1 = 1 : 3, \text{ or } 1\ to\ 3\ for,$$
$$\text{and} \quad f_1 : h_1 = 3 : 1, \text{ or } 3\ to\ 1\ against$$

For event E_2, next, $w_2 = 4$ as before, but $h_2 = 2$ (the count of the H_1T_2 and T_1H_2 possibilities). Hence, the *probability* of E_2 *happening* is

$$p_2 = h_2/w_2 = 2/4, \text{ or 2 chances out of 4}$$

Equivalently, by reduction of the fraction to lowest terms, etc., this is

$$p_2 = 1/2, \text{ or } 1\ chance\ out\ of\ 2$$
$$= 0{\cdot}5, \text{ or } a\ 50\%\ chance$$

And since $f_2 = w_2 - h_2 = 4 - 2 = 2$, the *odds on E_2 happening* are

$$h_2 : f_2 = 2:2 = 1:1, \text{ or } even\ \text{(for } or \text{ against)}$$

For event E_3, again $w_3 = 4$, but (as for E_1) $h_3 = 1$ (this time the count of the T_1T_2 possibility). Hence, the *probability* of, and *the odds on,* E_3 *happening* are exactly *the same as for E_1 happening.*

For event E, however, although $w = 4$ again, now $h = 3$ (the count of the H_1H_2, H_1T_2, and T_1H_2 possibilities).

Hence, *the probability of E happening* is

$$p = h/w = 3/4, \text{ or } 3\ chances\ out\ of\ 4$$
$$= 0{\cdot}75, \text{ or } a\ 75\%\ chance$$

Alternatively, since $f = w - h = 4 - 3 = 1$, the *odds on E happening* are

$$h : f = 3 : 1, \text{ or } 3 \text{ to } 1 \text{ for}$$
and $$f : h = 1 : 3, \text{ or } 1 \text{ to } 3 \text{ against}$$

EXAMPLE 2: A pack of bridge cards is shuffled and split at random. What is the probability that the card thus exposed is: (a) of a red suit—hearts or diamonds? (b) a diamond? (c) a queen? (d) the queen of diamonds?

SOLUTION: Since there are $w = 52$ cards in the pack and $h_a = 26$ of these are of the 2 red suits, the probability of the exposed card having a red face is

$$p_a = h_a/w = 26/52 = 1/2, \quad \text{ANS. } (a)$$

Since $h_b = 13$ of the cards are diamonds, the probability of the exposed card being a diamond is

$$p_b = h_b/w = 13/52 = 1/4, \quad \text{ANS. } (b)$$

Since $h_c = 4$ of the cards are queens, the probability of the exposed card being a queen is

$$p_c = h_c/w = 4/52 = 1/13, \quad \text{ANS. } (c)$$

But since there is only $h_a = 1$ queen of diamonds in the pack, the probability of the exposed card being this one is

$$p_d = h_a/w = 1/52, \quad \text{ANS. } (d)$$

RELATIVE PROBABILITIES

If the *probabilities* of *events* E_1 and E_2 are p_1 and p_2 respectively, then by definition *event* E_1 is p_1/p_2 times as probable (likely to happen) as *event* E_2. From the findings in the solution of Example 2, for instance, we may say that when a card is drawn at random from a shuffled bridge pack, the event of its being a diamond is $(13/52)/(4/52) = 13/4 = 3\frac{1}{4}$ times as probable as the event of its being a queen; or, more briefly, such a card is $3\frac{1}{4}$ times more likely to be a diamond than it is to be a queen (although, of course, it may be both in the special case of the queen of diamonds).

To illustrate the *relative likelihoods of all the possible outcomes of a game*, it is sometimes convenient to prepare a table of all its basic events and their consequent probabilities.

In the game of dice, for instance, each of 2 (cubical) dice has its faces numbered from 1 to 6 by a corresponding number of spots, and when the dice come to rest after being rolled, the players add the 2 of these numbers faced upward. We have already seen that the total number of possible combinations of the faces of 2 dice is $6^2 = 36$ (see the preceding chapter, Example 15, p. 216). But the corresponding game events (sums of 2 numbers each from 1 to 6) can also be catalogued in a table constructed as follows: all the possible face numbers from 1 to 6 for an arbitrarily selected first die are written on successive lines down the left side; the same numbers for the other die are written at the heads of columns along the top; and the 36 different possible sums of one number from each are written at corresponding intersections of lines and columns, thus:

		(Face numbers of second die)					
+		1	2	3	4	5	6
(Face numbers of first die)	1	2	3	4	5	6	7
	2	3	4	5	6	7	8
	3	4	5	6	7	8	9
	4	5	6	7	8	9	10
	5	6	7	8	9	10	11
	6	7	8	9	10	11	12

(*Sums of both face numbers*)

By counting identical entries in the body of this table, we can now see at a glance in how many ways, out of $w = 36$, a pair of die faces can add up to each possible total. Hence, assuming that all $w = 36$ ways are *equally likely*—that the dice are not 'loaded', that is—to compute the corresponding probabilities we have only to substitute these values of h in the definition formulae.

EXAMPLE 3: What is the sum (of 2 face numbers) most likely to be rolled on a pair of dice? How does the probability of this sum being rolled compare with the probabilities of others?

SOLUTION: Let $h_1, h_2 \cdots h_{13}$ and $p_1, p_2, \cdots p_{18}$ be respectively the number of favourable happenings and the probabilities for the sums, 1, 2, $\cdots$ 13. Then by inspection of the above table and actual count we find

$$h_1 = h_{13} = 0, \quad h_2 = h_{12} = 1,$$
$$h_3 = h_{11} = 2, \quad h_4 = h_{10} = 3, \text{ etc.}$$

And,

$$p_1 = p_{13} = 0/36, \text{ (or 0)},$$
$$p_2 = p_{12} = 1/36,$$
$$p_3 = p_{11} = 2/36, \text{ (or 1/18)},$$
$$p_4 = p_{10} = 3/36, \text{ (or 1/12)},$$
$$p_5 = p_9 = 4/36, \text{ (or 1/9)},$$
$$p_6 = p_8 = 5/36,$$
$$p_7 = \quad\quad 6/36, \text{ (or 1/6)}.$$

Thus, we see that the most likely number to be rolled is *7*, with a probability of *6 chances out of 36*, or of *1 out of 6*. By comparison, *6* and *8* are only (5/36)/(6/36) = *5/6 as likely* to be rolled; *5* and *9* are only 4/6 = *2/3 as likely*; *4* and *10* are only 3/6 = *1/2 as likely*; *3* and *11* are only 2/6 = *1/3 as likely*; *2* and *12* are only *1/6 as likely*; and *1* and *13* are *impossible* (as are all numbers higher than 13). ANS.

EXAMPLE 4: On his first cast a dice player rolls a 10, which he regards as his 'lucky number'. To win, he must again roll this number, called 'his point', before he rolls a 7 instead. Another player now offers a 'side bet' on 'even money' terms that he will not make his point. Is this a fair wager?

SOLUTION: Definitely not! As we have already seen from the preceding solution, a *10* is only *half as likely* to be rolled as a *7*. Hence, a *fair bet would require the holder of the dice to be given 2-to-1 odds*. If he takes the bet on the superstitious theory that 10 is his 'lucky number', then so much the worse for him in the long run of several trials!

WARNING: If event E_1 is h_1/h_2 *times as probable* as *event* E_2 this means only that the *odds* are h_1/h_2 for E_1 happening *before* E_2. To compute the *probability* of event E, that E_1 will happen before E_2, we must recognize that E can happen in only $h = h_1$ out of

$$w = h_1 + h_2 \text{ possible ways.}$$

Hence the *probability* that E_1 will happen before E_2 is only

$$p = \frac{h_1}{h_1 + h_2}$$

In the preceding case, for instance, the probability that a *4* will be rolled before a *7* is NOT $3/6 = 1/2$, BUT . . .

$$p = \frac{3}{3 + 6} = \frac{3}{9} = \frac{1}{3}$$

or equivalently

$$p = \frac{1}{1 + 2} = \frac{1}{3}$$

Likewise, the probability of a *7* being rolled before a *4* is

$$q = \frac{2}{2 + 1} = \frac{2}{3}$$

Partial check:

$$p + q = \frac{1}{3} + \frac{2}{3} = \frac{3}{3} = 1$$

ADDITION OF PROBABILITIES

Any 2 events are by definition *mutually exclusive* if the happening of either *excludes* the possible happening of the other *on the same trial*.

For instance, the event e of a coin landing heads is *mutually exclusive* with the event e' of the *same* coin landing tails on the same trial (toss). Similarly, the 11 events $E_2, E_3, \ldots E_{12}$, of a pair of dice rolling sums of *2, 3, . . . 12*, respectively, are all *mutually exclusive with respect to each other* on the same trial (roll).

In other words, *mutually exclusive events* are *alternative possible outcomes* of the *same trial*.

From this definition, it follows that *if* E_1 and E_2 are any two *mutually exclusive events*, then E_1 can happen in h_1 *ways* and E_2 in h_2 *ways* out of the *same* w equally likely ways, but the h_1 ways of E_1 happening must all be *different from* the h_2 ways of E_2 happening. (This has been illustrated at length in the dice-sum table on p. 226.)

Suppose, then, that E is the event of any one of n mutually exclusive events, E_1, *or* E_2, *or* E_2 *or* . . . E_n. Then E can happen in any of the ways,

$$h = h_1 + h_2 + \ldots + h_n$$

out of the *same* w equally likely ways.

Hence,

$$\frac{h}{w} = \frac{h_1 + h_2 + \cdots h_n}{w}$$

$$= \frac{h_1}{w} + \frac{h_2}{w} + \cdots + \frac{h_n}{w}$$

But, by definition,

$$p = \frac{h}{w}, \quad p_1 = \frac{h_1}{w}, \quad \cdots p_n = \frac{h_n}{w}$$

Therefore, by substitution,

$$p = p_1 + p_2 + \cdots p_n$$

Stated verbally, this means that if E_1, E_2, $\cdots E_n$ are *n mutually exclusive events*, then the *probability* of E_1 or E_2 or $\cdots E_n$ happening is equal to the *sum of the probabilities* of each of these events happening separately.

In Example 1 above, for instance, the events E_1 (that two tossed coins land heads) and E_2 (that one lands heads and one tails) are *mutually exclusive*. Also, E (that at least one lands heads) is the event that *either* E_1 or E_2 happens. In the direct solution of that problem we found, moreover, that

$$p_1 = \tfrac{1}{4}, \quad p_2 = \tfrac{1}{2}, \quad p = \tfrac{3}{4}$$

This solution is now confirmed by the above formula, as follows:

$$p_1 + p_2 = \tfrac{1}{4} + \tfrac{1}{2} = \tfrac{3}{4} = p, \text{ as anticipated}$$

In the special case where E_1, E_2, $\cdots E_n$ are *all* the *mutually exclusive events which can possibly happen in the given w equally likely outcomes of the same trial*, then by the same reasoning,

$$p = p_1 + p_2 + \cdots p_n = 1$$

From Example 1 again, for instance, we also have the event E_3 (that 2 tossed coins both land tails). With E_1 and E_2, this completes the mutually exclusive possibilities for the outcome of the trial of tossing 2 coins. Also, $p_3 = \tfrac{1}{4}$, and

$$p = \tfrac{1}{4} + \tfrac{1}{2} + \tfrac{1}{4} = 1, \text{ as anticipated}$$

Or, from Example 3 concerning cast dice:

$$p = (1 + 2 + 3 + 4 + 5 + 6 + 5 + 4 + 3 + 2 + 1)/36$$
$$= 36/36 = 1, \text{ as anticipated}$$

The above probability addition formulae can often be applied to shorten *the computation of probabilities, or to find further probabilities in terms of those already computed.*

EXAMPLE 5: Using the data in the solution of Example 3, find the probability that a player will roll a *7 or* an *11* on his first cast of the dice.

SOLUTION: The probability is

$$p = p_7 + p_{11} = 6/36 + 2/36 = 8/36$$
$$= 2/9, \text{ or } 2 \text{ chances out of } 9, \quad \text{ANS.}$$

EXAMPLE 6: Find the corresponding probability that he will roll a *2* or a *3* or a *12* on his first cast.

SOLUTION: The probability is

$$p = p_2 + p_3 + p_{12} = 1/36 + 2/36 + 1/36$$
$$= 4/36 = 1/9, \text{ or } 1 \text{ chance out of } 9, \quad \text{ANS.}$$

Two events which are *not mutually exclusive* may be so because they are (partially) overlapping. In Example 2 above, for instance, the event E_b (that a card drawn from a pack at random be a *diamond*) and the event E_c (that the card be a *queen*) are not mutually exclusive, but are (*partially*) *overlapping* in the case of the event E_d (that the card be the *queen of diamonds*).

By the methods of reasoning already illustrated, we may show that if E_1 and E_2 are any 2 (*partially*) *overlapping* events with probabilities p_1 and p_2 respectively, and if $p(E_1 \text{ and } E_2)$ is the probability that *both* E_1 and E_2 happen, then the *probability* that *either* E_1 or E_2 happen is

$$p(E_1 \text{ or } E_2) = p_1 + p_2 - p(E_1 \text{ and } E_2)$$

EXAMPLE 7: What is the probability that a card drawn at random from a bridge pack is *either* a diamond *or* a queen?

SOLUTION: From the solution of Example 2 we already know that the probability of such a card being a diamond is $13/52 = 1/4$, that the probability of it being a queen is $4/52 = 1/13$, and that the probability of it being *both* a queen *and* a diamond is $1/52$. Hence, the probability of it being *either* a queen *or* a diamond is

$$p = \frac{13}{52} + \frac{4}{52} - \frac{1}{52} = \frac{13 + 4 - 1}{52}$$
$$= \frac{16}{52} = \frac{4}{13}, \quad \text{ANS.}$$

EXAMPLE 8: The probability that a 5-card hand be 'any straight' (including a 'straight flush') is $p_1 = 0.003940$. The probability that it be 'any flush' (including a 'straight flush') is $p_2 = 0.001980$. The probability that it be a 'straight flush' is $p_{12} = 0.000015$. (For descriptions of these hands, recall p. 219 above.) What is the probability that a hand be *either* a 'straight' *or* a 'flush'?

SOLUTION: As in the preceding solution,

$$
\begin{array}{rl}
p_1 & 0.003940 \\
+ p_2 = & +0.001980 \\
\hline
p_1 + p_2 = & 0.005920 \\
- p_{12} = & -0.000015 \\
\hline
p = & 0.005905, \quad \text{ANS.}
\end{array}
$$

MULTIPLICATION OF PROBABILITIES

Events which are *neither mutually exclusive* nor (*partially*) *overlapping are separate events.*

By definition, any *2* events e_1 and e_2 are *separate* if the set of w_1 ways in which

e_1 can happen or fail to happen are *completely distinct* from the w_2 ways in which e_2 can happen or fail to happen. For instance, the event e_1 that a *first* coin land heads is separate from the event e_2 that a (different) *second* coin land heads, because the $w_1 = 2$ ways in which the *first* can land either heads or tails are *completely distinct from* the $w_2 = 2$ ways in which the *second* coin can land either heads or tails.

Also by definition, if $e_1, e_2, \cdots e_n$ are n separate events and E is the event that e_1, *and* $e_2 \cdots$ *and* e_n *all happen* (either concurrently or in succession) *on the same trial*, then E is a multiple event with $e_1, e_2 \cdots e_n$ as its (separate) constituent events. For instance, the event E (that 2 coins land heads) is a *multiple* event consisting of the 2 separate *constituent* events, e_1 (that the *first* coin lands heads) and e_2 (that the *second* coin lands heads).

Now let E be the *multiple event* that n separate constituent events, e_1 and e_2 *and* $\cdots e_n$ all happen on the same trial, and let h_1 be the number of ways in which e_1 can happen out of a total of w_1 possible ways *distinct from* the w_2, $w_3, \cdots w_n$ distinct ways of $e_2, e_3, \cdots e_n$ happening or failing to happen, etc. Then, by inter-group combinations of the possibilities, E can happen in

$$h = h_1 h_2 \cdots h_n \text{ ways}$$

out of a total of

$$w = w_1 w_2 \cdots w_n \text{ ways}$$

Hence,

$$\frac{h}{w} = \frac{h_1 h_2 \cdots h_n}{w_1 w_2 \cdots w_n} = \frac{h_1}{w_1} \cdot \frac{h_2}{w_2} \cdots \frac{h_n}{w_n}$$

And so, if all these ways are equally likely, then by substitution of $p = h/w$, etc.,

$$p = p_1 p_2 \cdots p_n$$

Stated verbally, this means that the *probability* of the happening of a *multiple event* on *any given trial* is equal to the *product* of separate probabilities of its n separate constituent events.

Along with the preceding *addition formulae* for the probabilities of *mutually exclusive events* (E_1 or E_2 or $\cdots E_n$), this *multiplication formula* for the probabilities of *separate events* (e_1 and e_2 and $\cdots e_n$) can shorten the work of computing many probabilities.

EXAMPLE 9: What are the odds against a coin landing heads 8 times in a row?

SOLUTION: The landings of the *same* coin 8 different times (or of *different* coins at the *same* time) are *separate* trials. Moreover, for each the probability of the coin landing heads is *1/2* as we have already seen. Hence, the probability of the *multiple event* of the coin landing heads all 8 times is, by the *multiplication theorem*,

$$p = (\tfrac{1}{2})^8 = 1/256$$

Accordingly,

$$h = 1, \quad w = 256$$
$$f = w - h = 256 - 1 = 255$$

and the odds on the multiple event are

$$f : h = 255 \text{ to } 1 \textit{ against } \text{its happening,} \quad \text{Ans.}$$

EXAMPLE 10: A coin has been tossed and fallen heads seven times in a row. An excited spectator, aware of the result of the previous solution, offers to bet 100 to 1 that it will not come heads again! What is the wisdom of his wager?

SOLUTION: The man is a fool who completely misses the point about separate events. His bet would have been a very advantageous one if made *before* the series of tosses began. But the solution of Example 9 is valid only because the *separate* probability of *each separate constituent event* is *exactly 1 chance out of 2*. And this holds for the eighth trial, the first trial, or the millionth trial, considered *separately* from the series in which it occurs.

Two separate events are said to be *independent* if the *happening* of *neither* affects the *probability* of the *happening* of the *other*. But 2 separate events are said to be *dependent* if the *happening* of *either* does *affect* the *probability* of the *happening* of the *other*. Numerical consequences of this distinction are illustrated in the following two examples:

EXAMPLE 11: One card is drawn from each of 2 different packs (or else a card is drawn from one pack and *returned* to it, after which another card is drawn from the *same* pack). What is the probability that both cards are aces?

SOLUTION: The 2 separate events, E_1 and E_2, of drawing an ace are here independent because the probability of each,

$$p_1 = p_2 = 4/52 = 1/13$$

is not affected by the occurrence of the other. Hence, the probability of both cards being aces is

$$p = p_1 p_2 = \tfrac{1}{13} \tfrac{1}{13} = 1/169, \quad \text{ANS.}$$

EXAMPLE 12: Two cards are drawn *in succession* from the *same* pack *without* the first being returned to it. What is the probability that both are aces?

SOLUTION: As before, the probability of event E_1, that the first card be an ace, is $p_1 = 1/13$. But now the event E_2, that the second card also be an ace, is *dependent upon* event E_1 having happened. For if an ace has already been drawn from the pack, then there are only $h_2 = 3$ aces left in a 'short pack': of $w_2 = 51$ cards. Hence, the probability of E_2 is

$$p_2 = h_2/w_2 = 3/51 = 1/17$$

and therefore the probability of both cards being aces is now only

$$p = p_1 p_3 = \tfrac{1}{13} \tfrac{1}{17} = 1/221, \quad \text{ANS.}$$

MISCONCEPTIONS AND SUPERSTITIONS

The *uncertainties of chance*, which is the subject matter of probability theory, is also the object of many hopes and fears. Understandably, therefore, the field is one in which wishful thinking often leads to *misconceptions*, or to *superstitions* based on misconceptions.

Two simple instances have already been pointed out in Examples 4 and 10 (pp. 226 and 231) above. Another now follows.

EXAMPLE 13: By the rules in the game of dice, the player who rolls the dice must match bets placed against him on an 'even money' basis. He can win either: (*a*) by rolling a 'natural'—7 or *11*— on his first cast; or else (*b*) by rolling a 'point'—*4, 5, 6, 8, 9,* or *10*—on his first cast and then 'making his point' by rolling the same number again before a 7. On the other hand, he can lose either (*c*) by rolling 'craps'—*2*, *3*, or *12*—on his first cast; or else (*d*) by rolling 7 before his 'point' if that is what he has rolled on his first cast.

Against the background of these rules, amateur dice players are known to 'feel' that it is 'luckier' to roll the dice themselves, even when they have no question about the honesty of their fellow players. But *professional* gambling-house operators always require the 'customer' to roll the dice rather than their own employee. Which policy is *mathematically* more advantageous?

SOLUTION: By applying the probability addition and multiplication formulae to the data of Examples 3 and 4 above (p. 226), we can now compute the probabilities of all the possible outcomes of this game—both favourable and unfavourable to the one who cast the dice. For instance, in the case with a 1/12 probability that the player rolls a *4* as his point, we know that he has only $1/(1 + 2) = 1/3$ chance thereafter of making this point (recall the WARNING on probabilities versus odds, p. 227). Hence, the probability of his winning in this instance is the *product* of the probabilities of *2 separate constituent events*,

$$\tfrac{1}{12} \cdot \tfrac{1}{3} = \tfrac{1}{36} = 2.78\%$$

Moreover, his probability of winning in the *mutually* exclusive event of his rolling a *10* as his point is the same. Hence, the probability of his winning in either of these cases is the *sum* of the probabilities of these two *mutually exclusive events*,

$$\tfrac{1}{36} + \tfrac{1}{36} = 2 \cdot \tfrac{1}{36} = \tfrac{1}{18} = 5.56\%$$

On the other hand, the probability of his *losing* in the cases of these same two mutually exclusive 'point' possibilities is the corresponding sum of two products,

$$2 \cdot \frac{1}{12} \cdot \frac{2}{2 + 1} = \frac{4}{36} = \frac{1}{9} = 11.11\%$$

Analysing all other possibilities in the same way, we obtain the table at the top of p. 233. There we see that the **one** who rolls the dice has a 49·29% *chance of winning* as opposed to a *50·71% chance of losing*. This means that, for a fair game, turns to roll the dice should be alternated among *all* the players. It also means that, even if there is no 'admission fee' or 'percentage charge for cashing chips', the *professional* gambling house *always wins in the long run*—for, in this case, the 'house' literally has the customer who rolls the dice working for it!

MATHEMATICAL EXPECTATION

As distinguished from *psychological* or *subjective* expectation, *mathematical expectation* is defined as the product, $p \cdot A$, of the probability p that a particular event will happen, and the amount A one will receive if it does happen.

Suppose, for instance, that a dice player has bet £5·00. When this amount is 'matched', it results in a 'pot' of $A = £10·00$ which the player will receive if he wins. From the following table we know that the probability of his winning is $p = 49·29\%$. Hence, his mathematical expectation is:

$$p \cdot A = 0·4929(£10·00) = £4·93$$

However, if he should roll a *4* or a *10* as his 'point', then the probability of his winning drops to $p = \tfrac{1}{3}$ and his mathematical expectation thereafter is only

$$£10·00/3 = £3·33$$

whereas that of the 'house' then increases from £5·07 to £6·67 to the nearest cent.

Table of Probabilities for Game Events in Dice

Possible Casts by Player	Probability Calculations	Percentage to Win	Chances to Lose
'Craps': 2, 3, or 12	$\frac{1+2+1}{36} = \frac{4}{36} = \frac{1}{9} =$		11·111
'Point' 4 or 10, made	$2 \cdot \frac{1}{12} \cdot \frac{1}{1+2} = \frac{2}{36} = \frac{1}{18} =$	5·556	
'Point' 4 or 10, lost	$2 \cdot \frac{1}{12} \cdot \frac{2}{2+1} = \frac{4}{36} = \frac{1}{9} =$		11·111
'Point' 5 or 9, made	$2 \cdot \frac{1}{9} \cdot \frac{4}{4+6} = \frac{8}{90} = \frac{4}{45} =$	8·889	
'Point' 5 or 9, lost	$2 \cdot \frac{1}{9} \cdot \frac{6}{6+4} = \frac{12}{90} = \frac{2}{15} =$		13·333
'Point' 6 or 8, made	$2 \cdot \frac{5}{36} \cdot \frac{5}{5+6} = \frac{50}{396} = \frac{25}{198} =$	12·626	
'Point' 6 or 8, lost	$2 \cdot \frac{5}{36} \cdot \frac{6}{6+5} = \frac{60}{396} = \frac{5}{33} =$		15·152
A 'natural': 7 or 11	$\frac{6+2}{36} = \frac{8}{36} = \frac{2}{9} =$	22·222	
	Total percentage chances:	49·29	50·71
	Partial Check: 49·29% + 50·71% =	100%	

This concept, to be applied later to the computation of insurance rates, may also help to analyse certain further misconceptions in the area of probability theory.

EXAMPLE 14: In a lottery 20 000 tickets are to be sold at 25p each, and the one prize is a car valued at £2500. In another lottery, 200 000 tickets are to be sold at £1 each, and the prizes are one of £50 000 and 25 of £1000. Mr. Sloe, who has never even heard of Adam Smith's comment that lotteries are the greatest single device ever invented by the ingenuity of man to separate a fool from his money, is thinking of making a £1 'investment' in either. He is tempted by the first lottery because he thinks it will give him '4 times as many *chances* (by which he means *tickets*) for the same money'. But he is also tempted by the second lottery because he thinks 'there is so much more prize money'. Compare his actual mathematical expectations in the two cases.

SOLUTION: In the first lottery, Sloe would have a 4/20 000 = 0·0002 probability of winning £2500. Hence, for his £1 *cost* of 4 tickets he would receive a *mathematical expectation* worth

$$V_1 = 0.0002(£2500) = £0.5$$

In the second lottery he would have a 1/200 000 = 0·000 005 probability of winning £50 000, plus a 25/200 000 = 0·000 125 probability of winning £1000.

Hence for the *same £1 cost* he would receive a *mathematical expectation* worth

$$V_2 = 0.000005(£50\,000) + 0.000125(£1000) = £0.375$$

Consequently, we might advise Sloe that neither purchase is an 'investment', and the second is by 25% an even worse 'speculation' than the first.

APPLYING COMBINATIONS FORMULAE

In the probability problems considered thus far it has been possible to *count* values of h, f, and w directly. In more complicated problems, however, these quantities are better *computed* by *formula* as in the preceding chapter.

EXAMPLE 15: What are the odds against the event of a card player being dealt a 2-pair hand—defined as one having a pair in each of 2 different ranks plus a fifth card in a still different rank?

SOLUTION: From Example 11 (p. 214) of the preceding chapter we know that a 5-card hand can be dealt in a total of $w = 2\,598\,960$ different ways. To compute h we can now reason that two different *ranks* may be combined in $C(13, 2)$ different ways; that in *each* of these cases 2 *pairs* of *different ranks* can occur in $C(4, 2)$ times $C(4, 2) = 6 . 6 = 36$ different ways; and that a *fifth* card from a *still different rank* can occur in *44* different ways, for a product of

$$h = C_2^{13} C_2^4 C_2^4 \, 44 = 78 . 36 . 44 = 123\,552$$

Hence,

$$f = w - h = 2\,598\,960 - 123\,552 = 2\,475\,408$$

and the *odds against* dealing a 2-pair hand are

$$f : h = 2\,475\,408 : 123\,552 = 20 \text{ to } 1, \quad \text{ANS.}$$

rounded off to the nearest whole number.

Since the various patterns of '5-card hands' illustrate very simply the most common types of combinations which are likely to occur in more technical

Table of Mathematical odds for 5-card Hands drawn in Poker

Hand	Computation of h	Value of h	Odds Against (*, approximate)
Straight Flush	$10C_1^4 = 10 . 4 =$	40	64 973 to 1
Four of a kind	$C_1^{13} C_4^4 C_1^{12} C_1^4 = 13 . 1 . 12 . 4 =$	624	4164 to 1
Full house	$C_2^{13} 2 C_2^4 C_3^4 = 78 . 2 . 6 . 4 =$	3744	693 to 1
Flush (non-straight)	$C_5^{13} C_1^4 - 40 = 5148 - 40 =$	5108	507 to 1*
Straight (non-flush)	$10(4^5) - 40 = 10\,240 - 40 =$	10 200	254 to 1*
Three of a kind	$C_1^{13} C_3^4 C_2^{12} 4^2 = 13 . 4 . 66 . 16 =$	54 912	46 to 1*
Two pairs	$C_2^{13} C_2^4 C_2^4 44 = 78 . 6 . 6 . 44 =$	123 552	20 to 1*
One pair	$C_1^{13} C_2^4 C_3^{12} 4^3 = 13 . 6 . 220 . 64 =$	1 098 240	1·4 to 1*
Other	$w -$ (all the above) $=$	1 302 540	1 to 1*
Any five cards	$w = C_5^{52} = 52!/5!47! =$	2 598 960	0·0 to 1

probability problems, a *comparative table* of their *computations* is appended on p. 234. For definitions of the patterns and more detailed explanations of the computations, also recall the solutions of Example 17 and Practice Exercise No. 82, problems 16 (*a*) through (*f*) of the preceding chapter.

EXAMPLE 16: What is the probability of dealing: (*a*) a bridge hand consisting of 13 cards all of the same suit; (*b*) 4 such hands from the same pack; (*c*) a bridge hand with a *4, 4, 3, 2* suit distribution?

SOLUTION: (*a*) One bridge hand of 13 cards can be dealt from a 52-card pack in any of

$$w_a = C_{13}^{52} = \frac{52!}{13!39!} = 635\ 013\ 559\ 600 \text{ ways}$$

Of these, only $h_a = 4$ have all 13 cards in the same suit. Hence, the probability of such a deal is

$$p_a = 4/w_a = 1/158\ 753\ 389\ 900, \quad \text{ANS. } (a)$$

(*b*) Since it makes no difference whether each of 4 hands is dealt, from a shuffled pack, 1 card at a time or all 13 cards at a time, let us for convenience assume the deal to be the latter. Then the 4 hands can be dealt consecutively to 4 players in

$$C_{13}^{52},\ C_{13}^{39},\ C_{13}^{26}, \text{ and } C_{13}^{13} \text{ ways}$$

respectively, and the product of these 4 quantities would give the number of ways in which bridge hands can be *permuted* among 4 players. Dividing this product by *factorial 4*, therefore, we find the total number of *combinations* of such hands to be

$$w_b = C_{13}^{52}C_{13}^{39}C_{13}^{26}C_{13}^{13}/4!$$

$$= \frac{52!}{13!39!} \cdot \frac{39!}{13!26!} \cdot \frac{26!}{13!13!} \cdot \frac{13!}{13!} \cdot \frac{1}{4!}$$

$$= \frac{52!}{(13!)^4 4!}$$

when like factors are cancelled from numerator and denominator. This last quantity has been machine computed to be *2235 followed by 24 additional* digits. Since $h_b = 1$, therefore, the required probability is

$$p_b = h_b/w_b = 1/2235 \text{ trillions of millions}, \quad \text{ANS. } (b)$$

(*c*) There are $C(4, 2) = 6$ ways of choosing the 2 suits in which (the same numbers of) 4 cards are to be dealt, and $2C_2^2 = 2 \cdot 1 = 2$ different ways of next choosing the 2 suits in which (the *different* numbers of) *3* and *2* cards are to be dealt. Hence,

$$h_c = 6C_4^{13}C_4^{13}2C_3^{13}C_2^{13}$$

$$= 12 \cdot \frac{13!}{4!9!} \cdot \frac{13!}{4!9!} \cdot \frac{13!}{3!10!} \cdot \frac{13!}{2!11!}$$

$$= \frac{(13!)^4}{11!10!(9!)^2(4!)^2} = \frac{13^4 12^4 11^3 10^2}{4^2 3^2 2^2}$$

$$= 13^4 12^2 11^3 5^2 = 136\ 852\ 887\ 600.$$

Divided by w_a (rounded off to 635 thousand million) from above, this gives us the required probability as

$$p_c = h_c/w = 0 \cdot 2155$$

or *somewhat better than 1 chance in 5*,   Ans. (c)

'CONTINUOUS PROBABILITY'

In all preceding parts of this chapter we have this far considered only events *E* for which it was possible *to count* or *to compute* a definite number of ways *h* in which *E* could happen out of an equally definite number of ways *w* in which *E* could either happen or fail to happen. These are variously called *discontinuous, arithmetic*, or *finite events*. And by, a somewhat inappropriate transfer of adjectives, the corresponding measures of the likelihoods of their happening are called *discontinuous, arithmetic*, or *finite probabilities*.

But suppose we are told that a 'stick' is broken 'anywhere at random', and we are asked to compute the probability of its being broken closer to its midpoint than to either end. Since the stick can be broken in an *infinite (indefinitely large) number* of points, we cannot count, or otherwise compute, any *definite* values for either *h* or *w*. In such a case the possibilities are said to be *continuous, geometric*, or *infinite* events. And obviously our previously stated definitions of probability and improbability do not, in their original form, apply to such events.

However, for convenience in referring to its points and to segments of its length, let us suppose the 'stick' in the preceding instance to be a common *12-cm* ruler. Obviously such a stick will be broken closer to the *6-cm* mark of its mid-point anywhere between its *3-cm* and *9-cm* marks. If, temporarily to *simplify* the problem therefore, we were to consider it breakable only at its *full-cm* marks, then we should have $w = 11$ (the count of the *1-cm*, *2-cm*, $\cdots$ *11-cm* marks), and $h = 5$ (the count of the *4-cm*, *5-cm*, $\cdots$ *8-cm* marks). Hence in this simplified case,

$$p = h/w = 5/11 = 0 \cdot 4545 \cdots$$

Or if, coming a little closer to the original problem's condition, we were to consider the ruler to be breakable at any of its *millimetre marks*, then we should have $w = 119$, $h = 59$,

$$p = \frac{h}{w} = \frac{59}{119} = 0 \cdot 4958$$

And by continuing this process of considering the ruler to be breakable in more and more points, we begin to suspect that, as can actually be proven in calculus, the value of $p = h/w$ comes closer and closer to $\frac{1}{2} = 0 \cdot 5$, which is the *ratio* of the *length* of the middle 6 cm of the ruler (between the *3-cm* mark and *9-cm* marks) to its entire length.

In such a case, therefore, we define the—so called—*continuous, geometric*, or *infinite probability* of the *continuous, geometric*, or *infinitely varied event* as the *limit p* to which the *ratio h/w* comes closer and closer as the number of possible cases increases indefinitely. You will find this reasoning more fully developed later in *calculus*. But in a *literally geometric case*, such as that just considered, we may take this *limit* to be *the ratio of the corresponding geometric lengths, areas, volumes, angles, or even time intervals* involved in the statement of the problem. In the present illustration, for instance,

$$p = (6 \text{ cm})/(12 \text{ cm}) = \tfrac{1}{2}, \quad \text{ANS.}$$

EXAMPLE 17: If a stick of length L is broken anywhere at random, what is the probability that one piece is more than twice as long as the other?

SOLUTION: One piece will be more than twice as long as the other only if the break occurs at a point less than $L/3$ distant from either end—that is, in one of the two segments marked more heavily in the accompanying diagram. Hence,

$$p = (L/3 + L/3)/L = \tfrac{2}{3}L/L = 2/3, \quad \text{ANS.}$$

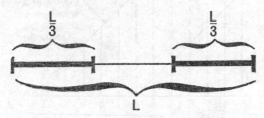

EXAMPLE 18: If a 12-cm ruler is broken in any 2 points at random, what is the probability that a triangle can be formed with the 3 segments?

SOLUTION: Let x be the length of the left-hand segment, and let y by the length of the middle segment, so that $12 - x - y$ is the length of the third segment. Then, since any side of a triangle must be less than the sum of the other 2 sides the condition of this problem will be satisfied only if x, y, and $12 - x - y$ are *all less than 6*:

$$x < 6, y < 6, 12 - x - y < 6$$

Now, to construct a *geometric diagram* which expresses these algebraic requirements, measure all possible lengths of x along a *horizontal 12-cm scale* from O to L, and measure corresponding lengths of y and of $12 - x$ along a *vertical 12-cm scale* from O to M, perpendicular to OL at O, as in the accompanying Fig. 6. Then, corresponding to the point A for $x = 1$ cm on OL, we have an *11-cm* vertical line AN representing $12 - x = 12 - 1 = 11$, perpendicular to OL at A; and beginning from A we can find points on AN corresponding to any possible values of y. Of the latter, however, only those between $y = 5$ and $y = 6$, dotted in Fig. 6, satisfy the above inequalities. Likewise, corresponding to the point B for $x = 2$ cm on OL, we have 10-cm. vertical line BP representing $12 - x = 12 - 2$; but on this line only those points representing values of y from $y = 4$ cm to $y = 6$ cm, dotted in the diagram, correspond to values of y, which satisfy the above equalities when $x = 2$, etc. *Moreover*, for all values of x on OL from D for $x = 6$ cm, to L for $x = 12$ cm, there is *no* corresponding value of y which satisfies the above inequalities, since x already violates the first inequality.

Thus, we see that for each possible pair of values of x and y there corresponds *some* point in the *large* triangle OLM with an *area* $= 12 \cdot 12/2 = 72$, and that for those values of x and y which satisfy the conditions of our problem the *points* all lie within the *smaller* triangle DEF with an *area* $= 6 \cdot 6/2 = 18$. Hence, our required probability is

$$p = 18/72$$
$$= 1/4, \text{ or one chance in four, } \quad \text{ANS.}$$

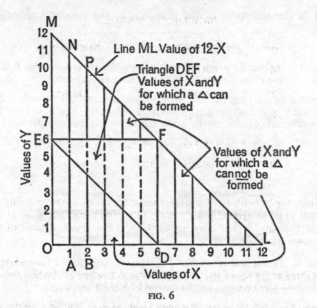

FIG. 6

The same reasoning would apply, of course, to a stick of any length. The only difference is that we should find it more awkward to refer repeatedly to different fractional parts of its length L.

STATISTICAL PROBABILITY

In some types of problems to which we may apply the concepts of probability theory, the values of h and f are not derived from theoretical assumptions, but are learned *statistically* from *experience*. Hence, the latter are distinguished as cases of *statistical, empirical,* or *inductive probability*, in contrast to the *non-statistical* or *deductive probabilities* with which we have hitherto been concerned here.

Below, for instance, is a *Mortality Table* which insurance companies use in computing premiums for life insurance. Obviously, there is no way of knowing in advance how long any particular 10-year-old child will live, and it is reasonable to assume that those who have serious organic defects, or who grow up to pursue dangerous occupations, may very well, on the average, die sooner than others. Nevertheless, this table has been compiled, and reduced to a *common denominator of 100 000*, from actual statistics concerning how long people do live. By it we see, for instance, that of $w = 100\,000$ people alive at age 10, only $h = 14\,474$ are still alive at age 80. Hence, we may say, in this *statistical sense*, that *the probability of a child of 10 living to attain the age of 80 is, on the average,*

$$p = h/w = 14\,474/100\,000$$
$$= 14\cdot5\% \text{ approximately}$$

MORTALITY TABLE

Age	Number Living	Age	Number Living
10	100 000	60	57 917
20	92 637	70	38 569
30	85 441	80	14 474
40	78 106	90	847
50	69 804	100	0

Of course, the last entry in this table means only that, corresponding to each person still alive at age 100, there were more than 200 000 alive at age 10. Otherwise, the entry would have rounded off to *1* or more, instead of to *zero*, as the nearest whole number.

EXAMPLE 19: To the nearest tenth of 1%, what is the *statistical probability* that a child living at the age of 10: (*a*) will *still* be living at age 60; (*b*) will *not* be living at age 60; (*c*) will *die between* the ages of 60 and 70?

SOLUTION: (*a*) From the above *Mortality Table*, of $w = 100\ 000$ children living at age 10, only $h = 57\ 917$ are still living at age 60. Hence, the probability of the latter event is

$$p = h/w$$
$$= 57\ 917/100\ 000 = 57 \cdot 9\%, \quad \text{ANS.} (a)$$

(*b*) From the same figures, the number not living at age 60 is

$$f = w - h = 100\ 000 - 57\ 917 = 42\ 083$$

Hence, the probability of such a child not living at age 60 is

$$q = f/w$$
$$= 42\ 083/100\ 000 = 42 \cdot 1\%, \quad \text{ANS.} (b)$$

Or, alternatively from answer (*a*),

$$q = 1 - p = 1 - 57 \cdot 9 = 42 \cdot 1\%, \quad \text{same ANS.}$$

(*c*) From the same *Table*, the number who die between ages 60 and 70 is

$$h = 57\ 917 - 38\ 569 = 19\ 348$$

Hence, the required statistical probability is

$$p = h/w$$
$$= 19\ 348/100\ 000 = 19 \cdot 3\%, \quad \text{ANS.} (c)$$

The amounts of most *insurance premiums* are based on *statistically determined mathematical expectations* plus pro-rated overhead costs and reserve or profit margins. Of this, the following is a greatly simplified illustration.

EXAMPLE 20: An insurance company knows from statistical studies that 45 out of every 10 000 houses in a particular area are destroyed annually by fire. A man in this area applies for a £2000 fire-insurance policy on his home. What must the company charge him annually for this *risk* in addition to its pro-rated other costs?

SOLUTION: The company should charge the man, for this part of its premium, the *mathematical expectation* which he would be purchasing. The amount is $A = $ £2000 and the *statistical probability* of *loss* in any year is

$$p = 45/10\ 000 = 0{\cdot}0045$$

Hence, the proper annual charge (for risk only) is the *mathematical expectation*,

$$V = p \cdot A = 0{\cdot}0045(£2000) = £9, \quad \text{Ans.}$$

Note: If your own fire-insurance premiums differ greatly from this rate you probably live in a very different statistical fire-experience area or are covered by a company with high overhead costs.

Practice Exercise No. 83

1. Letting the heads and tails faces of 3 coins be designated H_1T_1, H_2T_2, H_3T_3, make a systematic array of the $w = n^k = 2^3 = 8$ possible combinations in which they can fall face up when tossed. Then find the corresponding values of h and f for the following events:

> E_1—all 3 land heads
> E_2—only 2 land heads
> E_3—only 1 lands heads
> E_4—none lands heads
> E_5—at least 2 land heads

2. Using your answers to Question 1, find the probabilities p and q of these same events happening and not happening.

3. Which of the events defined in Question 1 are mutually exclusive? Verify your answer by applying probability addition formulae.

4. For events E_1, E_2, and E_5, defined as in Question 1, compute p_5 in terms of p_1 and p_2.

5. For the events in Question 1, what are the odds: (a) for E_1 to happen; (b) against E_2 to happen; (c) for E_5 to happen?

6. If a shuffled pack of cards is split at random, what is the probability that the exposed card is: (a) a 'face card'—*king, queen,* or *jack,* in any suit; (b) a 'black card'—*spades* or *clubs*; (c) a 'black face card'? (d) *either* a 'face card' *or* a 'black card'?

7. What is the probability that, if 5 coins are tossed, all will land with the same face—heads or tails—up?

8. Each of 2 boxes contains 10 balls which are identical except that, in each case, 4 are red and 6 are white. What is the probability that, if 2 balls are selected at random from each box, all 4 turn out to be red?

9. What is the probability in the preceding question if all 4 balls are drawn from the *same* box without any being returned?

10. Referring to the *Table* on p. 234, what is the probability of a player's being dealt an 'opening hand'—defined as one which has a pair of jacks or any higher combination?

11. A common (Monte Carlo) type of roulette wheel has 37 sectors numbered from 0 to 36, the *zero* being a 'free house number'. A player receives $36.00 for each dollar he bets on any given number if the wheel stops spinning with a small ball coming to rest on that number. On the sound, but here irrelevant, theory that 'the house always wins', Mr. Sloe decides to place a $10.00 bet on the 'house number'

zero, rather than on *13*, which he regards as 'unlucky'. Discuss his mathematical expectations in both cases.

12. From the solution of Example 16 (p. 235) of the text, we know that the probability of a *4, 4, 3, 2* bridge-hand distribution is $p_1 = 0.2155$. The corresponding probabilities of *4, 3, 3, 3* and *4, 4, 4, 1* bridge-hand distributions are $p_2 = 0.1054$ and $p_3 = 0.0299$, respectively. What is the probability of drawing a hand with a distribution of at least 5 cards in at least one suit?

13. One or more balls are drawn at random from a bag containing 7. What is the probability that an *even number* are drawn?

14. Due to a purely accidental failure of its power source, an electric clock has stopped. What is the (continuous) probability: (*a*) that the hour hand has stopped between the 12- and 1-hour marks; (*b*) that both hands have stopped there?

15. What is the (statistical) probability, to the nearest tenth of 1%, that of any 2 given persons alive at age 10 both will be alive at age 50?

SUMMARY

In the *theory of probability* we give *exact numerical measures* to our common-sense notions of *likelihood, unlikelihood, certainty*, and *impossibility*.

Approaching the subject first from a *non-statistical, deductive point of view*, we begin with the *hypothetical assumption* that a particular occurrence can happen in a definite number *w of equally likely ways*. This corresponds to such postulates in geometry, for instance, as the assumption that a straight-line segment, or an angle, can be duplicated any definite number of times.

Then if an *event* E can *happen* in *h* of these ways and fail to happen in $f = w - h$ of these ways we define 4 *basic ratios* as follows:

$p = h/w$—the *probability* of E *happening*

$q = f/w$—the *probability* of E *not happening*

$h:f$—the *odds for* E to happen

$f:h$—the *odds against* E to happen

From these it follows that:

$$0 \leq p \leq 1, \quad 0 \leq q \leq 1, \quad p + q = 1$$

0 is the *measure* of *impossibility*
1 is the *measure* of *certainty*

$$h + f = w, \quad h = w - f, \quad f = w - h$$

Of the above ratios, *p* and *q* may also be stated in terms of *h* or *f* chances out of *w*, in terms of *decimal fractions*, or in terms of equivalent *percentage chances*.

If the probability of event E_1 is h_1/w and the probability of event E_2 is h_2/w, then the *odds* for E_1 to happen before E_2 are $h_1:h_2$; and the probability of E_1 *happening before* E_2 is $h_1/(h_1 + h_2)$.

This last statement implies the concept of *mutually exclusive events*. By definition, events E_1 and E_2 are *mutually exclusive* if they are *alternative possible outcomes of the same trial;* or *they are partially overlapping events* if *some* of the ways in which E_1 can happen are *identical with some* of the ways in which E_2 can happen *as outcomes of the same trial*.

Let $p(E_1$ or $E_2)$ be the probability that *either* of *two mutually exclusive or partially overlapping* events E_1 or E_2 happen, and let $p(E_1$ and $E_2)$ be the probability that *both of the two partially overlapping events* E_1 and E_2 happen. Then the *basic probability addition formula* is

$$p(E_1 \text{ or } E_2) = p_1 + p_2 - p(E_1 \text{ and } E_2)$$

with the last term $= 0$ in the case when E_1 and E_2 are not *overlapping* but *mutually exclusive*.

In the *special case* when E_1, E_2, $\cdots E_n$ are *all the possible alternative outcomes of the same trial*, then the preceding formula becomes

$$p(E_1, \text{ or } E_2, \text{ or } \cdots E_n) = p_1 + p_2 + \cdots p_n = 1$$

On the other hand, events E_1 and E_2 are *separate* events if they are *possible outcomes of different trials*—*independent* if the outcome of *neither* affects the *probability* of the outcome of the *other*, or *dependent* if the outcome of *either* does *affect* the *probability* of the outcome of the *other*.

Also, when E_1, E_2, $\cdots E_n$ are *n independent events*, then E is by definition the *multiple event* that E_1, and E_2, and $\cdots E_n$ *all* happen—its *probability* being given by the *multiplication formula*

$$p(E_1, \text{ and } E_2, \text{ and } \cdots E_n) = p_1 p_2 \cdots p_n$$

In *simple* probability problems, $h, f,$ and w may be *counted*. In more *complicated* probability problems they are better *computed* by the combinations and permutations formulae summarized at the end of the preceding chapter.

When it is *not possible* either to *count* or to *compute* values of $h, f,$ and w, because they are *infinite (indefinitely large)*, the methods of *calculus* are needed for a complete analysis. But in simple cases we can regard the so-called '*continuous*', '*infinite*', or '*geometric*' *probabilities* of corresponding events as equal to the *ratios* of certain geometric quantities or time intervals.

Finally, when values of $h, f,$ and w are found *empirically* by *statistical surveys* rather than deductively by theoretical assumptions and computations based upon such assumptions, then we *define the same ratios* of these quantities as *statistical* or *empirical probabilities* and *odds*, applying to them all the same addition and multiplication theorems.

Whether the probability of an event is deductive and discontinuous, deductive and continuous, or statistical, however, we define one's *mathematical expectation V* with regard to it by the formula

$$V = p \cdot A$$

in which p is *the event's probability* and A is *the amount one may expect to receive if it happens*.

Finally, since the *uncertainties* of *chance* give rise to much *wishful thinking* in this area, certain resulting *superstitions* and *misconceptions* have also been analysed in this chapter.

TEST NO. 3

FINAL TEST

1. A brick wall tumbled down, leaving a height of 2·4 metres standing. You were told that $\frac{5}{8}$ of the wall had fallen. How many metres of brick would you have to add to rebuild the wall?
   (A) 1·5 m ...... (B) 3 m ...... (C) 3·6 m ...... (D) 4 m ......

2. Two cargoes together have a mass of 1800 kg. The lighter cargo is only half as heavy as the other cargo. What is the mass of the heavier cargo?
   (A) 1000 kg ...... (B) 1200 kg ...... (C) 1400 kg ...... (D) 1600 kg ......

3. 12% of one man's money equals 18% of the amount another man has. The poorer of the two men has \$300. How much does the richer man have?
   (A) \$350 ...... (B) \$374 ...... (C) \$400 ...... (D) \$450 ......

4. How much would you have to lend for $1\frac{1}{2}$ years at 4% to get back £424 in all?
   (A) £328·56 ...... (B) £407·04 ...... (C) £400 ...... (D) £390 ......

5. A retailer wishes to make 20% on shoes. At what price must he buy them in order to sell them at £4·50 a pair?
   (A) £2·50 ...... (B) £4·25 ...... (C) £3·50 ...... (D) £3·75 ......

6. A supply pipe with a capacity of 8 litres per minute can fill a tank in 18 h. What capacity pipe would be needed to fill the tank in 10 h?
   (A) $4\frac{4}{9}$ ...... (B) $14\frac{2}{5}$ ...... (C) $22\frac{1}{2}$ ...... (D) 24 ......

7. The formula for the length of the sides of a right triangle is $c^2 = a^2 + b^2$. If $c = 15$ and $a = 12$, what does b equal?
   (A) 9 ...... (B) 10 ...... (C) 11 ...... (D) 12 ......

8. What will be the diameter of a wheel whose area is 264 cm^2?
   (A) 15·25 cm ...... (B) 18·33 cm ...... (C) 21·66 cm ...... (D) 24·50 cm ......

9. The radius of a circular room having a tile floor is 2·1 m. You wish to use the tiling in a rectangular room of the same area. The latter room is to be 1·4 m wide. What will be its length?
   (A) 2·1 m ...... (B) 6·3 m ...... (C) 9·9 m ...... (D) 12 m ......

10. A man has six times as many 5p coins as he has 10p coins. The value of his total money is £4·80. How many 5p coins has he?
    (A) 60 ...... (B) 68 ...... (C) 72 ...... (D) 80 ......

11. What is the cube root of 262 144?
    (A) 512 ...... (B) 64 ...... (C) 256 ...... (D) 32 ......

12. What is the volume of a can 4 cm in diameter and 6 cm high?
    (A) $301\frac{5}{7}$ cm^3 ...... (B) $150\frac{6}{7}$ cm^3 ...... (C) $75\frac{3}{7}$ cm^3 ...... (D) $13\frac{5}{7}$ cm^3 ......

13. Which of the following is one of the factors of $15x^2 - 21xy - 18y^2$?
    (A) $5x - 3y$ ...... (B) $5x + 6y$ ...... (C) $3x - 3y$ ...... (D) $3x - 6y$ ......

14. What is the sum of the angles of a hexagon?
    (A) 360° ...... (B) 720° ...... (C) 540° ...... (D) 180° ......

15. If $\sin 22° = 0{\cdot}3746$, $\tan 32° = 0{\cdot}6249$, $\sin 39° = 0{\cdot}6293$ and $\sec 52° = 1{\cdot}6243$, at what angle of depression would an aeroplane 1 km high sight a target that was $1\frac{3}{5}$ km from a point directly below?
    (A) 22° ...... (B) 32° ...... (C) 39° ...... (D) 52° ......

ANSWERS

TEST NO. 1

1. If $\frac{2}{3}$ broke, then $\frac{1}{3}$ was left; 6 m $= \frac{3}{3}$. $\therefore \frac{1}{3} = 2$ m; and $\frac{2}{3} = 4$ m.
2. $\frac{7}{7} + \frac{2}{7} = 117$. If $\frac{9}{7} = 117$ then $\frac{1}{7} = 13$, and $\frac{7}{7} = 7 \times 13$ or 91.
3. $132\% = 33$. $\therefore 1\% = 33 \div 132 = 0.25$, and $100\% = 0.25 \times 100$ or 25.
4. Since boat is slower it takes $2\frac{1}{2} \times 10$ or $\frac{50}{2} = 25$ h.
5. At 2 drums daily we use 40 drums for 20 days. To make 40 drums last 30 days you must use $40 \div 30$ or $1\frac{1}{3}$ drums daily. $2 - 1\frac{1}{3} = \frac{2}{3}$.
6. £1·00 at 6% for $1\frac{1}{3}$ years would cost £1·08. $\therefore$ the number of pounds that would cost £432 is $432 \div 1.08 = 400$.
7. $C = \frac{5}{9}(113 - 32) = \frac{5}{9} \times 81 = 45$.
8. Let $x =$ number of 10p. $\therefore 36 - x =$ number of 5p.
$$10x + 5(36 - x) = 330$$
$$\therefore 5x + 180 = 330$$
$$x = 30.$$
9. $9 : 5 :: x : 15, 5x = 135, x = \frac{135}{5} = 27$.
10. Let $B =$ body length and $T =$ tail length.
$$T = 6 + \frac{1}{2}B$$
$$B = 6 + T \text{ or } T = B - 6$$
$$6 + \frac{1}{2}B = B - 6 \text{ for } T = T$$
$$B - \frac{1}{2}B = 6 + 6$$
$$\frac{3}{4}B = 12$$
$$B = 12 \times \frac{4}{3} = 16$$
$$6 + 16 + (6 + 4) = 32$$

TEST NO. 2

A

1. 15	2. 20	3. 18	4. 29	5. 32	6. 42	7. 83
8. 39	9. 29	10. 51	11. 66	12. 92	13. 76	14. 103
15. 166	16. 185	17. 235	18. 363	19. 533	20. 460	21. 674
22. 964	23. 1241	24. 1473	25. 2251			

B

26. 15	27. 18	28. 19	29. 28	30. 33	31. 41	32. 73
33. 37	34. 32	35. 61	36. 79	37. 92	38. 96	39. 113
40. 176	41. 185	42. 245	43. 343	44. 533	45. 480	46. 674
47. 944	48. 1271	49. 1473	50. 2192			

C

51. 18	52. 38	53. 47	54. 58	55. 42	56. 54	57. 83
58. 49	59. 32	60. 81	61. 81	62. 102	63. 106	64. 133
65. 154	66. 185	67. 215	68. 343	69. 420	70. 591	71. 694
72. 914	73. 1251	74. 1473	75. 2281			

TEST NO. 3

1. If $\frac{1}{3}$ fell, $\frac{2}{3}$ was left, and that was equal to 2·4 m. $\therefore \frac{1}{3} = 2.4 \div 3 = 0.8$, and $5 \times 0.8 = 4$ m.
2. Let $x =$ mass of heavier. $\therefore$ lighter $= \frac{1}{2}x, x + \frac{1}{2}x = 1\frac{1}{2}x = 1800$.
$x = \frac{1800}{\frac{3}{2}} = 1800 \times \frac{2}{3} = 1200$.

3. $100\% = 300$, then $18\% = \$54$. If $12\% = 54$, $100\% = \frac{100}{12} \times 54 = 450$.

4. £1 at 4% for $1\frac{1}{2}$ years amounts to £1·06. ∴ the sum that will amount to £424 is $424 \div 1·06$ or £400.

5. Let $x = $ cost. ∴ $x + \frac{1}{5}x = \frac{6}{5}x = 4·50$.

$6x = 4·50 \times 5$, $6x = 22·50$, $x = \dfrac{22·50}{6} = £3·75$.

6. $\frac{18}{10} = \frac{x}{8}$ (inverse proportion), $10x = 144$, $x = \frac{144}{10} = 14\frac{2}{5}$.

7. If $c^2 = a^2 + b^2$, $b^2 = c^2 - a^2$, and $b = \sqrt{c^2 - a^2} = \sqrt{225 - 144} = \sqrt{81} = 9$.

8. $A = \pi r^2$. ∴ $r^2 = \dfrac{A}{\pi}$, and $r = \sqrt{\dfrac{A}{\pi}} =$

$\sqrt{264 \times \frac{7}{22}} = \sqrt{84} = 9·165$

$D = 2r = 18·33$

9. Area $\bigcirc = \pi r^2$, $A = \frac{22}{7} \times 4·41 = 13·86$ m²

Area rectangle $= b \times h$. ∴ $\dfrac{13·86}{1·4} = 9·9$.

10. Let $n = $ number of 10p. ∴ $6n = $ number of 5p. $0·10n + 0·05(6n) = 4·80$.

$0·10n + 0·30n = 4·80$, $0·40n = 4·80$, $n = \dfrac{4·80}{0·40} = 12$.

If $n = 12$, then $6n = 72$.

11. $64 \times 64 \times 64 = 262\,144$.

12. $2^2 \times 3\frac{1}{2} \times 6 = 75\frac{3}{7}$ cm³.

13. $(5x + 3y)(3x - 6y) = 15x^2 - 21xy - 18y^2$.

14. Divide into 6 equilateral $\triangle$s. Each of the 6 $\angle$s of the hexagon equals 2 $\angle$s of an equilateral $\triangle$. $6 \times 2 \times 60° = 720°$. Or add all the angles of the $\angle$s formed (12 rt. $\angle$s) and subtract the 4 rt. $\angle$s at the centre; 12 rt. $\angle$s $-$ 4 rt. $\angle$s $= 8$ rt. $\angle$s $= 720°$.

15. Form a $\triangle$ with side $a \perp$ from plane to ground, b thence to target, and c thence to plane. Then $\angle$ at target $=$ angle of depression at plane (alt. int. $\angle$ of $\parallel$ lines). Tan of $\angle$ at target $= a : b = 1 : 1\frac{3}{5} = 0·625 = \tan 32°$ within 0·0001.

Exercise No. 1

	Col. I	Col. II	Col. III	Col. IV
A–B	37 249	57 365	41 757	22 728
B–C	45 262	39 969	48 211	19 858
C–D	37 339	38 753	43 447	21 680
D–E	38 886	41 657	38 601	35 667
A–C	82 511	97 334	89 968	42 586
B–D	82 601	78 722	91 658	41 538
C–E	76 225	80 410	82 048	57 347
A–D	119 850	136 087	133 415	64 266
B–E	121 487	120 379	130 259	77 205
A–E	158 736	177 744	172 016	99 933

Exercise No. 2

1. 389.   2. 4968.   3. 3484.   4. 41 482.   5. 37 789.   6. 452 884.

Exercise No. 3

1. 1504.   2. 5289.   3. 9464.   4. 4557.   5. 3264.

6. 993 016. To solve this, reverse the example, multiplying 893 by 12 and then by 11.

7. 537 633. Note that $36 = 4 \times 9$. $1457 \times 9 = 13\,113$. This $\times 4 = 52\,452$.

8. 2304. 4×4 combined with $8 \times 8 = 1664$. To this add twice 8×40.

9. 4399. 8×5 combined with $3 \times 3 = 4009$. To this add $3 \times (8 + 5)$ with 0 after it.

10. 13 225. Note ending in 5. Combine 11×12 with 5×5

11. 5 524 582. Use 11 and 12 as multipliers.

12. 2 439 720. 64 = 8 × 8. After multiplying by 8, multiply this figure by 8.

13. 7569. 8 × 8 combined with 7 × 7 = 6449. To this add twice 7 × 80.

14. 4416. 9 × 4 combined with 6 × 6 = 3636. To this add 6 × (9 + 4) with 0 after it.

15. 326 019. 997 is near 1000. 327 000 − 3 × 327 gives the answer.

Exercise No. 4

1. 382.   2. 384.   3. $405\frac{27}{63}$.   4. 534.   5. 645.   6. 843.   7. $917\frac{19}{46}$.
8. 903.   9. 593.

Exercise No. 5

1. 8.   2. 6.   3. 0.   4. 5.   5. 1.   6. 0.   7. 8.   8. 2.   9. 6.
10. 0.   11. 3.   12. 4.   13. 0.   14. 8.   15. 3.

Exercise No. 6

1. Right.   2. Right.   3. Wrong.   4. Right.   5. Wrong.
6. Right.   7. Right.   8. Right.   9. Wrong.   10. Wrong.
11. Right.   12. Wrong.   13. Right.   14. Right.   15. Right.
16. Wrong.   17. Right.   18. Wrong.   19. Right. $\frac{1}{4} = \frac{24}{96}$.   20. Wrong.
21. Right.   22. Right.   23. Right.   24. Wrong.

Exercise No. 7

1. 178 322. You should use the horizontal method and make only one addition for the numbers in the thousands.

2. 12 months in 1 year. ∴ 12 × 152 = £1824. *Note:* The sign ∴ means *therefore.*

3. Distance marched last day is 48 minus 12 + 9 + 7 + 9 = 48 − 37 = 11.

4. If 6 cost £1·5, one cost £0·25; 3·75 ÷ 0·25 = 15.

5. 450 ÷ 15 = 30 m as the distance travelled in 1 second.
   30 ÷ 3 = 10 m, the distance travelled in $\frac{1}{3}$ second.

6. 80 + 90 + 70 + 60 + 50 = 350; 350 ÷ 5 = 70.

7. Total volume of hangar is 100 × 50 × 10 = 50 000 m³. If 1000 m³ costs £25 per season then 50 000 m³ costs £25 × 50 = £1250.

8. If 50 kg covers 10 m² then 5 kg covers 1 m². Area of rectangle is 250 m². ∴ 1250 kg of cement required.

9. Area left over is 142 − (22 + 30 + 14 + 16) = 142 − 82 = 60 km².

10. Difference per hour = 200. Difference per day = 10 × 200 = 2000. Difference for 30 days is 2000 × 30 = 60 000.

11. 200 × 6 = 1200 km covered in 6 days + 2 days delay. 12 − 8 = 4 days left to cover remainder of distance, or 2400 − 1200 = 1200 km. 1200 ÷ 4 = 300 km per day average needed.

12. Payment on car £104; Send home £260; Insurance £48. Remainder is £408, i.e. £34 per month.

Exercise No. 8

1. 4.   2. 12.   3. 7.   4. 3.   5. 16.   6. 12.   7. 9.
8. 18.   9. 33.   10. 32.   11. 8.   12. 9.   13. 24.   14. 16.
15. 3.

Exercise No. 9

1. $\frac{3}{8}$.   2. $\frac{2}{3}$.   3. $\frac{2}{3}$.   4. $\frac{3}{4}$.   5. $\frac{3}{8}$.   6. $\frac{4}{11}$.   7. $\frac{5}{6}$.
8. $\frac{1}{4}$.   9. $\frac{5}{8}$.   10. $\frac{9}{28}$.   11. $\frac{5}{9}$.   12. $\frac{1}{4}$.   13. $\frac{5}{6}$.   14. $\frac{30}{61}$.
15. $\frac{1}{7}$.

Exercise No. 10

1. $\frac{2}{8}$.   2. $\frac{4}{12}$.   3. $\frac{8}{20}$.   4. $\frac{36}{81}$.   5. $\frac{24}{48}$.   6. $\frac{14}{49}$.   7. $\frac{8}{64}$.
8. $\frac{30}{78}$.   9. $\frac{9}{24}$.   10. $\frac{27}{45}$.   11. $\frac{8}{36}$.   12. $\frac{44}{60}$.   13. $\frac{42}{75}$.   14. $\frac{72}{88}$.
15. $\frac{40}{96}$.   16. $\frac{28}{68}$.

Exercise No. 11

1. $2\frac{2}{5}$.   2. 2.   3. $1\frac{7}{12}$.   4. $8\frac{2}{3}$.   5. $6\frac{1}{2}$.   6. $1\frac{1}{2}$.   7. $2\frac{2}{7}$.
8. $4\frac{2}{3}$.   9. $4\frac{3}{4}$.   10. 3.   11. 2.   12. 16.

Exercise No. 12

1. $\frac{11}{4}$.   2. $\frac{13}{4}$.   3. $\frac{24}{5}$.   4. $\frac{28}{5}$.   5. $\frac{38}{3}$.   6. $\frac{75}{4}$.   7. $\frac{139}{7}$.
8. $\frac{97}{6}$.   9. $\frac{86}{7}$.   10. $\frac{94}{7}$.   11. $\frac{71}{5}$.   12. $\frac{112}{5}$.

Exercise No. 13

1. $1\frac{5}{8}$.   2. $\frac{2}{9}$.   3. $\frac{11}{40}$.   4. $1\frac{13}{18}$.   5. $\frac{1}{2}$.   6. $\frac{1}{8}$.   7. $9\frac{1}{4}$.
8. $7\frac{1}{4}$.   9. $23\frac{17}{18}$.   10. $6\frac{13}{24}$.   11. $1\frac{7}{12}$.   12. $6\frac{13}{24}$.

Exercise No. 14

1. $\frac{9}{35}$.   2. $\frac{1}{4}$.   3. $\frac{1}{9}$.   4. $\frac{3}{10}$.   5. $4\frac{1}{2}$.   6. $\frac{7}{96}$.   7. $\frac{25}{34}$.
8. $\frac{1}{6}$.   9. 36.   10. 9.   11. $1\frac{1}{3}$.   12. $4\frac{4}{7}$.   13. $1\frac{1}{4}$.   14. $4\frac{3}{8}$.
15. 14.   16. 6.   17. $6\frac{4}{5}$.   18. 40.   19. $1\frac{1}{4}$.   20. $\frac{2}{3}$.   21. 28.
22. $2\frac{1}{5}$.

Exercise No. 15

1. $46\frac{7}{8}$.   2. $88\frac{1}{8}$.   3. $80\frac{1}{4}$.   4. $71\frac{1}{8}$.   5. $94\frac{6}{9}$.   6. $37\frac{8}{11}$.
7. $141\frac{1}{8}$.   8. $41\frac{7}{8}$.   9. $57\frac{3}{8}$.   10. 7650.   11. $2\frac{18}{35}$.   12. $2\frac{17}{86}$.
13. $3\frac{11}{108}$.   14. $2\frac{91}{110}$.   15. $6\frac{17}{30}$.   16. $3\frac{51}{79}$.   17. $6\frac{162}{167}$.   18. $23\frac{9}{11}$.
19. $5\frac{51}{148}$.   20. $7\frac{14}{17}$.

Exercise No. 15a

1. $\frac{8}{11}$.   2. $\frac{3}{80}$.   3. $\frac{16}{21}$.   4. $\frac{9}{14}$.   5. 200.   6. 160.   7. 154 8. $1\frac{1}{8}$.

Exercise No. 16

1. $23\frac{1}{2}$ divided by $\frac{1}{32} = \frac{51}{2} \times \frac{32}{1} = 816$.

2. $\frac{1}{16} =$ the amount he can do in 1 day. $\therefore$ the amount of work he can do in $\frac{1}{2}$ day $= \frac{1}{2} \times \frac{1}{16} = \frac{1}{32}$.

3. The top and bottom wings placed together measure $\frac{5}{8}$ m in height. $\frac{5}{8} \times \frac{3}{4} + \frac{5}{8}$ $\times 1\frac{3}{4} = \frac{5}{8} \times \frac{5}{2} = \frac{25}{16} = 1\frac{9}{16}$.

4. If $\frac{1}{3} = 10$, then 1 whole $= \frac{3}{3}$ or 3 times 10, which is 30. Or invert and multiply: if $\frac{1}{3} = 10$, then $\frac{3}{1} \times 10 = 30$.

5. If the output of the faster machine is 1, then the slower machine produces $2 \times \frac{1}{3}$ or $\frac{2}{3}$. The combined output is $1 + \frac{2}{3}$ or $\frac{5}{3}$. If $\frac{5}{3} = 600$, $\frac{3}{3} = 600 \times \frac{3}{5} = 360$.

6. If $\frac{2}{3}$ takes 5 h, $\frac{1}{3}$ will take half that time or $2\frac{1}{2}$ h, and $\frac{3}{3}$ or the whole job will take 3 times $2\frac{1}{2}$ or $7\frac{1}{2}$ h. Or—by inverting and multiplying: if $\frac{2}{3} = 5$, then $\frac{3}{3} = \frac{3}{2} \times 5 = \frac{15}{2}$ or $7\frac{1}{2}$.

7. If second turns out $\frac{5}{6}$, then he loses $\frac{1}{6}$. If first turns out $\frac{1}{2}$ of $\frac{5}{6}$ he turns out $\frac{5}{12}$ and loses $\frac{7}{12}$. $\therefore \frac{1}{6} + \frac{7}{12}$ or $\frac{9}{12} = \frac{3}{4}$. $\frac{3}{4}$ of the output of one machine equals $\frac{5}{6}$ of the combined output.

8. 2nd flight $= \frac{1}{2}$ of 432 or 216, 3rd flight $= \frac{1}{3}$ of 432 or 144, 4th flight $= \frac{1}{2}$ of $216 + 144 = 180$ km.

9. $\frac{1}{4} + \frac{3}{8} = \frac{5}{8}$ of the job completed in $24 + 31$ days or 55 days. If $\frac{5}{8} = 55$ days, then $\frac{1}{8} = \frac{1}{5}$ of 55 or 11 days, and $\frac{8}{8} = 88$ days. Or—by inverting, if $\frac{5}{8} = 55$, then $\frac{8}{8} = \frac{8}{5} \times 55 = 88$ days.

10. Both travel for 5 hours. $\therefore 62\frac{3}{5} \times 5 = 313$, and $69\frac{4}{5} \times 5 = 349$. $3000 - (313 + 349) = 2338$ km, the distance apart at the end of 5 hours.

Note: Easier methods for doing examples of this type are explained in the chapters on ratios and equations.

Exercise No. 17

1. $\frac{1}{100}$. 2. $\frac{1}{2}$. 3. $\frac{5}{8}$. 4. $2\frac{1}{10}$. 5. $23\frac{9}{20}$.
6. $\frac{1}{1250}$. 7. $\frac{38}{625}$. 8. $\frac{2341}{10000}$. 9. $\frac{4329}{100000}$. 10. $18\frac{1}{50}$.
11. 0·3. 12. 0·05. 13. 0·321. 14. 12·01. 15. 124·0003.
16. 18·7. 17. 0·3. 18. 1·45. 19. 22·3. 20. 4·33.

Exercise No. 18

1. 0·5. 2. 0·75. 3. 0·375. 4. 0·313. 5. 0·563. 6. 0·531.
7. 0·875. 8. 0·875. 9. 0·688. 10. 0·875.

Exercise No. 19

1. 0·77. 2. 25·03. 3. 641·099. 4. 22·165. 5. 38·89.
6. 12·42. 7. 0·004 5. 8. 0·869. 9. 0·802 10. 0·08.
11. 63·554. 12. 10·439. 13. 51·292. 14. 44·345 6. 15. 57·358 3
16. 4·512 35.

Exercise No. 20

1. 74. 2. 9·36. 3. 32·40. 4. 30·602.
5. 0·414 4. 6. 0·067 648. 7. 0·980 76. 8. 0·004 284.
9. 0·001 803. 10. 0·000 089 6. 11. 32·67. 12. 0·326 7.
13. 2·862 68. 14. 0·407 7. 15. 6·008 8. 16. 87.
17. 0·069. 18. 9560. 19. 4·53. 20. 4069.
21. 0·94. 22. 92. 23. 749. 24. 5·347 9.
25. 0·492 568. 26. 0·024 965 3. 27. 0·059 08. 28. 0·000 071 56.
29. 0·495 674. 30. 0·000 000 386 49.

Exercise No. 21

1. 0·17. 2. 0·05. 3. 0·6. 4. 4·73. 5. 420.
6. 3·7. 7. 0·047. 8. 0·67. 9. 404·286. 10. 36·818.

Exercise No. 22

1. £46. 2. £215. 3. $222·50. 4. R2385. 5. £14 665. 6. 800.
7. 750. 8. 1333⅓. 9. 120. 10. 1¾.

Exercise No. 23

1. Amount of sand is 0·84 of the quantity of mortar. ∴ 0·84 × 250 = 210 kg.
2. Since each side is 0·04 × 2 m less than the outside length, the total difference in length for the four sides is 0·08 × 4 m = 0·32; 8·32 m − 0·32 = 8 m.
3. 0·675 + 0·000 7 = 0·675 7 mm.
4. If £0·20 is the amount received from one dozen, then £44·20 divided by 0·2 = 221 dozen.
5. Length of one section is 158·72 ÷ 16 or 9·92 m. Six sections have a length 9·92 × 6 = 59·52 m.
6. The average of the four readings is 2576 m ÷ 4 = 0·644 m.
7. 33·75 − 0·09 = 33·66 mm.

Exercise No. 24

1. $\frac{1}{100}$. 2. $\frac{1}{50}$. 3. $\frac{1}{25}$. 4. $\frac{7}{100}$. 5. $\frac{1}{200}$.
6. $\frac{1}{16}$. 7. $\frac{1}{15}$. 8. $\frac{3}{40}$. 9. $\frac{1\cdot4}{300}$. 10. $\frac{1}{400}$.
11. $\frac{3}{200}$. 12. $\frac{1}{30}$.

Exercise No. 25

1. 186.   2. 12.   3. 10.   4. £50.   5. $\frac{19}{20}$.

6. 25%.   7. 50.   8. 7%.   9. 50%.   10. 75%.

11. 48.   12. 50.   13. 320.   14. 800.   15. 96.

16. $20\,000 \times \frac{4}{5} = 16\,000$, $20\,000 - 16\,000 = 4000$.

17. $0.08 \times 275 = 22$.

18. $45\% = \frac{19}{20}$, $75 \times \frac{19}{20} = 33\frac{3}{4}$, $75 - 33\frac{3}{4} = 41\frac{1}{4}$, or 55% of 75 = 41·25.

19. The number = 100%. ∴ 175% = 140, and 1% = 140 ÷ 175 or 0·8, and 100% = 100 × 0·8 or 80.

20. If 5·25 = 17·5%, 1% = 5·25 ÷ 17·5 or 0·3, and 100% = 100 × 0·3 = £30.

21. $9000 - 6750 = 2250$. $\frac{2250}{9000} = \frac{1}{4}$ or 25%.

22. $28 - 3\frac{1}{2} = 24\frac{1}{2}$, $\frac{24\frac{1}{2}}{28} = \frac{49}{2} \times \frac{1}{28} = \frac{7}{8}$ or $87\frac{1}{2}\%$.

23. If 8% = 144, then 1% = 144 ÷ 8 or 18, and 100% = 18 × 100 or 1800.

24. $140 \times \frac{1}{20} = 7$, $140 - 7 = 133$ volts.

25. 100% or total output of slower man equals 1500. ∴ 50% = 750. Since $33\frac{1}{3}\%$ of faster man = 50% of slower man, his $33\frac{1}{3}\%$ = 750. ∴ his total output is 100% or 3 times 750 or 2250.

Exercise No. 26

1. Selling price is £30 × 0·75 = £22·5.

2. Discount is $\frac{2}{3}$ × £54 = £36.

3. List price of paper is £60 × 0·4.
   Price paid is £60 × 0·4 × 0·75 × 0·975
   $$= £60 \times 0.4 \times \frac{3}{4} \times \frac{39}{40}$$
   $$= £17.55.$$

4. Final price is $£10 \times \frac{60}{100} \times \frac{90}{100} \times \frac{80}{100}$
   $$= £4.32.$$

5. Premium paid is £40 × 0·5 × 0·825
   $$= £16.50.$$

Exercise No. 27

1. £25 000 × 0·015 = £375.

2. £165 × 100 = £16 500, selling price of 100 sets; £16 500 × 0·16 = £2640, total commission.

3. If 5% = £400, then 100% equals $\frac{100}{5} \times 400 = \frac{40\,000}{5}$ = £8000.

4. $37\frac{1}{2}\%$ = £600. Since $37\frac{1}{2} = \frac{3}{8}$, then $\frac{8}{3} \times 600 = \frac{4800}{3}$ = £1600.

Exercise No. 28

1. Interest for one year on £188·60 at $4\frac{1}{2}\%$ = £188·60 × 0·045 = £8·49.

2. Interest on £1850 at 4% for 1 year = £1850 × 0·04 = £74·00; for $2\frac{1}{2}$ years interest = £74 × $\frac{5}{2}$ = £185.

3. For one year £275 × 0·045 = £12·375. For 3 months interest = £12·375 × $\frac{1}{4}$ = £3·09.

4. $\frac{2}{3}$ of £60 000 = £40 000; £40 000 × 0·04 = £1600; £1600 × 2 = £3200; $\frac{1}{3}$ of £60 000 = £20 000; £20 000 × 0·05 = £1000; £1000 × 2 = £2000; £3200 + £2000 = £5200.

5. If £67·50 = interest for 4½ years, then 67·50 ÷ $\frac{9}{2}$ = £15 of interest for 1 year; £15 on £300 is 15 ÷ 300 or 0·05 = 5%.

6. This means that rate of interest will yield £150 in 14 years. £150 ÷ 14 = £10·714, amount interest in 1 year $\frac{10·714}{150}$ = 7$\frac{1}{7}$%.

7. £36 for 3 months would be 4 × 36 or £144 for 1 year. If £144 equals 3%, 100% = £144 × $\frac{100}{3}$ = $\frac{£14\,400}{3}$ = £4800.

8. £800 at 6% for 1 year yields £48; $\frac{80}{48}$ = 1$\frac{32}{48}$ = 1$\frac{2}{3}$.

9. (i) £112·55(1), (ii) £1407·10, (iii) £12·25(04).

10. 1000 ÷ 1·47746 = £676·84.

11. $\frac{57 \times 400}{100}$ = £228

12. Rateable value = £24 000 × 100 = £2 400 000. Sum raised is £24 000 × 2 = £48 000.

Exercise No. 29

1. 20% = $\frac{1}{5}$, $\frac{1}{5}$ × 5 = 1c profit on one. 5c + 1c = 6 cents, selling price of one. 6 × 12 = 72 cents, selling price per doz.

2. 10 000 × 2·50 = £25 000, total cost, 25 000 × 0·06 = £1500 profit. £25 000 + £1500 = £26 500, total selling price.

3. $\frac{2}{5}$ of 10 = 4. £30 × 4 = £120, selling price of 4 books. 10 − 4 = 6. £25 × 6 = £150, selling price of 6 books.

£150 + £120 = £270. £270 − £200 = £70; $\frac{£70}{£200}$ = 35%.

4. 100% = cost of the boat; ∴ 125% or $\frac{5}{4}$ = selling price or £500; if $\frac{5}{4}$ = 500, $\frac{1}{4}$ = £500 ÷ 5 or £100, and $\frac{4}{4}$ = £400.

5. If 15% = £75, then 100% = $\frac{100}{15}$ × £75 = £500.

6. If $\frac{3}{8}$ = £90, then $\frac{8}{8}$ = $\frac{8}{3}$ × 90 or £240, total cost. Loss of £90 makes £150, selling price.

7. If loss = 12½%, then £259 = 100% − 12½ or 87½% = $\frac{7}{8}$. If £259 = $\frac{7}{8}$, then $\frac{8}{8}$ = $\frac{8}{7}$ × 259 = £296.

8. If mark-up was 24%, then 124% = £806, and 100% = $\frac{100}{124}$ × £806 = £650, cost.

9. £10 000 × 0·18 = £1800 profit; £10 000 × 0·18 = £1800 loss. £1800 minus £1800 = 0.

10. If rate of profit is 20%, and selling price is £3600, then £3600 = 120% or $\frac{6}{5}$, and $\frac{1}{5}$ = $\frac{5}{6}$ × £3600 or £3000. £3600 − £3000 = £600 profit. If rate of loss is 20%, and selling price is £3600, then £3600 = 80% or $\frac{4}{5}$, and $\frac{5}{5}$ = $\frac{5}{4}$ × £3600 or £4500; £4500 − £3600 = £900 loss. £900 − £600 = £300 total loss.

11. If the selling price includes 37½% profit, the cost must represent 62½% of the selling price. 62½% is $\frac{5}{8}$; $\frac{5}{8}$ of £4 is £2·50.

12. $1000 represents 100%, of which 12½% is profit and 87½% is cost. 87½% of $1000 is $875.

13. Since there is a profit, this represents a smaller proportion of the selling price than it does of the cost. Hence we divide 15% by 100% + 15%. $\frac{0·15}{1·15}$ = 0·1304 = 13·04%.

14. Where there is a loss it represents a larger percentage of the selling price than it does of the cost. Hence we divide 15% loss on selling price by 100% + 15%. $\frac{0·15}{1·15}$ = 0·1304 = 13·04%.

15. A pays £170 × $\frac{5}{8}$ = £106·25. B pays £150 × $\frac{31}{40}$ = £116·25, ∴ B pays £10 more than A.

Exercise No. 30

1. 1500 mm.
2. 1530 m.
3. 1530 g.
4. $60 \times 60 = 3600$ s.
5. 75 min.
6. $45 \times 60 = 2700'$.
7. $17 \times 24 = 408$ h.
8. 398 m².
9. $2 \cdot 3 \times 100 \times 100 = 23\,000$ m².
10. $238 \cdot 4 \div 100 = 2 \cdot 384$ km².
11. $4 \cdot 6 \times 100 = 460$ ha.
12. $100\,000 \div 1000 = 100$ m³.
13. 3·5 right angles.
14. 1·75 m.
15. 5·6 t.
16. $2347 \div 100 = 23 \cdot 47$ a.
17. $56\,722 \div 10\,000 = 5 \cdot 6722$ ha.

Exercise No. 30a

1. 109° 23′ 58″.
2. 6° 50′ 53″.
3. 45 ha 64 a 60 m².
4. 2 km² 78 ha 75 a.
5. One sixth of the time is 2 h 43 min. Journey now takes 13 h 35 min.
6. 2 km² 62 ha 50 a.
7. 12·6 t.

Exercise No. 30b

1. 6° 41′ 21″.
2. 149° 14′ 13″.
3. 2 h 9 min 37 sec.
4. 8 h 0 min 28 sec.

Exercise No. 30c

1. $15·60.
2. £0·40, $A25.
3. £4166·67.

Exercise No. 31

1. A space 8 m square contains 8×8 or 64 m². The difference between this and 8 m² is 56 m².
2. 8000 m².
3. 45 m @ £3 = £135.
4. $3 \cdot 6 \times 1 \cdot 6 \times 100 \times £500 = £288\,000$.
5. The bricks would naturally run the long way of the path. $250 \times 50 = 12\,500$.
6. $£75 \div 45 = £1\frac{2}{3}$.

Exercise No. 32

1. 2·184.
2. 0·36.
3. 28.
4. 10^9.
5. 10^3.
6. 1.

Exercise No. 33

1. 3·746 23.
2. 4253.
3. 0·000 085 46.
4. 473 860.
5. 3 560 000.
6. 0·000 374 658.
7. 3·426.

Exercise No. 34

1. 951·972 acres.
2. 95·070 miles.
3. 166·447 lb.
4. 79·553 m.
5. 63·677 ha. (All to three decimal places.)
6. 226·8 g.

Exercise No. 35

1. $\frac{24}{32} = \frac{3}{4}$.
2. $63 : 56 = \frac{63}{56} = 1\frac{7}{56} = 1\frac{1}{8}$.
3. $\dfrac{10}{5 \times 1000} = \dfrac{1}{500}$.
4. 2 parts + 3 parts = 5 parts. Since 2 parts are water, then $\frac{2}{5}$ of total is water, or 40%.
5. 6 parts tin + 19 parts copper = 25 parts to make bronze. The amount of tin in 500 g of bronze is $\frac{6}{25} \times 500$ or 120 g.

6. $5 + 14 + 21 = 40$ parts $=$ the total of £2000. $\frac{21}{40} \times 2000 = £1050$ as the largest share; $\frac{5}{40} \times 2000 = £250$ as the smallest share; £1050 $-$ £250 $=$ £800 difference.

Exercise No. 36

1. 6.   2. 12.   3. 6.   4. 2.   5. 21.   6. 4.   7. 3.
8. 125.   9. 3.   10. 18.

Exercise No. 37

1. $\frac{18}{27} = \frac{20}{x}$ Direct prop. $18x = 540$, $x = \frac{540}{18} = 30$.

2. $\frac{9}{30} = \frac{2}{x}$ Direct prop. $9x = 60$, $x = \frac{60}{9} = 6\frac{2}{3}$.

3. $\frac{5}{20} = \frac{90}{x}$ Direct prop. $5x = 1800$, $x = \frac{1800}{5} = 360$.

4. $\frac{112}{240} = \frac{28}{x}$ Inverse prop. $112x = 6720$, $x = \frac{6720}{112} = 60$.

5. $\frac{26}{20} = \frac{x}{35}$ Inverse prop. $20x = 910$, $x = \frac{910}{20} = 45\frac{1}{2}$.

6. $\frac{220}{x} = \frac{2}{8}$ Direct prop. $2x = 1760$, $x = \frac{1760}{2} = 880$.

Exercise No. 38

1. 23.   2. -36.   3. -7.   4. $-20d$.   5. $4b$.
6. -3.   7. $12x$.   8. $5a + 2b$.   9. $15a - 2b$.   10. $15a + 3b - 5$.

Exercise No. 39

1. 28.   2. -9.   3. -24.   4. 66.   5. 130.   6. $-40ab$.
7. 3.   8. -3.   9. -25.

Exercise No. 40

1. -32.   2. 216.   3. 72.   4. -144.   5. -3.   6. 4.
7. $2\frac{4}{5}$.   8. -12.

Exercise No. 41

1. $x + y = 5$.   2. $x + y + z = 9$.   3. $2x + 2y = 10$.
4. $x + y - z = 1$.   5. $x^2 + y^2 = 13$.   6. $y - 3 = 0$.
7. $2xz = 16$.

Exercise No. 42

1. 23.   2. 110.   3. 13.   4. 20.   5. 12.   6. 44.
7. 3.   8. 55.   9. 1260.   10. 3.

Exercise No. 43

1. $p = 2l + 2w$.   2. $d = rt$.   3. $P = va$.

4. $I = PRT$.   5. $A = \dfrac{W}{V}$.   6. $P = M - O$.

7. $d = 4 \cdot 9t^2$.   8. $A = S^2$.   9. $C = \frac{5}{9}(F - 32°)$.

10. $R = \dfrac{N}{T}$.

Exercise No. 44

1. $p = 5$.
2. $n = 12\frac{1}{2}$.
3. $x = 28$.
4. $c = 6$.
5. $y = 4$.
6. $n = 36$.
7. $a = 48$.
8. $b = Wc$.
9. $A = \dfrac{W}{V}$.
10. $W = \dfrac{P}{AH}$.

Exercise No. 45

1. Since $\dfrac{D}{d} = \dfrac{r}{R}$ then $rd = DR$, and $r = \dfrac{DR}{d}$; substituting, $r = \dfrac{18 \times 100}{6} = \dfrac{1800}{6} = 300$.

2. Since $DR = dr$, and D is the unknown, then $D = \dfrac{dr}{R}$; substituting, $D = \dfrac{9 \times 256}{144} = \dfrac{2304}{144} = 16$.

3. Let n represent the number. Then $3n + 2n = 90$, $5n = 90$, $n = \dfrac{90}{5} = 18$.

4. Let n represent the smaller number. Then $n + 7n = 32$, $8n = 32$, $n = \frac{32}{8} = 4$, and $7n = 7 \times 4$ or 28.

5. Let $x =$ the amount the first unit gets. Then the second unit receives $2x$, and the third unit receives $x + 2x$ or $3x$. The total $x + 2x + 3x = 6x = 600$, $x = 100$, and $2x = 200$.

6. Rate $\times$ Time = Distance; $R \times t = D$. They both travel the same amount of time. Let t equal the time they travel. Then $300t + 200t = 3000$, $500t = 3000$, $t = \dfrac{3000}{500} = 6$ h. $300 \times 6 = 1800$ km.

7. $R \times t = D$. Let $R =$ rate of slower soldier. Then $2R =$ rate of faster one. $10R + 20R = 24$ km, $30R = 24$, $R = \frac{24}{30}$ or $\frac{4}{5}$, and $2R = \frac{8}{5}$ or $1\frac{3}{5}$ km/h.

8.
$$2n + 3n = 200,$$
$$\therefore n = 40.$$
$$4n = 160.$$

9. Let T and t represent number of teeth in large and small gears and let R and r represent r.p.m.

$$\text{Then } \frac{T}{t} = \frac{r}{R}, \text{ and } \frac{72}{48} = \frac{160}{R}, \ 72R = 7680,$$
$$R = \frac{7680}{72} = 106\frac{2}{3}.$$

10. Mass $\times$ Distance = Mass $\times$ Distance. $\therefore 120 \times \frac{9}{2} = x \times 5$. $5x = 540$, $x = 108$.

Exercise No. 46

1. 8.
2. 10.
3. 9.
4. 3.
5. 5.
6. 12.
7. 10.
8. 1.
9. 0·2.
10. 0·3.
11. 1·2.
12. 0·05.

Exercise No. 47

1. 73.
2. 35.
3. 54·2.
4. 17·68 (2 d.p.)
5. 20·69.
6. 26.
7. 43.
8. 56.
9. 85.
10. 97.

Exercise No. 48

1. 36.
2. 729.
3. 5.
4. $\frac{1}{64}$.
5. 186 624.
6. 32 768.
7. 9.
8. 5.

9. 43 000 000. 10. 620 000. 11. $\frac{4}{9}$. 12. 14·6969.

13. 0·000 000 28. 14. 0·0025. 15. 122 000 000.

Exercise No. 49

1. 135. 2. 223. 3. 343. 4. 739. 5. 487.
6. $\frac{10}{12}$. 7. $\frac{14}{5}$. 8. $\frac{124}{49}$. 9. $\frac{65}{20}$. 10. $\frac{1068}{400}$.
11. $\frac{5}{6}$. 12. $\frac{31}{36}$. 13. $\frac{63}{72}$. 14. $\frac{7}{4}$. 15. $\frac{6}{8}$.
16. 9. 17. 7. 18. $2\frac{1}{3}$. 19. $3\frac{1}{5}$. 20. $1\frac{1}{2}$.
21. 441. 22. 529. 23. 1089. 24. 1369. 25. 1521.
26. 1225. 27. 4225. 28. 9025. 29. 11025. 30. 42025.
31. 841. 32. 1521. 33. 9801. 34. 784. 35. 1444.
36. 399. 37. 896. 38. 1591. 39. 2484. 40. 3456.

Exercise No. 50

1. $7abc(ac^2 - 4)$.
2. $5acd(3a + 4c - 3d)$.
3. $(2x + 3y)(2x + 3y)$.
4. $(3ab - 4ac)(3ab - 4ac)$.
5. $(3ax + 4ay)(3ax - 4ay)$.
6. $(7x^2 + 4y)(7x^2 - 4y)$.
7. $(3x - y + 2z)(x + 3y)$.
8. $(a^2 + a + 1)(a^2 - a + 1)$.
9. $(x + 7)(x + 3)$.
10. $(x - 15)(x - 3)$.
11. $(x + 9)(x - 4)$.
12. $(x - 16)(x + 3)$.
13. $(x - 11y)(x - 3y)$.
14. $(3x + 9)(2x + 1)$.
15. $(5x - 7)(3x + 3)$.
16. $(4x + 13)(3x - 3)$.

Exercise No. 51

1. Let x = no. of min. spaces passed over by min. hand; y = no. passed by hr. hand. $12y = x$, $y = x - 60$. Subtracting, $11y = x - (x - 60) = 60$; $y = 5\frac{5}{11}$. $5\frac{5}{11}$ min. spaces past 12 o'clock gives $1·05\frac{5}{11}$ o'clock.

2. Try elimination by substitution. x = part inv. at 5%; y = part inv. at 6%. $0·05x + 0·06y = £1220$; $x + y = £22\,000$. Multiplying $0·05x$ etc. by 20 we get $x + 1·20y = £24\,400$, from which $x = £24\,400 - 1·20y$. Substituting this value in other equation, $£24\,400 - 1·20y + y = £22\,000$; $0·20y = £2400$; $y = £12\,000$; hence $x = £10\,000$.

3. Try elimination by comparison. a = Jack's age; b = Joe's age. $a = 2b$; $a - 20 = 4(b - 20)$; $a = 4(b - 20) + 20$; hence $2b = 4b - 80 + 20$; $2b - 4b = -60$; $b = 30$ years for Joe; $2 \times 30 = 60$ years for Jack.

4. a = first; b = second. $a + \dfrac{b}{2} = 35$; $\dfrac{a}{2} + b = 40$. Multiply first equation by 2, $2a + b = 70$. Subtracting second equation from this $1\frac{1}{2}a = 30$; $a = 20$; $a + \dfrac{b}{2} = 35$; $\dfrac{b}{2} = 35 - 20$; $\dfrac{b}{2} = 15$; $b = 30$.

5. $a + \dfrac{b}{3} = £1700$; $\dfrac{a}{4} + b = £1800$. Multiplying first equation by 3, $3a + b = £5100$. Subtracting second equation, $2\frac{3}{4}a = £3300$; $a = £1200$. Substituting in first equation $£1200 + \dfrac{b}{3} = £1700$; $\dfrac{b}{3} = £1700 - £1200 = £500$; $b = £1500$.

6. $\dfrac{a}{2} + \dfrac{b}{3} = 45$; $\dfrac{a}{5} + \dfrac{b}{2} = 40$. Multiplying both equations, $a + \dfrac{2b}{3} = 90$; $a + \dfrac{5b}{2} = 200$. Subtracting the first from the second $\dfrac{11b}{6} = 110$; $b = 60$. Substituting in first equation, $\dfrac{a}{2} + 20 = 45$; $\dfrac{a}{2} = 25$; $a = 50$.

7. a = A's profit; b = B's profit. $a + b$ = £153; $a - b$ = £45. Adding, $2a$ = £198; a = £99; b = 54. Dividing £918 in the proportions of 99 and 54, £918 ÷ 153 = 6; £99 × 6 = £594 for A; £54 × 6 = £324 for B.

8. Try substitution. $14A + 15M$ = $153; $6A - 4M$ = $3; $6A$ = $3 + 4M$, whence A = \$.50 + $\frac{2M}{3}$. Substituting in other equation $7 + \frac{28M}{3} + 15M$ = $153; $9\frac{1}{3}M + 15M$ = $153 - $7. $\frac{73M}{3}$ = $146; M = $6. Substituting in original equation, $6A - $24 = $3; 6A$ = $27; A = \$4.50.

9. There are several ways to solve problems like this. The method by simultaneous equations might be as follows. Select letters to represent values that do not change.

a = mass of copper *to be added*; b = mass of tin. $b = 80 - \frac{7b}{3} = 24$. $a = \frac{11b}{4} - \frac{7b}{3}$ $= 66 - 56 = 10$ kg.

10. Eliminate by comparison. $B + \frac{J}{8}$ = £1200; $\frac{B}{9} + J$ = £2500. $B = £1200 - \frac{J}{8}$; $B = £22\,500 - 9J$;   £1200 $- \frac{J}{8}$ = £22 500 $- 9J$; $9J - \frac{J}{8}$ = £22 500 $-$ £1200; $\frac{71J}{8}$ = £21 300; J = £2400; $B + \frac{J}{8}$ = £1200; $B +$ £300 = £1200; B = £900.

Exercise No. 52

1. $\frac{5xy}{4ab}$.

2. $\frac{x + a}{3(x - a)}$.

3. $x - a + \frac{3}{x - a}$.

4. $\frac{a^2}{a - x}$.

5. $\frac{x - y + c}{x - y}$.

6. $\frac{a(x + a)}{ab}, \frac{a^2}{ab}, \frac{b(a - x)}{ab}$.

7. $\frac{x(1 - x)^2}{(1 - x)^3}, \frac{x^2(1 - x)}{(1 - x)^3}, \frac{x^3}{(1 - x)^3}$.

8. x.

9. $\frac{4x^2 - 5x + 3}{(x - 1)^3}$.

10. $2a + \frac{3(a - b)}{c}$.

11. $\frac{a^2 + x^2}{a^2 - x^2}$.

12. $\frac{2(x + y)}{a}$.

13. $\frac{4x(x - 2)}{3}$.

14. $\frac{3}{4}$.

15. $x^2 - y^2$.

Exercise No. 53

1. 1.	2. 2.	3. 4.	4. 3.	5. −1.
6. 1.	7. 0.	8. −4.	9. −2.	10. 0.

Exercise No. 54

1. 2·5490.	2. 1·8808.	3. 0·9031.	4. 3·8025.	5. 0·5631.
6. 1̄·3692.	7. 3̄·5465.	8. 0·7810.	9. 4̄·7262.	10. 2·8279.

Exercise No. 55

1. 22 310.	2. 1068.	3. 486 300.	4. 156·2.	5. 8·252.
6. 79 270.	7. 456 500.	8. 4·396.	9. 3·238.	10. 84·72.

11. Log 5 = 0·6990, log 7 = 0·8451, antilog 1·8539 = 0·71. Prefix minus sign.

12. Log $17 = 1.2304$, log $32 = 1.5051$, antilog $\bar{1}.7253 = 0.53$. Prefix minus sign.
13. First perform addition. Log $9 = 0.9542$, log $4 = 0.6021$, antilog $0.3521 = 2.25$.
14. First perform addition. Log $15 = 1.1761$, log $7 = 0.8451$, antilog $0.3310 = 2.14$.
15. First perform subtraction. Log $4 = 0.6021$, log $3 = 0.4771$, antilog $0.1250 = 1.33$.
16. First perform subtraction. Log $11 = 1.0414$, log $9 = 0.9542$, antilog $0.0872 = 1.22$.
17. First multiply. Log $8 = 0.9031$, log $15 = 1.1761$, antilog $\bar{1}.7270 = 0.53$.
18. First multiply. Log $24 = 1.3802$, log $11 = 1.0414$, antilog $0.3388 = 2.18$.

19. $\dfrac{7 \div 3}{4} = \dfrac{7}{12}$. Log $7 = .8451$, log $12 = 1.0792$, antilog $\bar{1}.7659 = 0.58$.

20. $\dfrac{16}{18 \div 5} = \dfrac{80}{18} = \dfrac{40}{9}$. Log $40 = 1.6021$, log $9 = 0.9542$, antilog $0.6479 = 4.44$.

Exercise No. 56

1. -496.6. 2. -496.6. 3. 229.3 4. -3.037. 5. -2.047. 6. -4.102.

Exercise No. 57

1. 0.01. 2. no error. 3. no error. 5. £2750 by both methods.

Exercise No. 58

1. Approx. $\dfrac{3 \times 20 \times 350}{700 \times 40}$, 0.75; 0.6708. 2. Approx. $\dfrac{70 \times 0.1 \times 5}{140 \times 0.4}$, $\dfrac{5}{8}$; 0.6876.

3. Approx. $\dfrac{10 \times 0.8 \times 50}{10 \times 0.16 \times 5}$, 50; 49.57. 4. Approx. $\dfrac{100 \times 100 \times 80}{0.2 \times 80 \times 10}$, 5000; 6410.

Exercise No. 59

1. $14, 20; 65$.   2. $928, 912; 792$.   3. $7, 9; 41$.   4. $18, 6; -54$.
5. $16, 21; 101$.   6. $-4, -10, -112$.

Exercise No. 60

1. $7, 9, 11, 13, 15, \ldots; t_{30}$. 2. $0, 3, 6, 9, 12, \ldots; t_{15}$. 3. $100, 95, 90, 85, 80, \ldots; t_{12}$.
4. $11, 10, 9, 8, 7, \ldots; t_{12}$. 5. $44, 40, 36, 32, 28, \ldots; t_9$.
6. $-44, -40, -36, -32, -28, \ldots; t_{15}$.

Exercise No. 61

1. $6, 12, 24, 48, 96, 192, \ldots; t_6$.
2. $6, 18, 54, 162, 486, \ldots; t_7$. 3. $486, 162, 54, 18, 6, \ldots; t_9$.
4. $64, 32, 16, 8, 4, 2, 1, \ldots; t_8$. 5. $3.4, 0.34, 0.034, 0.0034, 0.00034, \ldots; t_6$.
6. $1/64, 1/32, 1/16, 1/8, 1/4, \ldots; t_{10}$.

Exercise No. 62

1. $59, 320$. 2. $10, 630$. 3. $2, 2, 156$. 4. 10100. 5. 10000. 6. £231.

Exercise No. 63

1. 2×3^n.   2. $16/2^n$.   3. $18 \times 2^{n-1}$ or 9×2^n.
4. 2^{n-1}.   5. $3 \times 2^{n-1}$.   6. $2 \times 5^{n-1}$.

Exercise No. 64

1. 4, 12, 36, 108, 324; 484.   2. 288, 144, 72, 36, 18; 558.
3. 2, 480, 930.   4. $t_8 = 11/7$; 30811.
5. −25.   6. 1, −3, 9, −27, 81; 61.

Exercise No. 65

1. 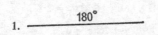   2.

3. 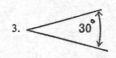   4.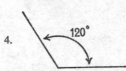

5. 30°.   6. 50°.   7. 80°.   8. 90°.   9. 130°.   10. 100°.   11. 50°.

Exercise No. 66

1. $\angle 2 = 30°$. $\angle s$ that coincide are $=$.
2. $\angle ABD = 22° 30'$. A bisector divides an $\angle$ in half.
3. (a) $\angle 1 = \angle 3$⎫ (c) Relationship unknown.
   (b) $\angle 2 = \angle 5$⎭ Ax. 1, p. 134.
4. $\angle 1$ and $\angle 2$, $\angle 2$ and $\angle 3$, $\angle 3$ and $\angle 4$, $\angle 4$ and $\angle 5$, $\angle 5$ and $\angle 6$, $\angle 6$ and $\angle 7$, $\angle 7$ and $\angle 1$.
5. $\angle 1$ and $\angle 5$, $\angle 3$ and $\angle 0$.
6. $\angle 2 = 50°$, $\angle 4 = 30°$, $\angle 5 = 50°$, $\angle 6 = 100°$.
7. $\angle AOC = 80°$, $\angle AOD = 180°$, $\angle BOE = 180°$, $\angle FOB = 130°$.
8. (a) 67° 30', (b) 60°, (c) 45°, (d) 30°, (e) 22° 30'.
9. (a) 22°, (b) 45°, (c) 35°, (d) 58°, (e) 85°, (f) 56° 30'.
10. (a) 155°, (b) 55°, (c) 136°, (d) 92°, (e) 105° 30', (f) 101° 30'.

Exercise No. 67

1. (a) alt. int., (b) alt. int., (c) corr., (d) alt. ext., (e) corr.
2. $\angle 1 = 50°$, $\angle 2 = 130°$, $\angle 4 = 130°$.
3. $\angle 6 = 140°$, $\angle 7 = 140°$, $\angle 8 = 40°$.
4. Two lines $\perp$ to a third line are $\parallel$.
5. (a) alt. int. $\angle s$ are $=$.   (b) corr. $\angle s$ are $=$.   (c) alt. ext. $\angle s$ are $=$.
6. If a pair of alt. int. $\angle s$ are $=$ the lines are $\parallel$.
7. $\angle 3$ is sup. to 115°. $\therefore$ $\angle 3 = 65°$, making corr. $\angle s =$.
8. 70°.

9. Extend AB to D and construct $\angle BDE =$ to 60°.
Then $DE \parallel BC$ because corr. $\angle s$ are $=$.

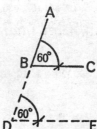

10. If the triangle is moved along the edge of the T-square into any two different positions, then lines drawn along side *a* will be ∥ to each other, and lines drawn along side *b* will also be ∥ to each other.

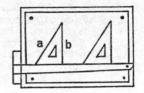

Exercise No. 68

1. $b^2 = c^2 - a^2, b = \sqrt{17^2 - 15^2}$
   $\therefore h = 8,$

2. $A = \dfrac{bh}{2}; \dfrac{18(63)}{2} = 567 \text{ m}^2.$

3. Perimeter = sum of 3 sides. In isosceles △ 2 sides are equal.
   $75 + 75 + 50 = 200; 200 \times £2 = £400.$

4. $c^2 = a^2 + b^2.$
   $c = \sqrt{90^2 + 90^2} =$
   $\sqrt{8100 + 8100} =$
   $\sqrt{16\,200} = 127 \cdot 27.$

5. Area of $A = \frac{1}{2}bh; \dfrac{20(10)}{2} = 100 \text{ m}^2.$

6. $a^2 = c^2 - b^2, a = \sqrt{(c)^2 - (b)^2} =$
   $\sqrt{(26)^2 - (10)^2} =$
   $\sqrt{676 - 100} = \sqrt{576} = 24.$

7. Let l = length of line, h = height of cliff, and b = distance from cliff at base. Then $l = \sqrt{h^2 + b^2} = \sqrt{1600 + 81} = \sqrt{1681} = 41.$

8. Form the triangle and make the necessary deduction afterwards. Let h = height of mast + elevation = $160 + 20 = 180$; l = line from top of mast to opposite shore; d = horizontal distance from mast to opposite shore. Then $d = \sqrt{l^2 - h^2} = \sqrt{500^2 - 180^2} = \sqrt{250\,000 - 32\,400} = \sqrt{217\,600} = 466 \cdot 47$. Subtracting 100 m leaves 366·47 as the width of the river.

Exercise No. 69

1. $AB = AD, \angle 1 = \angle 2$ and $AC = AC$ (by identity) $\therefore \triangle ABC \cong ADC$ by s.a.s. = s.a.s.

2. $AD = DC, BD = BD,$ and $\angle ADB = \angle CDB$ (all rt. ∠s are =)
   $\therefore \triangle ABD = \triangle CBD$ by s.a.s. = s.a.s.

3. $\angle 3 = \angle 5, BC = CD$ (bisected line) and $\angle 2 = \angle 6$ (vert. ∠s are =)
   $\therefore \triangle ABC \cong \triangle EDC$ by a.s.a. = a.s.a.

4. $AB = BD$, $EB = BC$, $\angle 1 = \angle 2$ (vert. $\angle s$)
$\triangle ABE \cong \triangle CBD$ by s.a.s = s.a.s.
$\angle 3 = \angle 4$ (corr. $\angle s$ of cong. $\triangle s$)
$\therefore AE \parallel CD$ (two lines are $\parallel$ if a pair of alt. int. $\angle s$ are =).

5. $AD = BC$, $AC = BD$, $AB = AB$ by identity
$\therefore \triangle BAD \cong \triangle CBA$ by s.s.s. = s.s.s.
$\therefore \angle 1 = \angle 2$ (corr. $\angle s$ of cong. $\triangle s$ are =).

6. $AB = CB$, $AD = CD$, $DB = DB$ by identity
$\therefore \triangle ABD \cong CBD$ by s.s.s. = s.s.s.
$\therefore \angle 5 = \angle 6$ (corr. $\angle s$ of cong. $\triangle s$)
$\therefore \angle 7 = \angle 8$ (supp. of = $\angle s$ are =)
$DE = DE$ by identity
$\therefore \triangle ADE \cong \triangle CDE$ by s.a.s. = s.a.s.
$\therefore \angle 1 = \angle 2$ (corr. $\angle s$ of cong. $\triangle s$).

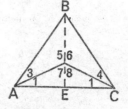

7. $AB = EF$, $\angle A = \angle F$ (alt. int. $\angle s$) and $\angle C = \angle D$ (alt. int. $\angle s$)
$\therefore \triangle ABC \cong \triangle DEF$ by s.a.a. = s.a.a.
$\therefore BC = DE$ (corr. sides of cong. $\triangle s$).

Exercise No. 70

1. $45°$.
2. $360°$ (any quad. can be divided into 2 $\triangle s$).
3. $120°$ ($\angle s$ of equilateral $\triangle = 60°$, and ext. $\angle$ = sum of 2 int. $\angle s$).
4. $20°$ and $70°$ (acute $\angle s$ of a rt. $\triangle$ are comp. $\therefore 9x = 90°$, $x = 10°$).
5. $52°$ (supp. $116° = 64°$; base $\angle s$ of isos. $\triangle$ are =
$\therefore 180 - (64° + 64°) = 52°$.
6. $32\frac{8}{11}°$, $49\frac{1}{11}°$, $98\frac{2}{11}°$ (Let x = angle; then $x + \frac{1}{2}x + \frac{1}{3}x = 180°$ and $x = 98\frac{2}{11}°$.)
7. Ratio is $1 : 1$ or equal.
8. $\angle 2$ supp. $\angle 1$ and $\angle 3$ supp. $\angle 4$
$\therefore \angle 2 = \angle 3$ (Ax. 1, p. 134)
$\therefore AB = BC$ and $\triangle ABC$ is isos. (if 2 $\angle s$ of a $\triangle$ are = the sides opp. are = and the $\triangle$ is isos.).
9. In $\triangle ABC$ $\angle 1 = \angle 2$ (base $\angle s$ of an isos. $\triangle$ are =)
$\angle 1 = \angle 3$ (corr. $\angle s$ of $\parallel$ lines are =)
$\therefore \angle 3 = \angle 2$ (Ax. 1)
$\therefore DA = DE$ (if two $\angle s$ of a $\triangle$ are = , the sides opp. are =).

Exercise No. 71

1. Perimeter = sum of 4 sides, and the opposite sides of a rectangle are equal.
$\therefore P = 50 + 50 + 90 + 90 = 280$.
2. Area of a rectangle = $l \times w$. $(30 + 12) \times (40 + 12) = 2184$.
3. Area of a square = S^2. $75 \times 75 = 5625 \times 0\cdot20 = \$1125$.

4. $4^2 = 16$; $1024 \div 16 = 64$.

5. Diag. of a rectangle makes 2 rt. angles. $c^2 = a^2 + b^2$.
$\therefore c = \sqrt{a^2 + b^2}$,
$c = \sqrt{(88)^2 + (66)^2} = \sqrt{4356 + 7744} = 110$.

6. Side of a square is equal to the square root of the area. $S = \sqrt{A} = \sqrt{288}$. Diagonal makes a rt. triangle in which $c^2 = a^2 + b^2$ or $c = \sqrt{288 + 288} = \sqrt{576} = 24$. By formula, diag. of a sq. $= \sqrt{2A}$ or $\sqrt{2 \times 288} = \sqrt{576} = 24$.

7. Perimeter = sum of 4 sides. If area = 81, side = $\sqrt{81}$ or 9; $9 \times 4 = 36$; $36 \times 6p = £2\cdot16$.

8. Area of trapezium $= \dfrac{B+b}{2} \times h = \dfrac{60+30}{2}$ $\times 15 = 45 \times 15 = 675$ m².

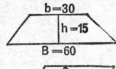
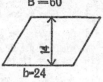

9. Area of a parallelogram equals base times height. $A = bh$. $\therefore A = 24 \times 14 = 336$ m².

10. To find unknown sections use formula for area of rt. triangle. $c^2 = a^2 + x^2$, or $x = \sqrt{c^2 - a^2} = \sqrt{(10)^2 - (8)^2} = \sqrt{100 - 64} = \sqrt{36} = 6$.

Area $= \dfrac{B+b}{2} \times h$; $\dfrac{18+12}{2} \times 8 = 15 \times 8 = 120$.

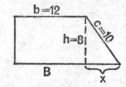

Exercise No. 72

1. $C = \pi d$. $C = \frac{22}{7} \times 28 = 88$ cm.
2. $C = 2\pi r$. $r = \dfrac{C}{2\pi}$, $r = \dfrac{110}{\frac{44}{7}} = 110 \times \dfrac{7}{44} = \dfrac{35}{2} = 17\frac{1}{2}$ mm.
3. $C = \pi d$. $C = \frac{22}{7} \times 4 = \frac{88}{7}$, $\frac{88}{7} \times 49 = 616$ cm.
4. $A = \pi r^2$, $D = 2r$. If $D = 1\cdot4$, then $r = 0\cdot7$; $A = \frac{22}{7} \times (0\cdot7)^2 = 1\cdot54$ m².

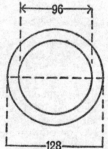

5. Area of ring $= \pi R^2 - \pi r^2$. $D = 128$, $d = 96$, $R = 64$, $r = 48$.

$A = \pi(64^2 - 48^2) = \frac{22}{7}(4096 - 2304) = \frac{22}{7} \times 1792$ $= 5632$; $5632 \times 0\cdot10 = £563\cdot20$.

6. $A = \pi r^2$, $r^2 = \dfrac{A}{\pi}$; $r = \sqrt{\dfrac{A}{\pi}} = \sqrt{\dfrac{50\frac{1}{4}}{\frac{22}{7}}} = \sqrt{50\frac{1}{4} \times \frac{7}{22}} = \sqrt{\frac{1407}{88}} = \sqrt{15\cdot99}$. Discarding the decimal, $r = 4$; $D = 2 \times 4$ or 8.

7. Since $A =$ constant times R^2, areas are to each other as the squares of their radii, or $A : a :: R^2 : r^2$. If $R = 2$ and $r = 1$, then $A = \pi \times (2)^2$, and $a = \pi(1)^2$, or 4 to 1. Answer is 4.

8. $R = \sqrt{(r^2 + r^2 + r^2 + r^2)}$
$= \sqrt{1\cdot4^2 + 1\cdot4^2 + 1\cdot4^2 + 1\cdot4^2}$
$= \sqrt{4(1\cdot96)} = 2\cdot8$.

9. Area $= \pi R^2 = \frac{22}{7} \times 2\cdot 8^2$. Cost is $\frac{22}{7} \times 2\cdot 8 \times 2\cdot 8 \times 25 = £616$.

10. Area of circle $= \pi R^2$. Side of equal square $=$
$\sqrt{\pi R^2} = \sqrt{\frac{22}{7} \times 28 \times 28} =$
$\sqrt{88 \times 28} = \sqrt{2464} = 49\cdot 6$ m.

Exercise No. 73

1. $\sin B = \dfrac{b}{c}$, $\cos B = \dfrac{a}{c}$, $\cot B = \dfrac{a}{b}$, $\sec B = \dfrac{c}{a}$, $\mathrm{cosec}\, B = \dfrac{c}{b}$.

2. $\tan A$. 3. $\cot A$. 4. $\sec A$. 5. $\mathrm{cosec}\, A$. 6. $\cos A = \frac{4}{5}$.
7. $\sin A = \frac{3}{5}$. 8. $\cos A = \frac{15}{17}$. 9. $\sec A = \frac{17}{15}$.
10. $\cos A = \frac{12}{13}$, $\tan A = \frac{5}{12}$, $\cot A = \frac{12}{5}$, $\sec A = \frac{13}{12}$, $\mathrm{cosec}\, A = \frac{13}{5}$.

Exercise No. 74

1. $\cos 64°$. 2. $\cot 47°$. 3. $\sin 65° 32'$. 4. $\tan 1° 10'$.
5. $\mathrm{cosec}\, 83° 50'$. 6. $\sec 12\frac{1}{2}°$.
7. $15°$ $(90° = 5A + A; \therefore 90° = 6A$, and $A = 15°)$.
8. $45°$ (reciprocals of the cofunctions are $=$, $\therefore \angle A = 45°$).
9. $45°$ $(90° - A = A; 90° = 2A; A = 45°)$.
10. $30°$ $(\cos A = \sin 90° - A$; since $\cos A = \sin 2A$, then $\sin 90° - A = \sin 2A$, $90° - A = 2A; 3A = 90°$ and $A = 30°)$.

Exercise No. 75

1. $0\cdot 1392$. 2. $0\cdot 6691$. 3. $0\cdot 8391$. 4. $0\cdot 5095$. 5. $1\cdot 079$.
6. $0\cdot 9063$. 7. $4\cdot 134$. 8. $0\cdot 9781$. 9. $0\cdot 3839$. 10. $6\cdot 3925$.
11. 4695. 12. $1\cdot 4826$. 13. $0\cdot 8480$. 14. $0\cdot 7071$. 15. $0\cdot 5000$
16. $15°$. 17. $35°$. 18. $60°$. 19. $70°$. 20. $10°$.

Exercise No. 76

1. $0\cdot 2672$. 2. $0\cdot 9014$. 3. $1\cdot 079$. 4, $0\cdot 7673$. 5. $1\cdot 3131$.
6. $5° 10'$. 7. $10° 50'$. 8. $65° 20'$. 9. $49° 40'$. 10. $25°$.

Exercise No. 77

1. $a = 54\cdot 5$. 2. $a = 3\cdot 42$. 3. $\angle A = 65° 33'$.
4. $c = 29\cdot 82$. 5. $\angle A = 46° 03'$.

6. $\dfrac{a}{c} = \sin A$, $c = \dfrac{a}{\sin A} = \dfrac{405}{0\cdot 2250} = 1800$ m.

7. $a = c \sin A = 150 \times 0\cdot 7660 = 114\cdot 9$ m.

8. $c = \dfrac{a}{\sin A} = \dfrac{1\cdot 2}{0\cdot 3090} = 3\cdot 883$ m.

9. $a = c \sin A = 20 \times 0\cdot 9455 = 18\cdot 91$ m.

10. $\dfrac{a}{c} = \sin A = \dfrac{54\cdot 5}{625} = 0\cdot 0872$ which is the sin of $5°$.

Exercise No. 78

1. $\angle A = 53° 8'$. 2. $b = 31\cdot 86$. 3. $b = 62\cdot 1$. 4. $\angle A = 60°$.
5. $c = 42$.

6. $\dfrac{b}{c} = \cos A = \dfrac{15}{17} = 0\cdot 8823$ which is the cos of $28° 05'$.

7. $\dfrac{b}{c} = \cos A$, $c = \dfrac{b}{\cos A} = \dfrac{681}{0\cdot 8910} = 764$.

8. $\angle B = 90° - 10° = 80°$, $\cos B = \dfrac{a}{c}$, $c = \dfrac{a}{\cos B} = \dfrac{125}{0\cdot 1736} = 720\cdot 0$.

9. $\cos B = \dfrac{a}{c}$, $a = c \cos B = 84 \times 0{\cdot}6428 = 53{\cdot}9952 = 54.$

10. $\cos A = \dfrac{b}{c} = \dfrac{16{\cdot}5}{100} = 0{\cdot}165$ which is the cos of 80° 30′.

Exercise No. 79

1. 36° 52′.   2. 64.   3. 45·04.   4. 18·19.   5. 15.

6. $\dfrac{a}{b} = \tan A$, $b = \frac{1}{2}$ of 280 = 140 m, $a = b \tan A = 140 \times 0{\cdot}9325 = 130{\cdot}55$ m.

7. $\tan B = \dfrac{b}{a}$, $\angle B = 90° - 10° = 80°$, $b = a \tan B = 24 \times 5{\cdot}6713 = 136{\cdot}1$ m.

8. $\tan A = \dfrac{a}{b}$, $b = \dfrac{a}{\tan A} = \dfrac{30}{0{\cdot}3640} = 82{\cdot}42$ m.

9. $\angle A = 90° - 15° = 75°$, $\angle A' = 90° - 14° = 76°$
   $CB = b \tan A = 10 \times 3{\cdot}7321 = 37{\cdot}321$ m.
   $CB' = b \tan A' = 10 \times 4{\cdot}0108 = 40{\cdot}108$ m.
   $CB' - CB = BB' = 40{\cdot}108 - 37{\cdot}321 = 2{\cdot}787$ m.

10. Let x = height of tower
    y = distance from nearer point to foot of tower.

   From $\triangle ACD$, $\dfrac{x}{30 + y} = \tan 30°$; $\tan 30° = \dfrac{1}{\sqrt{3}}$ ∴ $y = \sqrt{3}x - 30.$

   From $\triangle BCD$, $\dfrac{x}{y} = \tan 60°$, $\tan 60° = \sqrt{3}$ ∴ $y = \dfrac{x}{\sqrt{3}}.$

   Equating the values of y, $\sqrt{3}x - 30 = \dfrac{x}{\sqrt{3}}.$

   $2x - 30 \times 1{\cdot}732,\ x = 25{\cdot}98.$

Exercise No. 80

1. $a = 7$, $b = 8{\cdot}57.$
2. $\angle C = 69°$, $b = 58{\cdot}46$, $c = 58{\cdot}87.$
3. $\angle A = 76° 52′$, $\angle B = 35° 8′$, $c = 20{\cdot}95.$
4. $\angle A = 51° 24′$, $\angle B = 48° 49′$, $\angle C = 79° 47′.$
5. $\angle B = 12° 56′$, $\angle C = 146° 4′$, $c = 12{\cdot}43.$

6. $\angle ABC = 97° 44′.$
   $\angle BCA = 180° - (67° 31′ + 97° 44′) = 14° 45′.$

   $\dfrac{a}{c} = \dfrac{\sin A}{\sin C}$, $a = \dfrac{c \sin A}{\sin BCA} = \dfrac{1{\cdot}83 \times 0{\cdot}924}{0{\cdot}2546} = 6{\cdot}641.$

   $\sin 82° 16′ = \dfrac{x}{6{\cdot}641}$,

   $\qquad x = 0{\cdot}9909 \times 6{\cdot}641 = 6{\cdot}6$ km (1 d.p.).

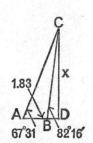

7. $\dfrac{b}{c} = \dfrac{\sin B}{\sin C}$, $\sin B = \dfrac{b \sin C}{c} = \dfrac{98{\cdot}5\,(\sin 64° 20′)}{146}$
   $= 0{\cdot}6081$ which is sin 37° 27′
   $\angle A = 180° - (64° 20′ + 37° 27′) = 78° 13′$
   $\dfrac{a}{b} = \dfrac{\sin A}{\sin B}$, $a = \dfrac{b \sin A}{\sin B} = \dfrac{98{\cdot}5 \times 0{\cdot}9790}{0{\cdot}6081}$
   $= 158{\cdot}6$ m.

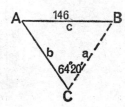

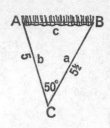

8. $A + B = 180° - 50° = 130°, \frac{1}{2}(A + B) = 65°$

$\tan \frac{1}{2}(A - B) = \dfrac{a - b}{a + b} \times \tan \frac{1}{2}(A + B) = \dfrac{5·5 - 5}{5·5 + 5} \times$

$2·1445 = 0·1021$ which is the tan of $5° 50'$

$\angle A = \frac{1}{2}(A + B) + \frac{1}{2}(A - B) = 70° 50'$

$\dfrac{c}{a} = \dfrac{\sin C}{\sin A}, c = \dfrac{a \sin C}{\sin A} = \dfrac{5·5 \times 0·7660}{0·9446} = 4·46$ km.

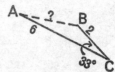

9. By cos law, $c = \sqrt{(a^2 + b^2 - 2ab \cos C)}$
$c = \sqrt{(2^2 + 6^2 - 2(2 \times 6) \cos 33°)}$
$= \sqrt{19·87}$
$= 4·46$ km.

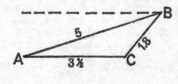

10. By cos law, $a^2 = b^2 + c^2 - 2bc \cos A$

$\therefore \cos A = \dfrac{b^2 + c^2 - a^2}{2bc}$

(a) $\cos A = \dfrac{3·5^2 + 5^2 - 1·8^2}{2(3·5 \times 5)} = 0·9717$

which is the cos of $13° 40'$

(b) alt. int. $\angle s$ of $\parallel$ lines are $=$;
$\therefore$ angle of depression $= 13° 40'$

Exercise No. 81

1.

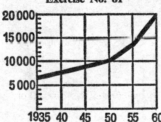

2.

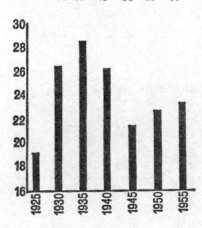

3.

timber	Price
1	$ ·08
15	1·20
25	2·00
35	2·80
45	3·60
55	4·40

Price of 36½ m = $2·92.

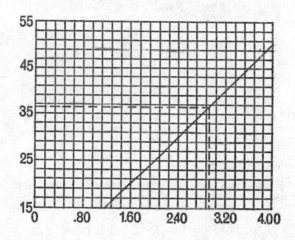

4. 6½ hr.

5. 3,200 m. Multiply the values shown on the graph by 100.

6.

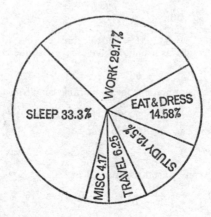

Exercise No. 82

1. $C(4, 3) = 4$ by count of

$$\overparen{A\ \&\ B\ \&\ C,}\quad \overparen{A\ \&\ B\ \&\ D,}$$

$$\overparen{A\ \&\ C\ \&\ D,}\quad \overparen{B\ \&\ C\ \&\ D.}$$

$P(4, 3) = 24$ by count of

ABC,	BAC,	CAB,	DAB,
ABD,	BAD,	CAD,	DAC,
ACB,	BCA,	CBA,	DBA,
ACD,	BCD,	CBD,	DBC,
ADB,	BDA,	CDA,	DCA,
ADC,	BDC,	CDB,	DCB.

2. $$C(4, 3) = \frac{4!}{3!(4-3)!} = \frac{4}{1!} = 4.$$

$$P(4, 3) = \frac{4!}{(4-3)!} = \frac{4!}{1!} = 4.3.2 = 24.$$

3. Since this is a question of grouping without regard to order, it is one of *combinations* with $n = 12, t = 5$, and

$$C(12, 5) = \frac{12!}{5!(12-5)!} = \frac{12!}{5!7!} = \frac{12.11.10.9.8}{5.4.3.2} = 792, \quad \text{Ans.}$$

4. Since different signals can be sent by different arrangements of the same combination of flags, these are questions of *permutations*.

(*a*) $\qquad\qquad P(5, 5) = 5! = 5.4.3.2 = 120, \quad \text{Ans. } (a).$

(*b*) $\qquad P(12, 5) = \frac{12!}{(12-5)!} = \frac{12!}{7!} = 12.11.10.9.8$

$$= 95\,040, \quad \text{Ans. } (b).$$

5. (*a*) $\quad C(n, t)\,P(t, t) = \frac{n!}{t!(n-t)!} \cdot t! = \frac{n!}{(n-t)!} = P(n, t)$, as required.

(*b*) With $n = 12$ and $t = 5$ as in the two preceding examples:

$$\begin{array}{rr} C(12, 5) = & 792 \\ P(5, 5) = & 120 \\ \hline & 15840 \\ & 792 \\ \hline \end{array}$$

$$P(12, 5) = 95040, \quad \text{CHECK.}$$

6. Regarding the couple as a unit, we have $n = 7$ in the circular-permutations formula

$$P_c(7, 7) = 6! = 6.5.4.3.2 = 720$$

But in each of these permutations the inseparable couple may sit on either side of each other. Hence,

$$2P_c = 2(720) = 1440, \quad \text{Ans.}$$

7. The 2 vowels can be arranged from 5 in any of

$$P(5, 2) = \frac{5!}{(5-2)!} = \frac{5!}{3!} = 5.4 = 20 \text{ ways}$$

The 3 other letters can be arranged from 21 in

$$P(21, 3) = \frac{21!}{(21-3)!} = 21.20.19 = 7980 \text{ ways.}$$

Hence, all 5 letters can be arranged in
$$P(5, 2)P(21, 3) = 20(7980)$$
$$= 159\,600 \text{ ways,} \quad \text{Ans.}$$

8. Here $n_1 = 3, n_2 = 6, n_3 = 4, n = 3 + 6 + 4 = 13$, and
$$P_a = \frac{n!}{n_1! n_2! n_3!} = \frac{13!}{3!6!4!} = \frac{13 \cdot 12 \cdot 11 \cdot 10 \cdot 9 \cdot 8 \cdot 7}{3 \cdot 2 \cdot 4 \cdot 3 \cdot 2} = 60\,060, \quad \text{Ans.}$$

9. (a) $N = 4(6) = 24$. (b) $N = 4(6)(3) = 72$.

10. $N = 2^5 = 32$.

11. $N = 6^4 = 1296$.

12. $N = 2(6)(4)(9) = 432$ models.

13. Excluding the one case in which *no dot* is raised,
$$N = 2^6 - 1 = 64 - 1 = 63, \quad \text{Ans.}$$

14. This question differs from question 4(a) in that duplications of flags are now possible. Hence, $n = k = 5$, and
$$N = n^k = 5^5 = 3{,}125, \quad \text{Ans.}$$

15. $\qquad M = 2^{16} - 1 = 65\,536 - 1 = 65\,535, \quad \text{Ans.}$

16. (a) We can combine any 2 ranks in
$$C_2^{13} = \frac{13 \cdot 12}{2} = 78 \text{ ways.}$$

And we can combine 'a pair' of one rank with '3 of a kind' of another rank in
$$2C_2^4 C_3^4 = 2 \cdot \frac{4 \cdot 3}{2} \cdot \frac{4 \cdot 3 \cdot 2}{3 \cdot 2} = 48 \text{ ways.}$$

Hence a 'full house' can be dealt in
$$C_2^{13} 2C_2^4 C_3^4 = 78(48) = 3744 \text{ ways,} \quad \text{Ans.}$$

(b) In any one suit we can combine 5 cards to form a 'flush' in C_5^{13} ways. Hence, with 4 suits in the pack, 'any flush' can be dealt in
$$4C_5^{13} = 4 \cdot \frac{13 \cdot 12 \cdot 11 \cdot 10 \cdot 9}{5 \cdot 4 \cdot 3 \cdot 2} = 5148 \text{ ways,} \quad \text{Ans.}$$

(c) 'Any straight' can be dealt with any one rank lowest (or highest) in
$$n^k = 4^5 = 1024 \text{ ways.}$$

Hence, it can be dealt with any one of 10 possible ranks lowest (or highest) in
$$10n^k = 10(1024) = 10\,240 \text{ ways,} \quad \text{Ans.}$$

(d) By combining the reasoning of (b) and (c), we find a 'straight flush' can be dealt in
$$4 \cdot 10 = 40 \text{ ways,} \quad \text{Ans.}$$

(e) $b - d = 5148 - 40 = 5108$ ordinary 'flushes'.

(f) $c - d = 10\,240 - 40 = 10\,200$ ordinary 'straights'.

Exercise No. 83

1. The required array and corresponding h values are:
$$h_5 = 4 \begin{cases} h_1 = 1 & H_1 H_2 H_3 \quad H_1 T_2 T_3 \\ h_2 = 3 \begin{cases} H_1 H_2 T_3 & T_1 H_2 T_3 \\ H_1 T_2 H_3 & T_1 T_2 H_3 \\ T_1 H_2 H_3 & T_1 T_2 T_3 \end{cases} h_4 = 1 \end{cases} h_3 = 3$$

Hence, the corresponding f values are:
$$f_1 = f_4 = 8 - 1 = 7,$$
$$f_2 = f_3 = 8 - 3 = 5,$$
$$f_5 = 8 - 4 = 4.$$

2.
$$p_1 = p_4 = \tfrac{1}{8} = 12{\cdot}5\%,$$
$$q_1 = q_4 = \tfrac{7}{8} = 87{\cdot}5\%,$$
$$p_2 = p_3 = \tfrac{3}{8} = 37{\cdot}5\%, \quad \text{ANS.}$$
$$q_2 = q_3 = \tfrac{5}{8} = 62{\cdot}5\%,$$
$$p_5 = q_5 = \tfrac{4}{8} = 50{\cdot}0\%.$$

3. Events E_1, E_2, E_3, and E_4 are all the mutually exclusive different outcomes possible when 3 coins are tossed. Hence,

$$p_1 + p_2 + p_3 + p_4 = \tfrac{1}{8} + \tfrac{3}{8} + \tfrac{3}{8} + \tfrac{1}{8} = \tfrac{8}{8} = 1, \quad \text{CHECK.}$$

Also, event E_5 is mutually exclusive with events E_3 and E_4, to exhaust the same possibilities with a different grouping. Hence,

$$p_5 + p_3 + p_4 = \tfrac{4}{8} + \tfrac{3}{8} + \tfrac{1}{8} = \tfrac{8}{8} = 1, \quad \text{CHECK.}$$

4. Since E_5 is the event that either of the mutually exclusive events, E_1 or E_2, happen

$$p_5 = p_1 + p_2 = \tfrac{1}{8} + \tfrac{3}{8} = \tfrac{4}{8} = 0{\cdot}5 \text{ as before.}$$

5. (a) The odds on E_1 happening are

$$h_1{:}f_1 = 1{:}7, \text{ or 1 to 7 for}$$

(b) The odds on E_2 happening are

$$f_2{:}h_2 = 5{:}3, \text{ or 5 to 3 against}$$

(c) The odds on E_5 happening are

$$h_5{:}f_5 = 4{:}4 = 1{:}1, \text{ or even either way.}$$

6. (a) For event E_a that the exposed card be a face card, $w = 52$, $h_a = 3 \cdot 4 = 12$, and

$$p_a = h_a/w = 12/52 = 3/13, \quad \text{ANS. } (a)$$

(b) For the event E_b that the exposed card be a black card, $h_b = 2 \cdot 13 = 26$, and

$$p_b = h_b/w = 26/52 = \tfrac{1}{2}, \quad \text{ANS. } (b)$$

(c) For the event E_c that the exposed card be a black face card, $h_c = 2 \cdot 3 = 6$, and

$$p_c = h_c/w = 6/52 = 3/26, \quad \text{ANS. } (c)$$

(d) Hence, the probability of the partially overlapping event E_d, that either E_a or E_b happen, is

$$p_d = p_a + p_b - p_c = \frac{12 + 26 - 6}{52} = \frac{32}{52} = \frac{8}{13}, \quad \text{ANS. } (d).$$

7. The coins can land in $w = n^k = 2^5 = 32$ ways. Of these, only $h = 2$ (1 all heads, and 1 all tails) comply with the condition that all land the same way. Hence, the required probability is

$$p = h/w = 2/32 = 1/16, \quad \text{ANS.}$$

Alternatively regarding the required landing as a multiple event, we can reason that the probability of the first coin landing either heads or tails is $p_1 = 2/2 = 1$ (for certainty), but that the probability of the other coins each separately landing the same way is 1/2 in each case. Hence, $p_2 = p_3 = p_4 = p_5 = 1/2$, and the probability of the specified multiple event is

$$p = p_1 p_2 p_3 p_4 p_5 = 1(\tfrac{1}{2})^4 = 1/2^4 = 1/16, \quad \text{SAME ANS.}$$

8. Any 2 balls from either box may be combined with any 2 balls in the other box in

$$w = C_2{}^{10} C_2{}^{10} = \left(\frac{10.9}{2}\right)^2 = 45^2 = 2025 \text{ ways.}$$

But 2 red balls from either may be combined with 2 red balls from the other in only

$$h = C_2{}^4 C_2{}^4 = \left(\frac{4.3}{2}\right)^2 = 6^2 = 36 \text{ ways.}$$

Hence, the required probability is

$$p = h/w = 36/2025 = 4/225, \quad \text{Ans.}$$

Alternatively, we may regard the drawing of all 4 balls as a multiple event consisting of the 4 separate drawings with the separate probabilities, $p_1 = p_3 = 4/10$ independent of each other, and $p_2 = p_4 = 3/9$ independent of each other but dependent upon p_1 and p_3 respectively. Then, by the multiplication theorem,

$$p = p_1 p_2 p_3 p_4 = \tfrac{4}{10} \cdot \tfrac{3}{9} \cdot \tfrac{4}{10} \cdot \tfrac{3}{9}$$
$$= (2.2)/(5.3.5.3) = 4/225, \quad \text{Same Ans.}$$

9. By the same reasoning, if all 4 balls are drawn from the same box, then

$$p = h/w = C_4{}^4/C_4{}^{10} = 1 \left/ \dfrac{\begin{matrix} 1 \\ 3 \ 2 \\ 10.9.8.7 \end{matrix}}{\begin{matrix} 4.3.2 \\ 1 \ 1 \ 1 \end{matrix}} \right. = 1/210, \quad \text{Ans.}$$

Or, alternatively, with each separate event dependent upon those preceding it,

$$p = p_1 p_2 p_3 p_4 = \tfrac{4}{10} \cdot \tfrac{3}{9} \cdot \tfrac{2}{8} \cdot \tfrac{1}{7} = \tfrac{1}{210}, \quad \text{Same Ans.}$$

10. Of the *1 098 240* ways of dealing a one-pair hand, 4/13 of these have pairs of jacks or better. We could add this number to all the numbers of still better hands above it in the *Table* to find $h =$ the total number of hands which have a pair of jacks or better. However, it is arithmetically simpler to find the number of hands which have single pairs less than jacks, or

$$1\ 098\ 240(9/13) = 760\ 320$$

and add to this the one number of hands in the *Table* still lower than these to find,

$$f = 760\ 320 + 1\ 302\ 540 = 2\ 062\ 860$$

Then, more quickly,

$$h = w - f = 2\ 598\ 960 - 2\ 062\ 860 = 536\ 100;$$

and the required probability is

$$p = h/w = 536\ 100/2\ 598\ 960 = 1 \text{ in } 4.8, \text{ approximately,} \quad \text{Ans.}$$

11. Regardless of the number on which Sloe bets, the mathematical expectation which he obtains for his \$10.00 is only

$$V = A \cdot p = \$360/37 = \$9.73.$$

In other words, all numbers are equally 'unlucky' for the player in the long run. The only reason the 'house' wins in the long run is that it never puts *any* money up on its 'free' number, zero. Moreover, it cannot afford dishonestly to have the wheel 'fixed' to come up more frequently on zero, or on any other number, because—quite aside from any question of ethics or good will—the kind of practised gambler who frequents such houses would soon detect the trend and 'break the bank' by placing large bets on the favoured outcomes. In other words, whether Sloe bets with or against it, the house will take his money from him most certainly in the long run by keeping the wheel 'honest' and Sloe naïvely hopeful. From the theory of probability we learn that, although there may be 'systems' which can beat the house that operates a dishonest or accidentally unbalanced wheel, there is no 'system' which will long win for anyone but the proprietor on an honest wheel in the long run!

12. The given possibilities are the only distributions with less than 5 cards in any suit. Hence,

$$p_1 = 0.2155$$
$$p_2 = 0.1054$$
$$p_3 = 0.0299$$

$$p_1 + p_2 + p_3 = 0.3508 = q,$$

and the required probability is

$$p = 1 - q = 1 - 0.3508 = 0.6492, \quad \text{Ans.}$$

13. One or more can be drawn in any of

$$w = M = n^k - 1 = 2^7 - 1 = 128 - 1 = 127$$

different combinations. Of these, combinations with the possible even numbers of 2, 4, or 6 balls are

$$h = C_2^7 + C_4^7 + C_6^7 = \frac{7!}{2!5!} + \frac{7!}{4!3!} + \frac{7!}{6!1!}$$

$$= \frac{7 \cdot 6}{2} + \frac{7 \cdot 6 \cdot 5}{3 \cdot 2} + \frac{7}{1} = 21 + 35 + 7 = 63.$$

Hence, the required probability is

$$p = h/w = 63/127, \quad \text{Ans.}$$

14. (a) The hour hand is between these marks for only 1 hour out of 12. Hence, the 'continuous' probability of event E_a is

$$p_a = 1 \text{ hour}/12 \text{ hours} = 1/12, \quad \text{Ans. } (a)$$

(b) However, both hands are between these marks for only 5 minutes out of each 12 hours = 720 minutes. Hence, the 'continuous' probability of event E_b is

$$p_b = 5 \text{ minutes}/720 \text{ minutes} = 1/144, \quad \text{Ans. } (b)$$

Alternatively, we can reason that the minute hand is between these marks for only 5 minutes out of 60, or for 1 hour out of 12. Hence, the probability of the independent separate event E_m that the minute hand stop between the marks is

$$p_m = p_a = \tfrac{1}{12}$$

And therefore the probability of the multiple event E_b is

$$p = p_a \cdot p_m = \tfrac{1}{12} \cdot \tfrac{1}{12} = \tfrac{1}{144}, \quad \text{Same Ans. } (b)$$

15. From the *Mortality Table* on p. 239, we learn that the statistical probability of one such person still being alive at age 50 is

$$p_1 = h/w = 69\,804/100\,000 = 0.698 = 69.8\%$$

Hence, the probability of the multiple statistical event that 2 such persons still be alive at age 50 is

$$p_2 = p_1 \cdot p_1 = (0.698)^2 = 0.487 = 48.7\%, \quad \text{Ans.}$$

Index